Lecture Notes in Mathematics

T0259915

Editors-in-Chief:
J.-M. Morel, Cachan
B. Teissier, Paris

For further volumes:
http://www.springer.com/series/304

"Lévy Matters" is a subseries of the Springer Lecture Notes in Mathematics, devoted to the dissemination of important developments in the area of Stochastics that are rooted in the theory of Lévy processes. Each volume will contain state-of-the-art theoretical results as well as applications of this rapidly evolving field, with special emphasis on the case of discontinuous paths. Contributions to this series by leading experts will present or survey new and exciting areas of recent theoretical developments, or will focus on some of the more promising applications in related fields. In this way each volume will constitute a reference text that will serve PhD students, postdoctoral researchers and seasoned researchers alike.

Editors

Ole E. Barndorff-Nielsen
Thiele Centre for Applied Mathematics
    in Natural Science
Department of Mathematical Sciences
Aarhus University
8000 Aarhus C, Denmark
oebn@imf.au.dk

Jean Jacod
Institut de Mathématiques de Jussieu
CNRS-UMR 7586
Université Paris 6 - Pierre et Marie Curie
75252 Paris Cedex 05, France
jean.jacod@upmc.fr

Jean Bertoin
Institut für Mathematik
Universität Zürich
8057 Zürich, Switzerland
jean.bertoin@math.uzh.ch

Claudia Klüppelberg
Zentrum Mathematik
Technische Universität München
85747 Garching bei München, Germany
cklu@ma.tum.de

Managing Editors

Vicky Fasen
Institut für Stochastik
Karlsruher Institut für Technologie
76133 Karlsruhe, Germany
vicky.fasen@kit.edu

Robert Stelzer
Institute of Mathematical Finance
Ulm University
89081 Ulm, Germany
robert.stelzer@uni-ulm.de

The volumes in this subseries are published under the auspices of the Bernoulli Society.

Björn Böttcher • René Schilling • Jian Wang

Lévy Matters III

Lévy-Type Processes: Construction, Approximation and Sample Path Properties

 Springer

Björn Böttcher
Institut für Mathematische Stochastik
Technische Universität Dresden
Dresden, Germany

René Schilling
Institut für Mathematische Stochastik
Technische Universität Dresden
Dresden, Germany

Jian Wang
School of Mathematics and Computer
  Science
Fujian Normal University
Fuzhou, Fujian
People's Republic of China

ISBN 978-3-319-02683-1   ISBN 978-3-319-02684-8 (eBook)
DOI 10.1007/978-3-319-02684-8
Springer Cham Heidelberg New York Dordrecht London

Lecture Notes in Mathematics ISSN print edition: 0075-8434
                              ISSN electronic edition: 1617-9692

Library of Congress Control Number: 2013955905

Mathematics Subject Classification (2010): 60-02; 60J25; 60J35; 60G17; 35S05; 60J75; 60G48; 60G51;
              60H10; 47D03; 35S30

Printed on acid-free paper

Springer is part of Springer Science+Business Media (www.springer.com)

Preface to the Series
Lévy Matters

Over the past 10–15 years, we have seen a revival of general Lévy processes theory as well as a burst of new applications. In the past, Brownian motion or the Poisson process had been considered as appropriate models for most applications. Nowadays, the need for more realistic modelling of irregular behaviour of phenomena in nature and society like jumps, bursts and extremes has led to a renaissance of the theory of general Lévy processes. Theoretical and applied researchers in fields as diverse as quantum theory, statistical physics, meteorology, seismology, statistics, insurance, finance and telecommunication have realized the enormous flexibility of Lévy models in modelling jumps, tails, dependence and sample path behaviour. Lévy processes or Lévy-driven processes feature slow or rapid structural breaks, extremal behaviour, clustering and clumping of points.

Tools and techniques from related but distinct mathematical fields, such as point processes, stochastic integration, probability theory in abstract spaces and differential geometry, have contributed to a better understanding of Lévy jump processes.

As in many other fields, the enormous power of modern computers has also changed the view of Lévy processes. Simulation methods for paths of Lévy processes and realizations of their functionals have been developed. Monte Carlo simulation makes it possible to determine the distribution of functionals of sample paths of Lévy processes to a high level of accuracy.

This development of Lévy processes was accompanied and triggered by a series of Conferences on Lévy Processes: Theory and Applications. The First and Second Conferences were held in Aarhus (1999, 2002), the Third in Paris (2003), the Fourth in Manchester (2005) and the Fifth in Copenhagen (2007).

To show the broad spectrum of these conferences, the following topics are taken from the announcement of the Copenhagen conference:

- Structural results for Lévy processes: distribution and path properties
- Lévy trees, superprocesses and branching theory

- Fractal processes and fractal phenomena
- Stable and infinitely divisible processes and distributions
- Applications in finance, physics, biosciences and telecommunications
- Lévy processes on abstract structures
- Statistical, numerical and simulation aspects of Lévy processes
- Lévy and stable random fields

At the Conference on Lévy Processes: Theory and Applications in Copenhagen the idea was born to start a series of Lecture Notes on Lévy processes to bear witness of the exciting recent advances in the area of Lévy processes and their applications. Its goal is the dissemination of important developments in theory and applications. Each volume will describe state-of-the-art results of this rapidly evolving subject with special emphasis on the non-Brownian world. Leading experts will present new exciting fields, or surveys of recent developments, or focus on some of the most promising applications. Despite its special character, each article is written in an expository style, normally with an extensive bibliography at the end. In this way each article makes an invaluable comprehensive reference text. The intended audience are Ph.D. and postdoctoral students, or researchers, who want to learn about recent advances in the theory of Lévy processes and to get an overview of new applications in different fields.

Now, with the field in full flourish and with future interest definitely increasing it seemed reasonable to start a series of Lecture Notes in this area, whose individual volumes will appear over time under the common name "Lévy Matters," in tune with the developments in the field. "Lévy Matters" appears as a subseries of the Springer Lecture Notes in Mathematics, thus ensuring wide dissemination of the scientific material. The mainly expository articles should reflect the broadness of the area of Lévy processes.

We take the possibility to acknowledge the very positive collaboration with the relevant Springer staff and the editors of the LN series and the (anonymous) referees of the articles.

We hope that the readers of "Lévy Matters" enjoy learning about the high potential of Lévy processes in theory and applications. Researchers with ideas for contributions to further volumes in the Lévy Matters series are invited to contact any of the editors with proposals or suggestions.

Aarhus, Denmark   Ole E. Barndorff-Nielsen
Paris, France   Jean Bertoin and Jean Jacod
Munich, Germany   Claudia Klüppelberg
June 2010

A Short Biography of Paul Lévy

A volume of the series "Lévy Matters" would not be complete without a short sketch about the life and mathematical achievements of the mathematician whose name has been borrowed and used here. This is more a form of tribute to Paul Lévy, who not only invented what we call now Lévy processes, but also is in a sense the founder of the way we are now looking at stochastic processes, with emphasis on the path properties.

Paul Lévy was born in 1886 and lived until 1971. He studied at the Ecole Polytechnique in Paris and was soon appointed as professor of mathematics in the same institution, a position that he held from 1920 to 1959. He started his career as an analyst, with 20 published papers between 1905 (he was then 19 years old) and 1914, and he became interested in probability by chance, so to speak, when asked to give a series of lectures on this topic in 1919 in that same school: this was the starting point of an astounding series of contributions in this field, in parallel with a continuing activity in functional analysis.

Very briefly, one can mention that he is the mathematician who introduced characteristic functions in full generality, proving in particular the characterization theorem and the first "Lévy's theorem" about convergence. This naturally led him to study more deeply the convergence in law with its metric and also to consider sums of independent variables, a hot topic at the time: Paul Lévy proved a form of the 0-1 law, as well as many other results, for series of independent variables. He also introduced stable and quasi-stable distributions and unravelled their weak and/or strong domains of attractions, simultaneously with Feller.

Then we arrive at the book *Théorie de l'addition des variables aléatoires*, published in 1937, and in which he summaries his findings about what he called "additive processes" (the homogeneous additive processes are now called Lévy processes, but he did not restrict his attention to the homogeneous case). This book contains a host of new ideas and new concepts: the decomposition into the sum of jumps at fixed times and the rest of the process; the Poissonian structure of the jumps for an additive process without fixed times of discontinuities; the "compensation" of those jumps so that one is able to sum up all of them; the fact that the remaining continuous part is Gaussian. As a consequence, he implicitly gave the formula providing the form of all additive processes without fixed discontinuities, now called the Lévy–Itô formula, and he proved the Lévy–Khintchine formula for the characteristic functions of all infinitely divisible distributions. But, as fundamental as all those results are, this book contains more: new methods, like martingales which, although not given a name, are used in a fundamental way; and also a new way of looking at processes, which is the "pathwise" way: he was certainly the first to understand the importance of looking at and describing the paths of a stochastic process, instead of considering that everything is encapsulated into the distribution of the processes.

This is of course not the end of the story. Paul Lévy undertook a very deep analysis of Brownian motion, culminating in his book *Processus stochastiques et mouvement Brownien* in 1948, completed by a second edition in 1965. This is a remarkable achievement, in the spirit of path properties, and again it contains so many deep results: the Lévy modulus of continuity, the Hausdorff dimension of the path, the multiple points and the Lévy characterization theorem. He introduced local time and proved the arc-sine law. He was also the first to consider genuine stochastic integrals, with the area formula. In this topic again, his ideas have been the origin of a huge amount of subsequent work, which is still going on. It also laid some of the basis for the fine study of Markov processes, like the local time again, or the new concept of instantaneous state. He also initiated the topic of multi-parameter stochastic processes, introducing in particular the multi-parameter Brownian motion.

As should be quite clear, the account given here does not describe the whole of Paul Lévy's mathematical achievements, and one can consult for many more details the first paper (by Michel Loève) published in the first issue of the *Annals of Probability* (1973). It also does not account for the humanity and gentleness of the person Paul Lévy. But I would like to end this short exposition of Paul Lévy's work by hoping that this series will contribute to fulfilling the program, which he initiated.

Paris, France   Jean Jacod

Preface

Behind every decent Markov process there is a family of Lévy processes. Indeed, let $(X_t)_{t \geq 0}$ be a Markov process with state space $\mathbb{R}^d$ and assume, for the moment, that the limit

$$\lim_{t \to 0} \frac{1 - \mathbb{E}^x\, e^{i\xi \cdot (X_t - x)}}{t} = q(x, \xi) \quad \forall x, \xi \in \mathbb{R}^d \tag{$\star$}$$

exists such that the function $\xi \mapsto q(x, \xi)$ is continuous. We will see below that this is enough to guarantee that $q(x, \cdot)$ is, for each $x \in \mathbb{R}^d$, the characteristic exponent of a Lévy process; as such, it enjoys a Lévy–Khintchine representation

$$q(x, \xi) = -il(x) \cdot \xi + \frac{1}{2}\xi \cdot Q(x)\xi + \int_{\mathbb{R}^d \setminus \{0\}} \left(1 - e^{iy \cdot \xi} + i\xi \cdot y \mathbb{1}_{[-1,1]}(|y|)\right) N(x, dy)$$

where $(l(x), Q(x), N(x, dy))$ is for every fixed $x \in \mathbb{R}^d$ a Lévy triplet. The function $q : \mathbb{R}^d \times \mathbb{R}^d \to \mathbb{C}$ is called the **symbol of the process**. The processes which admit a symbol behave locally like a Lévy process, and their infinitesimal generators resemble the generators of Lévy processes *with variable, i.e. x-dependent, coefficients*—this justifies the name **Lévy-type processes**. This guides us to the main topics of the present tract:

Characterization: For which Markov processes does the limit ($\star$) exist?

Construction: Is there a Lévy-type process with a given symbol $q(x, \xi)$? Is there a 1-to-1 correspondence between symbols and processes?

Sample paths: Can we use the symbol $q(x, \xi)$ in order to describe the sample path behaviour of the process?

Approximation: Is it possible to use $q(x, \xi)$ to approximate and to simulate the process?

Let us put this point of view into perspective by considering first of all some d-dimensional Lévy process $(X_t)_{t \geq 0}$. Being a (strong) Markov process, $(X_t)_{t \geq 0}$ can be described by the transition function $p_t(x, dy) = \mathbb{P}^x(X_t \in dy) = \mathbb{P}(X_t + x \in dy)$ which, in turn, is uniquely characterized by the characteristic function

$$\mathbb{E}^x e^{i\xi \cdot X_t} = \int_{\mathbb{R}^d} e^{i\xi \cdot y} \, p_t(x, dy) = e^{i\xi \cdot x} e^{-t\psi(\xi)} \tag{1}$$

and the characteristic exponent ψ. Thus,

$$1 - \int_{\mathbb{R}^d} e^{i\xi \cdot (y-x)} \, p_t(x, dy) = t\psi(\xi) + o(t) \quad \text{as } t \to 0 \tag{2}$$

and, with some effort, we can derive from this the Lévy–Khintchine representation of the exponent ψ

$$\psi(\xi) = -il \cdot \xi + \frac{1}{2}\xi \cdot Q\xi + \int_{\mathbb{R}^d \setminus \{0\}} \left(1 - e^{iy \cdot \xi} + i\xi \cdot y \mathbb{1}_{[-1,1]}(|y|)\right) \nu(dy) \tag{3}$$

where (l, Q, ν) is the Lévy triplet. The key observation is that the family of measures $t^{-1} p_t(x, B + x) = t^{-1} \mathbb{P}(X_t \in B)$ converges[1] to the Lévy measure $\nu(B)$ as $t \to 0$ for all Borel sets $B \subset \mathbb{R}^d \setminus \{0\}$ satisfying $\nu(\bar{B} \setminus B^\circ) = 0$ and $0 \notin \bar{B}$.

In our calculation there is only one place where we used Lévy processes: The second equality sign in (1) which is the consequence of the translation invariance (spatial homogeneity) and infinite divisibility of a Lévy process. If we do away with it, and if we only assume that $(X_t)_{t \geq 0}$ is strong Markov with transition function $(p_t(x, dy))_{t \geq 0, x \in \mathbb{R}^d}$, we still have that

$$\lambda_t(x, \xi) := \mathbb{E}^x e^{i\xi \cdot (X_t - x)} = \int_{\mathbb{R}^d} e^{i\xi \cdot (y-x)} \, p_t(x, dy). \tag{1'}$$

Assume we *knew* that $t^{-1} p_t(x, B + x)$ has, as $t \to 0$, for every $x \in \mathbb{R}^d$ and suitable Borel sets $B \subset \mathbb{R}^d \setminus \{0\}$, a limit $N(x, B)$ which is a kernel on $\mathbb{R}^d \times \mathcal{B}(\mathbb{R}^d \setminus \{0\})$. Then we would get, as in (2),

$$1 - \lambda_t(x, \xi) = t q(x, \xi) + o(t) \quad \text{as } t \to 0. \tag{2'}$$

But, what can be said about $q(x, \xi)$?

With some elementary harmonic analysis this can be worked out. Since $\xi \mapsto \lambda_t(x, \xi)$ is a characteristic function, it is continuous and positive definite (see p. 41 for the definition); and since $1 \geq \lambda_t(x, 0) \geq |\lambda_t(x, \xi)|$, we get from this that

$$\sum_{j,k=1}^{n} \left(\lambda_t(x, \xi_j - \xi_k) - 1\right) \mu_j \bar{\mu}_k \geq 0 \tag{4}$$

---

[1] The classical proofs use here Lévy's continuity theorem or the Helly–Bray theorem.

for all $n \geqslant 0$, $\xi_1, \ldots, \xi_n \in \mathbb{R}^d$ and $\mu_1, \ldots, \mu_n \in \mathbb{C}$ with $\sum_{j=1}^{n} \mu_j = 0$. This means that $\xi \mapsto \lambda_t(x, \xi) - 1$ is continuous and *conditionally positive definite*[2] (because of the *condition* $\sum_j \mu_j = 0$, cf. p. 42). The important point is now that every continuous and conditionally positive definite function enjoys a Lévy–Khintchine representation.

Obviously, inequality (4) remains valid if we divide by t and let $t \to 0$, so

$$\lim_{t \to 0} \frac{1 - \mathbb{E}^x \, e^{i\xi \cdot (X_t - x)}}{t} = q(x, \xi) \quad \forall x, \xi \in \mathbb{R}^d \qquad (\star)$$

defines a conditionally positive definite function $\xi \mapsto -q(x, \xi)$. If it is also continuous, then it has for every fixed $x \in \mathbb{R}^d$ a Lévy–Khintchine representation, and each $q(x, \cdot)$ is the characteristic exponent of a Lévy process. We will call the function $q(x, \xi)$ the **symbol** of the process $(X_t)_{t \geqslant 0}$. In this sense it is correct to say that *behind every decent Markov process* $(X_t)_{t \geqslant 0}$ *there is a family of Lévy processes* $(L_t^{(x)})_{t \geqslant 0, x \in \mathbb{R}^d}$ whose characteristic exponents are given by $(\star)$, and we are back at the point where we started our discussion.

Sufficient conditions for the limit $(\star)$ to exist are best described by a list of Lévy-type processes: Lévy processes, of course, (cf. Sect. 2.1), any Feller process whose infinitesimal generator has a sufficiently rich domain (Sects. 2.3 and 2.4), many Lévy-driven stochastic differential equations (Sect. 3.2), or temporally homogeneous Markovian jump-diffusion semimartingales (Sect. 2.5) provided that their extended generator contains sufficiently many functions. As it turns out, the symbol $q(x, \xi)$ encodes, via its Lévy–Khintchine representation and the (necessarily) x-dependent Lévy triplet, the semimartingale characteristics of the stochastic process; moreover, it yields a simple representation of the infinitesimal generator as a pseudo-differential operator.

We are not aware of necessary conditions such that $(\star)$ defines a negative definite symbol, although the class of temporally homogeneous Markovian jump-diffusion semimartingales looks pretty much to be the largest class of decent strong Markov processes admitting a symbol.

Most of our results hold for any *decent* strong Markov process admitting a symbol $q(x, \xi)$, but we restrict our attention to Feller processes where *decency* comes from the natural assumption that the compactly supported smooth functions $C_c^{\infty}(\mathbb{R}^d)$ are contained in the domain of the infinitesimal generator. The key results in this direction are our short proof of the Courrège–von Waldenfels theorem (Theorem 2.21) and the probabilistic formula for the symbol, Theorems 2.36 and 2.44.

Let us briefly explain how the material is organized. The *Primer on Feller Semigroups*, Chap. 1, is included in order to make the material accessible for the novice and also to serve as a reference. For the more experienced reader, the ideal point of departure should be Sect. 2.1 on Lévy processes which leads directly

---

[2] Also known as *negative definite*, and we will prefer this notion in the sequel, cf. Sect. 2.2

to the characterization of Feller processes. Among the central results of Chap. 2 is the characterization of the generators as pseudo-differential operators and the fact that Feller processes are semimartingales: In both cases the symbol $q(x, \xi)$ and its x-dependent Lévy triplet are instrumental. Chapter 3 is devoted to various construction methods for Feller processes. This is probably the most technical part of our treatise since techniques from different areas of mathematics come to bearing; it is already difficult to describe the results, to present complete proofs in this essay is near impossible. Nevertheless we tried to describe the ideas how things fit together, and we hope that the interested reader follows up on the references provided. Perturbations and time-changes for Feller processes are briefly discussed in Chap. 4. In particular, we obtain conditions such that the Feller property is preserved under these transformations. From Chap. 5 onwards, things become more probabilistic: Now we show how to use the symbol $q(x, \xi)$ in order to describe the behaviour of the sample paths of a Feller process. For Lévy processes this approach has a long tradition starting with the papers by Blumenthal–Getoor [31,32] in the early sixties; a survey is given in Fristedt [112]. The principal tool for these investigations is probability estimates for the running maximum of the process in terms of the symbol (Sect. 5.1). Using these estimates we can define Blumenthal–Getoor–Pruitt indices for Feller processes which, in turn, allow us to find bounds for the Hausdorff dimension of the sample paths, describe the (polynomial) short- and long-time asymptotics of the paths, their p-variation, their Besov regularity, etc. Returning to the level of transition semigroups we then investigate global properties (in the sense of Fukushima et al.) in Chap. 6. We focus on functional inequalities and their stability under subordination and on coupling methods; the latter are explained in detail for Lévy- and linear Ornstein–Uhlenbeck processes. The classical topics of transience and recurrence are discussed from the perspective of Meyn and Tweedie, with an emphasis on stable-like processes. In the final Chap. 7 we show how the viewpoint of a Feller process being locally Lévy can be used to approximate the sample paths of Feller processes. This allows us, for the first time, to simulate Feller processes with unbounded coefficients. We close this treatise with a list of open problems which we think are important for the further development of the subject.

We cannot cover all aspects of Lévy-type processes in this survey. Notable omissions are probabilistic potential theory, the general theory of Dirichlet forms, heat kernel estimates and processes on domains. Our choice of material was, of course, influenced by personal liking, by our own research interests and by the desire to have a clear focus. Some topics, e.g. probabilistic potential theory and Dirichlet forms, are more naturally set in the wider framework of general Markov processes and there are, indeed, monographs which we think are hard to match: In potential theory there are Chung's books [67,68] (for Feller processes), Sharpe [298] (for general Markov processes), Port–Stone [238] and Bertoin [27] (for Brownian motion and Lévy processes) and for Dirichlet forms there is Fukushima et al. [115] (for the symmetric case), and Ma–Röckner [210] (for the non-symmetric case). Heat kernel estimates are usually discussed in an L^2-framework, cf. Chen [60] for an excellent survey, or for various perturbations of stable Lévy processes (also on domains), e.g. as in Chen–Kim–Song [65, 66]. An interesting geometric approach has recently been proposed by [168]. Finally, processes on domains with general Wentzell boundary conditions [352] for the generator are still a problem: While some progress has been made in the one-dimensional case (cf. Mandl [216], Langer and co-workers [199, 200]), the multidimensional case is wide open, and the best treatment is Taira [313].

A few words on the style of this treatise are in order. Some time ago, we have been invited to contribute a survey paper to the *Lévy Matters* subseries of the Springer Lecture Notes in Mathematics, updating the earlier paper *Lévy-type processes and pseudo-differential operators* by N. Jacob and one of the present authors. Soon, however, it became clear that the developments in the past decade have been quite substantial while, on the other hand, much of the material is scattered throughout the literature and that a comprehensive treatise on Feller processes is missing. With this essay we try to fill this gap, providing a reliable source for reference (especially for those elusive *folklore* results), making a technically demanding area easily accessible to future generations of researchers and, at the same time, giving a snapshot of the state-of-the-art of the subject. Just as one would expect in a survey, we do not always (want to) give detailed proofs, but we provide precise references whenever we omit proofs or give only a rough outline of the argument (sometimes also sailing under the nickname "proof"). On the other hand, quite a few theorems are new or contain substantial improvements of known results, and in all those cases we do include full proofs or describe the necessary changes to the literature. We hope that the exposition is useful for and accessible to anyone with a working knowledge of Lévy- or continuous-time Markov processes and some basic functional analysis.

It is a pleasure to acknowledge the support of quite a few people. Niels Jacob has our best thanks, his ideas run through the whole text, and we shall think it a success if it pleases him.

Without the named (and, as we fear, often unnamed) contributions of our co-authors and fellow scientists such a survey would not have been possible; we are grateful that we can present and build on their results. Anite Behme, Xiaoping Chen, Katharina Fischer, Julian Hollender, Victorya Knopova, Franziska Kühn, Huaiqian Li, Felix Lindner, Michael Schwarzenberger and Nenghui Zhu read substantial portions of various β-versions of this survey, pointed out many mistakes and inconsistencies, and helped us to improve the text; the examples involving affine processes were drafted by Michael Schwarzenberger.

Special thanks go to Claudia Klüppelberg and the editors of the *Lévy Matters* series for the invitation to write and their constant encouragement to finish this piece.

Finally, we thank our friends and families who—we are pretty sure of it—are more than happy that this work has come to an end.

Dresden, Germany   Björn Böttcher and René Schilling
Fuzhou, China   Jian Wang
April 2013

Contents

Summary of Notation

This list is intended to aid cross-referencing, so notation that is specific to a single section is generally not listed. Some symbols are used locally, without ambiguity, in senses other than those given below; numbers following an entry are page numbers.

Unless otherwise stated, functions are real-valued, and binary operations between functions such as $f \pm g$, $f \cdot g$, $f \wedge g$, $f \vee g$, comparisons $f \leq g$, $f < g$ or limiting relations $f_j \xrightarrow{j \to \infty} f$, $\lim_j f_j$, $\underline{\lim}_j f_j$, $\overline{\lim}_j f_j$, $\sup_j f_j$ or $\inf_j f_j$ are understood pointwise. "Positive" and "negative" always means "≥ 0" and "≤ 0".

General Notation: Analysis

$a \vee b, a \wedge b$	$\max(a, b), \min(a, b)$
a^+, a^-	$\max(a, 0), -\min(a, 0)$
$f \asymp g$	$\exists c \; \forall t \; : \; \frac{1}{c} g(t) \leq f(t) \leq c g(t)$
$\lfloor x \rfloor$	Largest integer $n \leq x$
$\lvert x \rvert$	Euclidean vector and matrix norm
$x \cdot y, \langle x, y \rangle$	Scalar product in $\mathbb{R}^d$, $\sum_{j=1}^{d} x_j y_j$
$\mathbb{1}_A$	$\mathbb{1}_A(x) = \begin{cases} 1, & x \in A \\ 0, & x \notin A \end{cases}$
$e_\xi(x)$	$e^{i \xi \cdot x}, x, \xi \in \mathbb{R}^d$
δ_x	Point mass at x
$\operatorname{supp} f$	Support, $\overline{\{f \neq 0\}}$
∇	Gradient $\left(\frac{\partial}{\partial x_1}, \dots \frac{\partial}{\partial x_d} \right)^\top$
∇^α	$\frac{\partial^{\alpha_1 + \dots + \alpha_d}}{\partial x_1^{\alpha_1} \dots \partial x_d^{\alpha_d}}$
$\mathcal{F}u, \hat{u}$	Fourier transform, 31
$\mathcal{F}^{-1}u, \check{u}$	Inverse Fourier transform, 31
bp-lim	Bounded pointwise (bp) convergence, 1
$(A, \mathcal{D}(A))$	Generator, 18, 23
$\hat{A}, \hat{A}_b$	Full generator, 25

$\psi(D)$	Fourier multiplier, 37, 51
$q(x, D)$	Pseudo-differential operator, 51

General Notation: Probability

$\sim$	"Is distributed as"
a.s.	Almost surely
$(X_t, \mathscr{F}_t)_{t \geq 0}$	Adapted process
(l, Q, v)	Lévy triplet, 33
χ	Truncation function, 33
τ_r^x	$\inf\{t > 0 : X_t \in \overline{\mathbb{B}}^c(x, r)\}$

Sets and σ-Algebras

A^c	Complement of the set A
A°	Open interior of the set A
$\overline{A}$	Closure of the set A
$\mathbb{B}(x, r)$	Open ball, centre x, radius r
$\overline{\mathbb{B}}(x, r)$	Closed ball, centre x, radius r
$\mathscr{B}(E)$	Borel sets of E
$\mathscr{F}_t^X$	$\sigma(X_s : s \leq t)$

Spaces of Measures and Functions

$\|u\|_{(k)}$	$\sum_{0 \leq |\alpha| \leq k} \|\nabla^\alpha u\|_\infty$
$\|\mu\|_{TV}$	Total variation norm
$B(E)$	Borel functions on E
$B_b(E)$	——, bounded
$C(E)$	Continuous functions on E
$C_b(E)$	——, bounded
$C_\infty(E)$	——, $\lim_{|x| \to \infty} u(x) = 0$
$C_c(E)$	——, compact support
$C^k(E)$	k times continuously diff'ble functions on E
$C_b^k(E)$	——, bounded (with their derivatives)
$C_\infty^k(E)$	——, 0 at infinity (with their derivatives)
$C_c^k(E)$	——, compact support
$L^p(E, \mu), L^p(\mu), L^p(E)$	L^p space w.r.t. the measure space $(E, \mathcal{F}, \mu)$
$\mathcal{M}(E)$	(Signed) Radon measures on E
$\mathcal{M}_b(E)$	——, with finite mass
$\mathcal{M}^+(E)$	——, positive
$\mathcal{M}^1(E)$	Probability measures on E
$\mathcal{S}(\mathbb{R}^d)$	Schwartz space of rapidly decreasing smooth functions

Chapter 1
A Primer on Feller Semigroups and Feller Processes

Throughout this chapter, E denotes a locally compact and separable space; later on we will restrict ourselves to the Euclidean space $\mathbb{R}^d$ and its subsets. By $C_\infty(E)$ we denote the space of continuous functions $u : E \to \mathbb{R}$ which **vanish at infinity**, i.e.

$$\forall \epsilon > 0 \quad \exists K \subset E \text{ compact} \quad \forall x \in K^c : |u(x)| \leqslant \epsilon. \qquad (1.1)$$

If $E = \mathbb{R}^d$, then (1.1) is the same as $\lim_{|x|\to\infty} u(x) = 0$; if $E = \mathbb{B}(z, r)$ is an open ball in $\mathbb{R}^d$, then (1.1) entails that (there is an extension of u such that) $u(x) = 0$ on the boundary $|x - z| = r$, and if E is a compact set, then $C_\infty(E) = C(E)$, i.e. the space of all continuous functions on E. Observe that

$$(C_\infty(E), \| \cdot \|_\infty), \quad \|u\|_\infty := \sup_{x \in E} |u(x)|, \qquad (1.2)$$

is a Banach space, and the space of compactly supported continuous functions $C_c(E)$ is a dense subspace. If E is not compact, we can use the one-point compactification E_∂ by adding the point ∂. Since the complements of compact sets $K \subset E$ form a neighbourhood base of the point ∂ at infinity, we can identify $C_\infty(E)$ with $\{u \in C(E_\partial) : u(\partial) = 0\}$.

The topological dual $C_\infty^*(E)$ of $C_\infty(E)$ consists of the bounded signed Radon measures $\mathcal{M}_b(E)$, i.e. the signed Borel measures μ on E with finite total mass $|\mu|(E) < \infty$. A sequence $(u_n)_{n \geqslant 1} \subset C_\infty(E)$ **converges weakly** to $u \in C_\infty(E)$, if $\lim_{n\to\infty} \int u_n \, d\mu = \int u \, d\mu$ for all $\mu \in \mathcal{M}_b(E)$. Weak convergence is the same as **bp (bounded pointwise) convergence**

$$\text{bp-} \lim_{n\to\infty} u_n = u \iff \sup_{n \geqslant 1} \|u_n\|_\infty < \infty \text{ and } \lim_{n\to\infty} u_n(x) = u(x) \quad \forall x \in E, \tag{1.3}$$

cf. Dunford–Schwartz [93, Corollary IV.6.4, p. 265]. Note that this only holds for sequences, cf. Ethier–Kurtz [100, Appendix 3, pp. 495–496].

B. Böttcher et al., *Lévy Matters III*, Lecture Notes in Mathematics 2099, DOI 10.1007/978-3-319-02684-8_1,
© Springer International Publishing Switzerland 2013

The norm topology on the Banach space $\mathcal{M}_b(E)$ is given by the total variation norm $\|\mu\|_{TV} := \mu^+(E) + \mu^-(E)$ where $\mu = \mu^+ - \mu^-$ is the Hahn–Jordan decomposition. More often, we use on $\mathcal{M}_b(E)$ the **weak-*** or **vague topology**, i.e.

$$\mu_n \xrightarrow[n \to \infty]{\text{vaguely}} \mu \iff \lim_{n \to \infty} \int u \, d\mu_n = \int u \, d\mu \quad \forall u \in C_\infty(E). \qquad (1.4)$$

If, in addition, $\lim_{n \to \infty} \mu_n^\pm(E) = \mu^\pm(E)$, then one speaks of **weak convergence (of measures)**, i.e.

$$\mu_n \xrightarrow[n \to \infty]{\text{weakly}} \mu \iff \lim_{n \to \infty} \int u \, d\mu_n = \int u \, d\mu \quad \forall u \in C_b(E). \qquad (1.5)$$

Mind that this is *not weak convergence in the topological sense*.

1.1   Feller Semigroups

There is no standard usage of the term **Feller semigroup** in the literature and *every author has his or her own definition of "Feller semigroup"* (Rogers and Williams [255, p. 241]). Therefore we take the opportunity to develop some of the core material in a consistent way.

Definition 1.1. Let $(T_t)_{t \geq 0}$ be a family of linear operators defined on the bounded Borel measurable functions $B_b(E)$. If

$$T_0 = \mathrm{id} \quad \text{and} \quad T_t T_s u = T_s T_t u = T_{t+s} u \quad \forall u \in B_b(E), \ s, t \geq 0$$

then $(T_t)_{t \geq 0}$ is said to be a **(one-parameter operator) semigroup**.

A **sub-Markov semigroup** is an operator semigroup $(T_t)_{t \geq 0}$ which is **positivity preserving**

$$T_t u \geq 0 \quad \forall u \in B_b(E), \ u \geq 0 \qquad (1.6)$$

and has the **sub-Markov property**

$$T_t u \leq 1 \quad \forall u \in B_b(E), \ u \leq 1. \qquad (1.7)$$

A **Markov semigroup** is a sub-Markov semigroup which is **conservative**, i.e. $T_t 1 = 1$.

Note that a sub-Markov semigroup is automatically **monotone**

$$T_t v \leq T_t w \quad \forall v, w \in B_b(E), \ v \leq w \qquad (1.8)$$

(take $u = w - v$ in (1.6)), it satisfies **Jensen's inequality**

$$\phi(T_t u) \leqslant T_t \phi(u) \quad \forall u \in B_b(E) \text{ and any convex } \phi : \mathbb{R} \to \mathbb{R}, \; \phi(0) = 0 \qquad (1.9)$$

(observe that a convex function with $\phi(0) = 0$ is the upper envelope of affine-linear functions $\ell(x) = ax + b$ where $a \in \mathbb{R}$ and $b \leqslant 0$ such that $\ell(x) \leqslant u(x)$ for all x, and use $\ell(T_t u) \leqslant T_t \ell(u) \leqslant T_t \phi(u)$, see e.g. [283, Theorem 12.14, p. 116] for the standard proof) and it is **contractive**

$$\|T_t u\|_\infty \leqslant \|u\|_\infty \quad \forall u \in B_b(E) \qquad (1.10)$$

(use $|T_t u| \leqslant T_t |u| \leqslant T_t \|u\|_\infty \leqslant \|u\|_\infty$).

Definition 1.2. A **Feller semigroup** is a sub-Markov semigroup $(T_t)_{t \geqslant 0}$ which satisfies the **Feller property**

$$T_t u \in C_\infty(E) \quad \forall u \in C_\infty(E), \; t > 0 \qquad (1.11)$$

and which is **strongly continuous** in the Banach space $C_\infty(E)$

$$\lim_{t \to 0} \|T_t u - u\|_\infty = 0 \quad \forall u \in C_\infty(E). \qquad (1.12)$$

One of the reasons to consider semigroups acting on spaces of continuous functions is the fact that such semigroups are integral operators with *pointwise everywhere* defined measure kernels, cf. the Riesz representation theorem, Theorem 1.5 below. This is particularly attractive for the study of stochastic processes where these kernels will serve as transition functions, cf. Sect. 1.2.

Example 1.3. Throughout the text we will use the following standard examples for Feller semigroups. For simplicity we consider only $E \subset \mathbb{R}^d$ and $u \in B_b(E)$.

a) (*Shift semigroup*) Let $\ell \in \mathbb{R}^d$. The shift semigroup is $T_t u(x) := u(x + t\ell)$, $t \geqslant 0$.

b) (*Poisson semigroup*) Let $\ell \in \mathbb{R}^d$ and $\lambda > 0$. The Poisson semigroup is defined as $T_t u(x) = \sum_{j=0}^\infty u(x + j\ell) \frac{(\lambda t)^j}{j!} e^{-t\lambda}$.

c) (*Heat/Brownian semigroup*) Let $g_t(x) = (2\pi t)^{-d/2} e^{-|x|^2/2t}$ be the heat kernel or normal distribution (mean zero, variance t) on $\mathbb{R}^d$. The heat or Brownian semigroup is $T_t u(x) = \int_{\mathbb{R}^d} u(y) g_t(y - x) \, dy$.

d) (*Symmetric stable semigroups*) Let $g_{t,\alpha}(x)$, $\alpha \in (0, 2]$ be the symmetric stable probability density. It is implicitly defined through the characteristic function (inverse Fourier transform)[1] $\mathcal{F}^{-1}[g_{t,\alpha}](\xi) := \int_{\mathbb{R}^d} e^{ix \cdot \xi} g_{t,\alpha}(x) \, dx = e^{-t|\xi|^\alpha}$. The symmetric α-stable semigroup is $T_t u(x) := \int_{\mathbb{R}^d} u(y) g_{t,\alpha}(y - x) \, dy$.

---

[1] See the beginning of Chap. 2 for the conventions for the Fourier transform and characteristic functions.

Only for $\alpha = 1$ and 2 the densities $g_{t,\alpha}$ are explicitly known: $\alpha = 1$ yields the Cauchy density $g_{t,1}(x) = t\Gamma\left(\frac{d+1}{2}\right)\pi^{-\frac{d+1}{2}}\left(t^2 + |x|^2\right)^{-\frac{d+1}{2}}$, and $\alpha = 2$ gives the heat semigroup $g_{t,2}(x) = g_{2t}(x) = (4\pi t)^{-\frac{d}{2}} e^{-|x|^2/4t}$ with twice the normal speed.

e) (*Convolution/Lévy semigroups*) Let $(\mu_t)_{t\geq 0}$ be a family of infinitely divisible probability measures on $\mathbb{R}^d$, i.e. for every $n \geq 2$ we can write μ_t as an n-fold convolution of the measures $\mu_{t/n}$. Moreover, assume that $t \mapsto \mu_t$ is continuous in the vague topology.

Then $T_t u(x) := \int_{\mathbb{R}^d} u(x+y)\, \mu_t(dy)$ is a semigroup of convolution operators. We will discuss the structure of these semigroups in Sect. 2.1 below.

Note that all previously defined semigroups fall in this category. Because of the structure of these semigroups, the Feller property is easily seen using the dominated convergence theorem. Strong continuity follows from the vague continuity of the family $(\mu_t)_{t\geq 0}$, see also Berg–Forst [24, Chap. II.§12, pp. 85–97] or [284, Proposition 7.3, pp. 87–89] for a probabilistic proof for the heat semigroup which carries over to general convolution semigroups.

f) (*Ornstein–Uhlenbeck semigroup*) Let $(\mu_t)_{t\geq 0}$ be a family of infinitely divisible probability measures on $\mathbb{R}^d$ such that $t \mapsto \mu_t$ is continuous in the vague topology, and $B \in \mathbb{R}^{d\times d}$.

Then, $T_t u(x) := \int_{\mathbb{R}^d} u(e^{tB}x + y)\, \mu_t(dy)$ defines the so-called Ornstein–Uhlenbeck semigroup. Note that this is a special case of the Mehler semigroup, see e.g. Bogachev et al. [36]. The strong continuity and the Feller property of the Ornstein–Uhlenbeck semigroup was proved in Sato–Yamazato [268, Theorem 3.1].

Further examples are *generalized, Lévy-driven Ornstein–Uhlenbeck semigroups* which have been studied by Behme–Lindner [18], see also Examples 1.17(f) and 3.34(b) below for details.

g) (*One-sided stable semigroups*) On $E = [0, \infty)$ one defines the density $p_{t,\alpha}(x)$, $t, x \geq 0$, $0 < \alpha < 1$ through the Laplace transform $\int_0^\infty e^{-sx} p_{t,\alpha}(x)\, dx = e^{-ts^\alpha}$. Then $T_t u(x) := \int_0^\infty u(x + y) p_{t,\alpha}(y)\, dy$ is the one-sided α-stable semigroup.

The Lévy density $p_{t,1/2}(x) = (4\pi)^{-1/2} t\, x^{-3/2} e^{-t^2/4x}\, \mathbb{1}_{(0,\infty)}(x)$ is the only density in this family for which a closed-form expression is known.

h) (*Diffusion semigroups*) Consider a second order partial differential operator in divergence form $L = \frac{1}{2}\nabla \cdot (Q(\cdot)\nabla)$ where $Q : \mathbb{R}^d \to \mathbb{R}^d \times \mathbb{R}^d$ is a measurable, symmetric matrix-valued function which is uniformly elliptic, i.e. there exist constants $0 < c \leq C < \infty$ such that

$$c|\xi|^2 \leq \langle Q(x)\xi, \xi\rangle \leq C|\xi|^2 \quad \forall \xi \in \mathbb{R}^d$$

and (for simplicity) $Q \in C_b^\infty(\mathbb{R}^d)$. It is well known that the initial value problem $\frac{d}{dt}u(t, x) = Lu(t, x)$, $u(0, x) = \phi(x)$ admits a fundamental solution $p(t, x, y)$ satisfying $p \in \bigcup_{n=1}^\infty C_b^\infty([1/n, n]\times\mathbb{R}^d\times\mathbb{R}^d; (0, \infty))$. The fundamental solution leads to a Feller semigroup $u(t, x) = T_t\phi(x) = \int_{\mathbb{R}^d} \phi(y)p(t, x, y)\, dy$, cf. Itô [151, Chap. 1] or Stroock [310]. In general, the explicit expression of $p(t, x, y)$

is not known; however, we have Aronson's estimates which allow us to compare $p(t, x, y)$ from above and below with the well-known fundamental solution of the heat equation, i.e. where the differential operator is $\frac{1}{4}q_j \Delta$, $q_j > 0$, $j = 1, 2$:

$$(q_1 \pi t)^{-d/2} \exp\left(-\frac{|x - y|^2}{q_1 t}\right) \leqslant p(t, x, y) \leqslant (q_2 \pi t)^{-d/2} \exp\left(-\frac{|x - y|^2}{q_2 t}\right)$$

with $q_j = q_j(c, C, d)$. This beautiful result is originally due to Aronson [6], see also Stroock [310].

i) (*Affine semigroups*) Consider on $E = \mathbb{R}_+^m \times \mathbb{R}^{d-m}$, $d \geqslant m \geqslant 0$, the semigroup $(T_t)_{t \geqslant 0}$ given by $T_t u(x) = \int_E u(y) p_t(x, dy)$ (with a suitable transition kernel $p_t(x, dy)$). Then $(T_t)_{t \geqslant 0}$ is called **affine**, if for every $t \in [0, \infty)$ the characteristic function (inverse Fourier transform) of the measure $p_t(x, \cdot)$ has exponential-affine dependence on x. In general, $p_t(x, \cdot)$ is not known explicitly; however, affine semigroups are characterized by the existence of functions

$$\phi : [0, \infty) \times i\mathbb{R}^d \to \mathbb{C}_- \quad \text{and} \quad \psi : [0, \infty) \times i\mathbb{R}^d \to \mathbb{C}_-^m \times i\mathbb{R}^{d-m}$$

(as usual, we write $\mathbb{C}_- = \{z \in \mathbb{C} : \operatorname{Re} z \leqslant 0\}$) such that for every $x \in E$ and for all $(t, \xi) \in [0, \infty) \times \mathbb{R}^d$

$$T_t e_\xi(x) = \int e^{iy \cdot \xi} p_t(x, dy) = e^{\phi(t, i\xi) + \sum_{j=1}^d x_j \psi_j(t, i\xi)} = e^{\phi(t, i\xi) + x \cdot \psi(t, i\xi)}$$

holds; in this generality, affine semigroups have been considered for the first time by Duffie–Filipović–Schachermayer [91, Sect. 2].

If the measures $p_s(x, \cdot)$ converge weakly (in the sense of measures) to $p_t(x, \cdot)$ as $s \to t$ for all $(t, x) \in [0, \infty) \times E$ or, equivalently, if the functions $\phi(t, i\xi)$ and $\psi(t, i\xi)$ are continuous in $t \in [0, \infty)$, for every $\xi \in \mathbb{R}^d$, then $(T_t)_{t \geqslant 0}$ is a Feller semigroup, cf. Keller-Ressel [175, Sect. 1.3, Theorem 1.1, p. 16] or Keller-Ressel–Schachermayer–Teichmann [176, Sect. 3, Theorem 3.5]. □

The Role of Strong Continuity. Using the linearity and contractivity (1.10) of a Feller semigroup, it is not hard to see that (1.12) is equivalent to

$$\lim_{s \to t} \|T_s u - T_t u\|_\infty = 0 \quad \forall u \in C_\infty(E), \ t \geqslant 0. \tag{1.12'}$$

In fact, we can even replace (1.12) by the notion of pointwise convergence.

Lemma 1.4. *Let $(T_t)_{t \geqslant 0}$ be a sub-Markov semigroup which satisfies the Feller property. Then each of the following conditions is equivalent to the strong continuity (1.12).*

$$\lim_{t \to 0} T_t u(x) = u(x) \quad \forall u \in C_\infty(E), \ x \in E; \tag{1.13}$$

$$[0, \infty) \times E \ni (t, x) \mapsto T_t u(x) \quad \text{is continuous for each } u \in C_\infty(E); \tag{1.14}$$

$$[0, \infty) \times E \times C_\infty(E) \ni (t, x, u) \mapsto T_t u(x) \quad \text{is continuous.} \tag{1.15}$$

Proof. It is enough to prove that $(1.12) \Rightarrow (1.15)$ and $(1.13) \Rightarrow (1.12)$. The first implication is a standard $\epsilon/3$-argument: Fix $(t, x, u) \in [0, \infty) \times E \times C_\infty(E)$ and pick any (s, y, v) in some $\epsilon/3$-neighbourhood. Then

$$|T_t u(x) - T_s v(y)| \leqslant |T_t u(x) - T_t u(y)| + |T_t u(y) - T_s u(y)| + |T_s u(y) - T_s v(y)|$$

$$\leqslant |T_t u(x) - T_t u(y)| + \|T_{|t-s|} u - u\|_\infty + \|u - v\|_\infty.$$

The second implication is less trivial. We follow the proof given in Dellacherie–Meyer [84, Théorème XIII.19, pp. 98–99], see also Revuz–Yor [250, Proposition III.2.4, p. 89]. Let $u \in C_\infty(E)$. Clearly, (1.13) entails $\lim_{t \to s} T_t u(x) = T_s u(x)$ for all $s > 0$ and $x \in E$. Therefore the integral

$$U_\alpha u(x) := \int_0^\infty e^{-\alpha s} T_s u(x) \, ds$$

defines a family of linear operators on $C_\infty(E)$ and, by dominated convergence and a simple change of variables, it is easy to see that $\lim_{\alpha \to \infty} (\alpha U_\alpha u(x) - u(x)) = 0$ for every $x \in E$. In fact, $(U_\alpha)_{\alpha > 0}$ is a resolvent satisfying the resolvent equation

$$U_\alpha u - U_\beta u = (\beta - \alpha) U_\alpha U_\beta u \quad \forall \alpha, \beta > 0. \tag{1.16}$$

Thus, the range $\mathcal{R} := U_\alpha C_\infty(E)$ does not depend on $\alpha > 0$. Once again by dominated convergence, we see for any $\rho \in \mathcal{M}_b^+(E)$

$$\int u(x) \, \rho(dx) = \lim_{\alpha \to \infty} \int \alpha U_\alpha u(x) \, \rho(dx).$$

If ρ is orthogonal to $\mathcal{R}$, this equality shows that $\int u \, d\rho = 0$ for all $u \in C_\infty(E)$, hence $\rho = 0$. This proves that $\mathcal{R}$ is dense in $C_\infty(E)$. Now we can use Fubini's theorem to deduce

$$T_t U_\alpha u(x) = e^{\alpha t} \int_t^\infty e^{-\alpha s} T_s u(x) \, ds$$

as well as

$$\|T_t U_\alpha u - U_\alpha u\|_\infty \leqslant (e^{\alpha t} - 1) \|U_\alpha u\|_\infty + t \|u\|_\infty.$$

This shows that $\lim_{t \to 0} \|T_t f - f\|_\infty$ for all $f \in \mathcal{R}$, and a standard density argument proves (1.12). $\qquad \square$

Lemma 1.4 may be a bit surprising as it allows to replace uniform convergence by pointwise convergence. This is a variation of a theme from the theory of operator semigroups which says that for contraction semigroups the notions of continuity in the norm topology and in the weak topology coincide, see e.g. [78, Proposition 1.23, p. 15]. For Feller semigroups, weak continuity means that $t \mapsto \int T_t u(x) \, \rho(dx)$ is continuous for all $\rho \in \mathcal{M}_b(E)$. Since (sequential) weak convergence is the same as bounded pointwise convergence, it is indeed enough to check that the function $t \mapsto \int T_t u(x) \, \delta_y(dx) = T_t u(y)$ is continuous for each $y \in E$.

Feller Semigroups Defined on $C_\infty(E)$. Sometimes a strongly continuous, positivity preserving, conservative semigroup $(T_t)_{t \geqslant 0}$ with $T_t : C_\infty(E) \to C_\infty(E)$ is called a Feller semigroup—although it is only defined on $C_\infty(E)$. Using a variant of the Riesz representation theorem, cf. Rudin [258, Theorem 6.19, p. 130], we can extend $(T_t)_{t \geqslant 0}$ onto $B_b(E)$.

Theorem 1.5 (Riesz). *Let $T_t : C_\infty(E) \to C_\infty(E)$, $t > 0$, be a family of positivity preserving linear operators. Then T_t is an integral operator of the form*

$$T_t u(x) = \int u(y) \, p_t(x, dy) \tag{1.17}$$

where $p_t(x, \cdot)$ is a uniquely defined positive Radon measure.

It is not hard to see that $p_t(x, dy)$ is a sub-probability measure, if $T_t u \leqslant 1$ whenever $u \leqslant 1$. Moreover, $(t, x) \mapsto p_t(x, B)$ is for every $B \in \mathcal{B}(E)$ measurable: If $B = U$ is an open set, this follows immediately from (1.15) since we can approximate $\mathbb{1}_U$ by an increasing sequence of positive C_∞-functions. For general $B \in \mathcal{B}(E)$ we use a Dynkin system or monotone class argument. If $(T_t)_{t \geqslant 0}$ is a semigroup, the kernels $p_t(x, dy)$ satisfy the **Chapman–Kolmogorov equations**

$$p_{s+t}(x, B) = \int p_t(y, B) \, p_s(x, dy) \quad \forall B \in \mathcal{B}(E), \ s, t > 0. \tag{1.18}$$

This shows that every Feller semigroup defined on $C_\infty(E)$ can be uniquely extended to a sub-Markov semigroup in the sense of Definition 1.1, i.e. it becomes a Feller semigroup in the sense of Definition 1.2.

Other Feller Properties. As already mentioned, there is no uniform agreement on what a "Feller semigroup" should be. Usually the question is on which space the semigroup should be defined. Let us review some common alternative definitions and give them distinguishing names.

Definition 1.6. A sub-Markov semigroup $(T_t)_{t \geqslant 0}$ is called a C_b**-Feller semigroup** if it enjoys the C_b**-Feller property**, i.e.

$$T_t u \in C_b(E) \quad \forall u \in C_b(E), \ t > 0 \tag{1.19}$$

and if $t \mapsto T_t u$ is continuous in the topology of locally uniform convergence in the space $C_b(E)$.

If E is compact, the notions of Feller- and C_b-Feller semigroups coincide. The correct choice of topology on $C_b(E)$ is a major issue. Although $(C_b(E), \|\cdot\|_\infty)$ is a perfectly good Banach space, the requirement of strong continuity is so strong that only few semigroups enjoy this property.

Example 1.7. a) Let $T_t u(x) = u(x + t\ell)$ be the shift-semigroup on $\mathbb{R}^d$ (Example 1.3(a)). Since $T_t u(x) - u(x) = u(x + t\ell) - u(x)$, strong continuity of the shift semigroup entails that u is uniformly continuous. This means that $(T_t)_{t \geq 0}$ is not strongly continuous on $(C_b(\mathbb{R}^d), \|\cdot\|_\infty)$.

b) Let $T_t u(x)$ be the Poisson semigroup on $\mathbb{R}^d$ (Example 1.3(b)). Then

$$|T_t u(x) - u(x)| = \left| \sum_{j=1}^\infty (u(x + j\ell) - u(x)) \frac{(\lambda t)^j}{j!} e^{-\lambda t} \right| \leq 2\|u\|_\infty (1 - e^{-\lambda t})$$

shows that $(T_t)_{t \geq 0}$ is strongly continuous on $C_b(\mathbb{R}^d)$ (and even on $B_b(\mathbb{R}^d)$).

c) Denote by $(A, \mathcal{D}(A))$ the generator of the Feller semigroup $(T_t)_{t \geq 0}$, cf. Sect. 1.4 below. If A is a bounded operator with respect to $\|\cdot\|_\infty$, then $(T_t)_{t \geq 0}$ is strongly continuous on $C_b(\mathbb{R}^d)$, and even on $B_b(\mathbb{R}^d)$. This follows from

$$|T_t u(x) - u(x)| = \left| \int_0^t A T_s u(x)\, ds \right| \leq t \|A\| \|u\|_\infty \quad \forall x \in \mathbb{R}^d,\ u \in C_\infty(\mathbb{R}^d)$$

cf. Lemma 1.26, and a standard extension argument for linear operators (the B.L.T. theorem, Reed–Simon [248, Theorem I.7, p. 9]). This shows that a (Feller) semigroup with bounded generator is continuous in the strong operator topology: $\|T_t - 1\| = \sup_{\|u\|_\infty \leq 1} \|T_t u - u\|_\infty \leq t \|A\|$. Conversely, any semigroup which is continuous in the strong operator topology has a bounded generator, cf. Pazy [236, Theorem 1.2, p. 2].

d) The heat semigroup (Example 1.3(c)) is not strongly continuous on $C_b(\mathbb{R})$. To see this, define a function $u \in C_b(\mathbb{R})$ by

$$u(x) := \sum_{n=2}^\infty u_n(x) \quad \text{and} \quad u_n(x) := \begin{cases} 0, & |x - n| \geq \frac{1}{n}, \\ n(x - n + \frac{1}{n}), & n - \frac{1}{n} < x \leq n, \\ n(n + \frac{1}{n} - x), & n \leq x < n + \frac{1}{n}. \end{cases}$$

Then we find for $x \in \mathbb{R}$, $t > 0$ and $\delta > 0$

$$|T_t u(x) - u(x)| \geq \left| \int_{|y| \leq \delta} (u(x + y) - u(x)) g_t(y)\, dy \right| - 2\|u\|_\infty \int_{|y| > \delta} g_t(y)\, dy.$$

Pick $x = n$ and $\delta = n^{-1}$. Then

$$\|T_t u - u\|_\infty \geq \int_{\frac{1}{2n} \leq |y| \leq \frac{1}{n}} (u(n) - u(n+y))g_t(y)\,dy - 2\int_{|y| > \frac{1}{n}} g_t(y)\,dy$$

$$\geq \frac{1}{2}\int_{\frac{1}{2n} \leq |y\sqrt{t}| \leq \frac{1}{n}} g_1(y)\,dy - 2\int_{|y\sqrt{t}| > \frac{1}{n}} g_1(y)\,dy.$$

Now we use $t = t_n := (4n^2)^{-1}$ and write $\Phi(x) = \int_{-\infty}^x g_1(y)\,dy$ for the normal cumulative distribution function. Then

$$\varlimsup_{t \to 0} \|T_t u - u\|_\infty \geq \varlimsup_{n \to \infty} \|T_{t_n} u - u\|_\infty \geq \tfrac{5}{2}\Phi(2) - \tfrac{1}{2}\Phi(1) - 2 \geq 0.01.$$

A similar calculation shows that $(T_t)_{t \geq 0}$ is actually strongly continuous for all uniformly continuous functions u.

e) Let $(T_t)_{t \geq 0}$ be an affine semigroup (Example 1.3(i)) on $E = \mathbb{R}_+^m \times \mathbb{R}^{d-m}$ which is a Feller semigroup, i.e. $p_s(x, \cdot)$ converges weakly to $p_t(x, \cdot)$ as $s \to t$ for all $(t, x) \in [0, \infty) \times E$. Then $(T_t)_{t \geq 0}$ is also a C_b-Feller semigroup. According to Theorem 1.9 below, this follows from

$$T_t 1(x) = T_t e_0(x) = e^{\phi(t,0) + x \cdot \psi(t,0)},$$

which shows that $x \mapsto T_t 1(x)$ is continuous and bounded. $\qquad\square$

If we replace uniform convergence by uniform convergence *on compact sets*, the restriction of a Feller semigroup to $C_b(E)$ will be continuous at $t = 0$, cf. [274, Lemma 3.1].

Lemma 1.8. *Let $(T_t)_{t \geq 0}$ be a Feller semigroup. Then $\lim_{t \to 0} T_t u(x) = u(x)$ locally uniformly in x for all $u \in C_b(E)$.*

The following criterion for a Feller semigroup to be a C_b-Feller semigroup is again taken from [274, Sect. 3].

Theorem 1.9. *Let $(T_t)_{t \geq 0}$ be a sub-Markov semigroup. Then*

$$\Big(T_t : C_\infty(E) \to C_\infty(E) \quad and \quad T_t 1 \in C_b(E)\Big) \implies T_t : C_b(E) \to C_b(E).$$

In particular, if $(T_t)_{t \geq 0}$ is a Feller semigroup with $T_t 1 \in C_b(E)$, then it is also a C_b-Feller semigroup.

A necessary and sufficient condition that a C_b-Feller semigroup is a Feller semigroup is given in the next theorem.

Theorem 1.10. *Let $(T_t)_{t \geq 0}$ be a C_b-Feller semigroup and $(p_t(x, dy))_{t > 0}$ the transition kernels, i.e. for any $t > 0$, $x \in E$ and $u \in C_b(E)$, $T_t u(x) = \int u(y)\, p_t(x, dy)$.*

Then, $(T_t)_{t>0}$ is a Feller semigroup if, and only if, for all $t > 0$ and any increasing sequence of bounded sets $B_n \in \mathcal{B}(E)$ with $\bigcup_{n \geq 1} B_n = E$ we have

$$\lim_{|x| \to \infty} p_t(x, B_n) = 0 \quad \forall n \geq 1. \tag{1.20}$$

Proof. Since E is locally compact and separable, E is σ-compact, and there exists a sequence of bounded (even compact) sets B_n increasing towards E. Assume that $(T_t)_{t \geq 0}$ has the C_b-Feller property.

By the definition of $C_\infty(\mathbb{R}^d)$, there is for every $\epsilon > 0$ some $N(\epsilon)$ such that for all $n \geq N(\epsilon)$ we have $|u| \mathbb{1}_{E \setminus B_n} \leq \epsilon$. Thus, for $x \in E$,

$$|T_t u(x)| \leq \int_{B_n} |u(y)| \, p_t(x, dy) + \int_{E \setminus B_n} |u(y)| \, p_t(x, dy)$$

$$\leq \|u\|_\infty p_t(x, B_n) + \epsilon.$$

Hence,

$$\lim_{|x| \to \infty} |T_t u(x)| \leq \|u\|_\infty \lim_{|x| \to \infty} p_t(x, B_n) + \epsilon = \epsilon.$$

Letting $\epsilon \to 0$ yields that $T_t u \in C_\infty(E)$. In order to see strong continuity on $C_\infty(E)$, we remark that $t \mapsto T_t u(x)$ is continuous for all $x \in E$ and $u \in C_\infty(E)$. Thus, by Lemma 1.4, we conclude that $(T_t)_{t \geq 0}$ is strongly continuous on $C_\infty(E)$, hence a Feller semigroup.

On the other hand, for any bounded set $B \in \mathcal{B}(E)$, there is some $u \in C_\infty(E)$ such that $u \geq 0$ and $u|_B \equiv 1$. Therefore,

$$T_t u(x) \geq \int_B u(y) \, p_t(x, dy) = p_t(x, B).$$

Since $(T_t)_{t \geq 0}$ is a Feller semigroup,

$$0 = \lim_{|x| \to \infty} |T_t u(x)| = \lim_{|x| \to \infty} T_t u(x) \geq \lim_{|x| \to \infty} p_t(x, B). \qquad \square$$

The criterion (1.20) ensuring the Feller property in Theorem 1.10 is not easy to check. If $E = \mathbb{R}^d$ we can use the structure of the infinitesimal generator to obtain a simpler condition; we postpone this to Theorem 2.49 in Sect. 2.5.

In potential theory one often requires the following strong Feller property.

Definition 1.11. A sub-Markov semigroup $(T_t)_{t \geq 0}$ is said to be a **strong Feller semigroup** if $T_t : B_b(E) \to C_b(E)$ for all $t > 0$.

Among other things, the strong Feller property ensures that α-excessive functions are lower semicontinuous, see Blumenthal–Getoor [33, (2.16), p. 77]; for a detailed discussion we also refer to Bliedtner–Hansen [30, Sect. V.3, pp. 175–184].

If $(T_t)_{t\geqslant0}$ is a strong Feller semigroup, then the operators $T_t : B_b(E) \to C_b(E)$, $t > 0$, are compact if we equip $B_b(E)$ with the topology of uniform convergence and $C_b(E)$ with the topology of locally uniform convergence, cf. Revuz [249, Proposition 1.5.8, Theorem 1.5.9, p. 37] or [290, Proposition 2.3]. The following result is from Bliedtner–Hansen [30, Proposition 2.10, p. 181].

Lemma 1.12. *Let $(T_t)_{t\geqslant0}$ be a sub-Markov semigroup on $B_b(E)$. Then the following assertions are equivalent.*

a) *$(T_t)_{t\geqslant0}$ is a strong Feller semigroup and for every $t > 0$ and $u \in C_c(E)$ it holds that $\lim_{s\downarrow t} T_s u = T_t u$ locally uniformly.*
b) *For every $u \in B_b(E)$ the function $(t, x) \mapsto T_t u(x)$ is continuous on $(0, \infty) \times E$.*

Example 1.13. a) The shift and the Poisson semigroups [Examples 1.3(a) and 1.3(b)] are not strongly Feller.
b) A convolution semigroup (Example 1.3(e)) is strongly Feller if, and only if, the convolution kernel $\mu_t(dy)$ is absolutely continuous with respect to Lebesgue measure. This result is due to Hawkes [132, Lemma 2.1, p. 338], see also Jacob [157, Lemmas 4.8.19, 4.8.20, pp. 438–439]. □

A strong Feller semigroup need not be C_b-Feller nor Feller. Conversely, the strong Feller property does not follow from the (C_b-)Feller property without further conditions. Typically one has to assume some kind of (uniform) absolute continuity property or some ultracontractivity property. The following results are adapted from [290, Sects. 2.1 and 2.2].

Theorem 1.14. *Let $(T_t)_{t\geqslant0}$ be a C_b-Feller semigroup with kernels $(p_t(x, dy))_{t\geqslant0}$. Then the following assertions are equivalent.*

a) *$(T_t)_{t\geqslant0}$ is a strong Feller semigroup.*
b) *There exists a probability measure $\mu \in \mathcal{M}^+(E)$ such that for every $t > 0$ the family $(p_t(x, dy))_{x\in E}$ is locally absolutely continuous with respect to μ, i.e. for any compact set $K \subset E$ it holds $\lim_{\delta\to0} \sup_{B\in\mathcal{B}(E),\,\mu(B)\leqslant\delta} \sup_{z\in K} p_t(z, B) = 0$.*

In particular, if $(T_t)_{t\geqslant0}$ is a C_b-Feller semigroup such that the representing kernels are of the form $p_t(x, dy) = p_t(x, y)\,\mu(dy)$ for some Radon measure $\mu \in \mathcal{M}^+(E)$ and a locally bounded density $(x, y) \mapsto p_t(x, y)$, then $(T_t)_{t\geqslant0}$ is a strong Feller semigroup.

Another criterion is based on ultracontractivity. Hoh remarked in [138, Theorem 8.9, p. 134], see also Jacob–Hoh [140, Theorem 2.1], that a Feller semigroup on $C_\infty(\mathbb{R}^d)$ which is ultracontractive, i.e.

$$\|T_t u\|_\infty \leqslant c_t \|u\|_{L^2(dx)} \quad \forall t > 0, \ u \in C_c(\mathbb{R}^d),$$

is already a strong Feller semigroup.

Using Orlicz spaces we can obtain a necessary and sufficient condition. Let us recall some facts about Orlicz space from Rao–Ren [247]. A positive function $\Phi : \mathbb{R} \to [0, \infty]$ is a **Young function** if it is convex, even, satisfies $\Phi(0) = 0$ and

$\lim_{x \to \infty} \Phi(x) = \infty$. Given a Young function Φ and a Radon measure $\mu \in \mathcal{M}^+(E)$, we define the **Orlicz space** as

$$\mathbb{L}^\Phi(\mu) = \left\{ f : E \to \mathbb{R} \text{ measurable and } \int \Phi(\alpha f)\, d\mu < \infty \text{ for some } \alpha > 0 \right\}.$$

The set $\mathbb{L}^\Phi(\mu)$ is a linear space. If $\Phi(x) = |x|^p$, $p \geq 1$, then $\mathbb{L}^\Phi(\mu)$ coincides with the usual Lebesgue space $L^p(\mu)$. The **Orlicz norm**

$$\|f\|_\Phi = \sup \left\{ \int |fg|\, d\mu \; : \; \int \Phi_c(g)\, d\mu \leq 1, \; g \in B_b(E) \right\},$$

where Φ_c is the Legendre transform of Φ, i.e. $\Phi_c(y) := \sup_{x \geq 0} \big(x|y| - \Phi(x) \big)$, turns $\mathbb{L}^\Phi(\mu)$ into a Banach space. We have, cf. [290, Theorem 2.8],

Theorem 1.15. *Let $(T_t)_{t \geq 0}$ be a C_b-Feller semigroup. Then the following assertions are equivalent.*

a) *$(T_t)_{t \geq 0}$ is a strong Feller semigroup.*
b) *For every $t > 0$ there exists a Radon measure $\mu_t \in \mathcal{M}^+(E)$ and some Young function $\Phi_t : \mathbb{R} \to [0, \infty)$ which is strictly increasing on $[0, \infty)$ such that for all compact sets $K \subset E$ and $u \in C_c(E)$*

$$\|\mathbb{1}_K T_t u\|_\infty \leq C(K, t)\|u\|_{\Phi_t}. \tag{1.21}$$

Proof. This is a variant of [290, Theorem 2.8]. Note that the definition of a C_b-Feller semigroup includes the condition [290, Theorem 2.8 (2)]. In order to see that (b) entails (a), one only needs that $x = 0$ is the only zero of the Young functions Φ_t; this is clearly ensured by the strong monotonicity of Φ_t. The proof of the converse is based on the de la Vallée–Poussin characterization of uniform integrability, cf. [283, Theorem 16.8(vii), p. 170], and the argument in [283] allows us to take $\Phi_t(x)$ strictly increasing on $[0, \infty)$. $\qquad\square$

One-Point Compactifications and Sub-Markovianity. The following technique allows us to restrict our attention to Markov semigroups, i.e. sub-Markov semigroups satisfying $T_t 1 = 1$. Let $(T_t)_{t \geq 0}$ be a Feller semigroup which is not necessarily conservative. Denote by E_∂ the one-point compactification of E and define

$$T_t^\partial u := u(\partial) + T_t(u - u(\partial)) \quad \forall u \in C_\infty(E_\partial) = C_b(E_\partial). \tag{1.22}$$

Then $(T_t^\partial)_{t \geq 0}$ is a *conservative* Feller semigroup. Without problems we see that the new semigroup inherits all relevant properties from $(T_t)_{t \geq 0}$. Only the positivity is not so obvious. This can be seen by functional-analytic arguments as in Ethier–Kurtz [100, Lemma 4.2.3, p. 166]; alternatively let $(p_t(x, dy))_{t \geq 0}$ be the kernels from the

Riesz representation of $(T_t)_{t\geq 0}$, cf. Theorem 1.5. Then the corresponding kernels for $(T_t^\partial)_{t\geq 0}$ are given by

$$
\begin{aligned}
p_t^\partial(x, \{\partial\}) &:= 1 - p_t(x, E), & t &\geq 0, \ x \in E, \\
p_t^\partial(\partial, B) &:= \delta_\partial(B), & t &\geq 0, \ B \in \mathscr{B}(E_\partial), \\
p_t^\partial(x, B) &:= p_t(x, B), & t &\geq 0, \ x \in E, \ B \in \mathscr{B}(E),
\end{aligned}
\tag{1.23}
$$

and the positivity of each T_t^∂ follows.

This means that we can restrict ourselves to conservative semigroups, if needed.

1.2 From Feller Processes to Feller Semigroups—and Back

Let $(\Omega, \mathscr{F}, \mathbb{P})$ be a probability space and assume that $(X_t, \mathscr{F}_t)_{t\geq 0}$ is a time-homogeneous Markov process with state space $(E, \mathscr{B}(E))$. As usual, we denote by $\mathbb{P}^x$ and $\mathbb{E}^x$ the probability measures $\mathbb{P}(\cdot \mid X_0 = x)$ and the corresponding expectation, respectively. From the Markov property one easily sees that

$$
\mathbb{E}^x u(X_t) := \int_E u(y) \, \mathbb{P}^x(X_t \in dy) \quad \forall u \in B_b(E), \ x \in E
\tag{1.24}
$$

defines a Markov semigroup.

We always require that $(X_t)_{t\geq 0}$ is **normal**, i.e. $\mathbb{P}^x(X_0 = x) = 1$ for all $x \in E$. Moreover, we assume for simplicity that the process has **infinite life-time**, i.e. $\mathbb{P}^x(X_t \in E) = 1$ for all $t > 0$ and $x \in E$, otherwise we would get a sub-Markov semigroup.

Definition 1.16. A **Feller process** is a time-homogeneous Markov process whose transition semigroup $T_t u(x) = \mathbb{E}^x u(X_t)$ is a Feller semigroup.

A function $p_t(x, B)$ defined on $[0, \infty) \times E \times \mathscr{B}(E)$ is a time-homogeneous **transition function** if

$$
p_t(x, \cdot) \text{ is a (sub-)probability measure on } E, \ t \geq 0, \ x \in E,
$$
$$
p_0(x, \cdot) = \delta_x(\cdot), \ x \in E,
$$
$$
p.(\cdot, B) \text{ is jointly measurable}, \ B \in \mathscr{B}(E),
\tag{1.25}
$$
$$
p_{t+s}(x, B) = \int p_s(y, B) \, p_t(x, dy), \ s, t \geq 0, \ B \in \mathscr{B}(E).
$$

Clearly, $p_t(x, B) = \mathbb{P}^x(X_t \in B) = \mathbb{P}(X_t \in B \mid X_0 = x)$ is such a transition function.

Conversely, assume that we start with a Feller semigroup $(T_t)_{t\geq 0}$. Using the Riesz representation theorem, Theorem 1.5, we can write T_t as an integral operator

$$T_t u(x) = \int u(y) \, p_t(x, dy) \tag{1.26}$$

where the family of kernels $(p_t(x, \cdot))_{t \geq 0, x \in E}$ is a uniquely defined transition function.

Using Kolmogorov's standard procedure we can construct a probability space $(\Omega, \mathscr{F}, \mathbb{P})$ and a Markov process $(X_t)_{t \geq 0}$ with state space E such that

$$\mathbb{P}^x(X_t \in B) = \mathbb{P}(X_t \in B \mid X_0 = x) = p_t(x, B) \quad \text{and} \quad \mathbb{E}^x u(X_t) = T_t u(x).$$

Example 1.17. The semigroups of Example 1.3 correspond to the following stochastic processes.

a) (*Shift semigroup*) $X_t = t\ell$ is a deterministic movement with speed $\ell \in \mathbb{R}^d$.
b) (*Poisson semigroup*) X_t is a Poisson process with intensity $\lambda > 0$ and jump height $\ell \in \mathbb{R}^d$: $\mathbb{P}(X_t = j\ell) = \frac{(\lambda t)^j}{j!} e^{-t\lambda}$, $j = 0, 1, 2, \ldots$.
   $(X_t)_{t \geq 0}$ is spatially homogeneous, i.e. $\mathbb{P}^x(X_t \in B) = \mathbb{P}(X_t + x \in B)$.
c) (*Heat/Brownian semigroup*) X_t is a d-dimensional standard Brownian motion, $\mathbb{P}^x(X_t \in dy) = g_t(x - y) \, dy$. $(X_t)_{t \geq 0}$ is spatially homogeneous.
d) (*Symmetric stable semigroups*) X_t is a rotationally symmetric α-stable Lévy process, $\mathbb{P}^x(X_t \in dy) = g_{t,\alpha}(x - y) \, dy$. If $\alpha = 1$, we get the Cauchy process. $(X_t)_{t \geq 0}$ is spatially homogeneous.
e) (*Convolution/Lévy semigroups*) X_t is a **Lévy process**, i.e. a stochastic process with values in $\mathbb{R}^d$ and with the following properties:
   stationary increments: $X_t - X_s \sim X_{t-s}$ for $0 \leq s < t$;
   independent increments: $(X_{t_j} - X_{t_{j-1}})_{j=1}^n$, $0 \leq t_0 < \cdots < t_n$ are independent random variables;
   stochastic continuity: $\lim_{h \to 0} \mathbb{P}(|X_h| > \epsilon) = 0$ for all $\epsilon > 0$.
   Note that the stationary increment property entails that $X_0 \sim \delta_0$ or $X_0 = 0$ a.s. Since the transition semigroup is a convolution operator, $(X_t)_{t \geq 0}$ is **spatially homogeneous**:

$$\mathbb{E}^x u(X_t) = T_t u(x) = \int u(x + y) \, \mu_t(dy) = \int u(x + y) \, \mathbb{P}^0(X_t \in dy)$$

$$= \mathbb{E}^0 u(X_t + x),$$

   i.e. $\mathbb{P}^x(X_t \in dy) = \mu_t(dy - x)$. All previously considered examples are Lévy processes.
f) (*Ornstein–Uhlenbeck semigroups*) Let $(Z_t)_{t \geq 0}$ be a Lévy process and $B \in \mathbb{R}^{d \times d}$. The process $X_t^x := e^{tB} x + \int_0^t e^{(t-s)B} \, dZ_s$, $x \in \mathbb{R}^d$, is a (Lévy-driven) Ornstein–Uhlenbeck process, which is the unique strong solution to the following stochastic differential equation:

$$dX_t = BX_t \, dt + dZ_t, \qquad X_0 = x \in \mathbb{R}^d.$$

A *generalized Ornstein–Uhlenbeck* process is the strong solution of the SDE

$$dV_t = V_{t-}\, dX_t^{(1)} + dX_t^{(2)}, \qquad V_0 = x \in \mathbb{R},$$

where $X_t = (X_t^{(1)}, X_t^{(2)})$, $t \geqslant 0$, is a two-dimensional Lévy process. Behme–Lindner [18, Theorem 3.1] show that $(V_t)_{t\geqslant 0}$ is a one-dimensional Feller process.

g) *(One-sided stable semigroups)* X_t is an α-stable subordinator, i.e. an increasing Lévy process with values in $[0, \infty)$.

h) *(Affine semigroups)* An affine process $(X_t)_{t\geqslant 0}$ is a Markov process that corresponds to the affine semigroup $(T_t)_{t\geqslant 0}$ of Example 1.3. Well-known examples are the Cox–Ingersoll–Ross process on $E = [0, \infty)$, the Ornstein–Uhlenbeck process on $E = \mathbb{R}^d$, the process of a Heston model on $E = [0, \infty) \times \mathbb{R}^d$ or the Wishart process on the more general state space of positive semidefinite d-dimensional matrices $E = S_d^+$. Note that the condition to be a Feller semigroup in Example 1.3 is equivalent to the stochastic continuity of the affine process, cf. (1.27).   □

We have seen that Feller processes and Feller semigroups are in one-to-one correspondence. Clearly, the semigroup property is equivalent to the Chapman–Kolmogorov equations of the transition function, hence the Markov property of the Feller process. The strong continuity is linked to stochastic continuity of the process. Recall that a Markov process $(X_t)_{t\geqslant 0}$ is **stochastically continuous**, if

$$\lim_{t \to 0} \mathbb{P}^x(X_t \in E \setminus U_x) = 0 \quad \forall x \in E, \ U_x \text{ open neighbourhood of } x. \quad (1.27)$$

If (1.27) holds uniformly for all x (in compact sets) we speak of **(local) uniform stochastic continuity**. For example, any Lévy process (Example 1.17(e)) is uniformly stochastically continuous, see Dynkin [97, Chap. II.§5, 2.23, p. 77].

Lemma 1.18. *Let $(X_t)_{t\geqslant 0}$ be a (temporally homogeneous) Markov process and $(T_t)_{t\geqslant 0}$ be the corresponding Markov semigroup; assume that each T_t has the Feller property, i.e. $T_t : C_\infty(\mathbb{R}^d) \to C_\infty(\mathbb{R}^d)$. Then the strong continuity of $(T_t)_{t\geqslant 0}$ entails that $(X_t)_{t\geqslant 0}$ is (locally uniformly) stochastically continuous. Conversely, if $(X_t)_{t\geqslant 0}$ is stochastically continuous, the semigroup $(T_t)_{t\geqslant 0}$ is on the space $C_\infty(\mathbb{R}^d)$ weakly, hence strongly continuous.*

If $(X_t)_{t\geqslant 0}$ is a Feller process then we denote by $\mathscr{F}_t^X = \sigma(X_s : s \leqslant t)$ its natural filtration. Using some standard martingale regularization arguments one proves, see e.g. Revuz–Yor [250, Theorem III.2.7, p. 81],

Theorem 1.19. *Let $(X_t)_{t\geqslant 0}$ be a Feller process. Then it has a **càdlàg modification**, that is there exists a Feller process $(\tilde{X}_t)_{t\geqslant 0}$ such that $\mathbb{P}^x(X_t = \tilde{X}_t) = 1$ for all $t \geqslant 0$ and $x \in E$, and $t \mapsto \tilde{X}_t(\omega)$ is for almost all ω right-continuous with finite left-hand limits (càdlàg[2]).*

---

[2]càdlàg is the acronym for the French *continue à droite et limitée à gauche*.

In particular, any Lévy process (Example 1.17(e)) has a càdlàg modification.

Often the natural filtration is too small. Usually one considers the right-continuous filtration $\mathscr{F}_t = \mathscr{F}_{t+}^X := \bigcap_{\epsilon>0} \mathscr{F}_{t+\epsilon}^X$, cf. Ethier–Kurtz [100, Theorem 4.2.7, p. 169], or even larger universal augmentations, cf. Revuz–Yor [250, Proposition III.2.10, p. 93], which are automatically right-continuous. For a Lévy process it is enough to augment $\mathscr{F}_t^X$ by all $\mathbb{P}$ null sets to get a right-continuous filtration, see Protter [243, Theorem I.31, p. 22]. We define

$$\tilde{\mathscr{F}}_t := \bigcap_{\mu\in\mathcal{M}^+(E), \mu(E)=1} \sigma(\mathscr{F}_t, \mathcal{N}^\mu) \tag{1.28}$$

where $\mathcal{N}^\mu$ is the family of the null sets corresponding to the initial distribution μ. Then $(\tilde{\mathscr{F}}_t)_{t\geq 0}$ is a complete and right-continuous filtration [159, Theorem 3.5.10, p. 101] and we have the following extension of Theorem 1.19, see, for example, Jacob [159, Theorem 3.5.14, p. 104].

Theorem 1.20. *Let $(\tilde{X}_t)_{t\geq 0}$ be the càdlàg modification of a Feller process and $(\tilde{\mathscr{F}}_t)_{t\geq 0}$ be the filtration constructed in (1.28). Then $((\tilde{X}_t)_{t\geq 0}, (\tilde{\mathscr{F}}_t)_{t\geq 0})$ is a strong Markov process.*

1.3 Resolvents

Let $(T_t)_{t\geq 0}$ be a Feller semigroup and denote by $(p_t(x,\cdot))_{t>0,x\in E}$ the transition function. By Fubini's theorem, the integral

$$U_\alpha u(x) := \int_0^\infty e^{-\alpha t} T_t u(x)\, dt, = \int_0^\infty \int e^{-\alpha t} u(y)\, p_t(x, dy)\, dt \tag{1.29}$$

exists for all $\alpha > 0$, $x \in E$, $u \in B_b(E)$, and is a linear map $U_\alpha : B_b(E) \to B_b(E)$.

Definition 1.21. *Let $(T_t)_{t\geq 0}$ be a Feller semigroup and $\alpha > 0$. The operator U_α given by (1.29) is the α-**potential operator** or **resolvent operator** at $\alpha > 0$.*

If we interpret U_α as the (vector-valued) Laplace transform of the Feller semigroup $(T_t)_{t\geq 0}$, it is not surprising that there is a one-to-one relationship between $(U_\alpha)_{\alpha>0}$ and $(T_t)_{t\geq 0}$.

Theorem 1.22. *Let $(T_t)_{t\geq 0}$ be a Feller semigroup. Then $(U_\alpha)_{\alpha>0}$ is a Feller contraction resolvent, i.e. for all $\alpha > 0$ the operators αU_α*

a) *are positivity preserving and sub-Markov: $0 \leq u \leq 1 \implies 0 \leq \alpha U_\alpha u \leq 1$;*
b) *satisfy the Feller property: $\alpha U_\alpha : C_\infty(E) \to C_\infty(E)$;*
c) *are strongly continuous on $C_\infty(E)$: $\lim_{\alpha\to\infty} \|\alpha U_\alpha u - u\|_\infty = 0$;*

d) *satisfy the **resolvent equation***

$$U_\alpha u - U_\beta u = (\beta - \alpha) U_\beta U_\alpha u \quad \forall \alpha, \beta > 0, \ u \in B_b(E); \qquad (1.30)$$

e) *satisfy the **inversion formula** (or **exponential formula**)*

$$T_t u = \lim_{n \to \infty} \left[\tfrac{n}{t} U_{n/t} \right]^n u \quad \forall u \in C_\infty(E), \ (\text{strong limit in } C_\infty(E)). \qquad (1.31)$$

Proof. The properties (a)–(c) and (1.30) follow at once from the integral representation (1.29), see e.g. [284, Proposition 7.13, p. 97]. The inversion formula (1.31) is the vector-valued real Post–Widder inversion formula for the Laplace transform. The (non-trivial) proof can be found in Pazy [236, Theorem 1.8.3, p. 33]. □

Since the formula (1.31) holds for general contraction semigroups, we can use it to deduce the following result.

Corollary 1.23. *Let $(T_t)_{t \geqslant 0}$ be a contraction semigroup on $B_b(E)$ and $(U_\alpha)_{\alpha > 0}$ the corresponding family of potential operators.*

a) *$(T_t)_{t \geqslant 0}$ is positivity preserving (sub-Markov, strongly continuous) if, and only if, $(\alpha U_\alpha)_{\alpha > 0}$ is positivity preserving (sub-Markov, strongly continuous);*
b) *$T_t : C_\infty(E) \to C_\infty(E)$ for all $t \geqslant 0$ if, and only if, $U_\alpha : C_\infty(E) \to C_\infty(E)$ for all $\alpha > 0$.*

Note that the analogue of property (b) for C_b also holds, but it fails for the strong Feller property: The shift semigroup $T_t u(x) = u(x + t\ell)$ is not a strong Feller semigroup while its resolvent $U_\alpha u(x) = \int_0^\infty e^{-t\alpha} u(x + t\ell) \, dt$ is a convolution operator which maps $B_b(E)$ to $C_b(E)$.

1.4 Generators of Feller Semigroups and Processes

Let $(T_t)_{t \geqslant 0}$ be a Feller semigroup. If we understand the semigroup property

$$T_{t+s} = T_t T_s \quad \text{and} \quad T_0 = \text{id}$$

as an operator-valued functional equation it is an educated guess to expect that T_t is some kind of exponential e^{tA} where A is a suitable operator. For matrix (semi-)groups this is an elementary exercise. Having in mind the classical functional equation, we know that we have to assume some kind of boundedness and continuity; in fact, strong continuity of $(T_t)_{t \geqslant 0}$ will be enough. The key issue is the question how to define T_t as an "exponential" if A is an unbounded operator. This problem was independently solved by Hille and Yosida in 1948, and we refer to any text on operator semigroups for a complete description, for example [78, 236, 354] or [150] for a probabilistic perspective. Here we concentrate on Feller semigroups as in [100] or [284].

Definition 1.24. A **Feller generator** or (**infinitesimal**) **generator** of a Feller semi-group $(T_t)_{t \geq 0}$ or a Feller process $(X_t)_{t \geq 0}$ is a linear operator $(A, \mathcal{D}(A))$ defined by

$$\mathcal{D}(A) := \left\{ u \in C_\infty(E) : \lim_{t \to 0} \frac{T_t u - u}{t} \text{ exists as uniform limit} \right\}, \tag{1.32}$$

$$Au := \lim_{t \to 0} \frac{T_t u - u}{t} \quad \forall u \in \mathcal{D}(A).$$

In general, $(A, \mathcal{D}(A))$ is an unbounded operator which is densely defined, i.e. $\mathcal{D}(A)$ is dense in $C_\infty(E)$, and **closed**

$$\left. \begin{aligned} (u_n)_{n \geq 1} \subset \mathcal{D}(A), \ \lim_{n \to \infty} u_n = u, \\ (Au_n)_{n \geq 1} \text{ is a Cauchy sequence} \end{aligned} \right\} \implies \left\{ \begin{aligned} & u \in \mathcal{D}(A) \text{ and} \\ & Au = \lim_{n \to \infty} Au_n. \end{aligned} \right. \tag{1.33}$$

Example 1.25. In general, it is difficult to determine the exact domain of the generator. For the semigroups from Example 1.3 we find

a) (*Shift semigroup*) $Au(x) = \ell \cdot \nabla u(x)$ where $\ell \in \mathbb{R}^d$ and ∇ is the d-dimensional gradient. We have $C_\infty^1(\mathbb{R}^d) \subset \mathcal{D}(A)$.
b) (*Poisson semigroup*) $Au(x) = \lambda(u(x + \ell) - u(x))$ with $\lambda > 0, \ell \in \mathbb{R}^d$. Since this is a bounded operator, $\mathcal{D}(A) = C_\infty(\mathbb{R}^d)$.
c) (*Heat/Brownian semigroup*) $Au(x) = \frac{1}{2}\Delta u(x)$ where Δ is the d-dimensional Laplacian on $\mathbb{R}^d$. It is easy to see that $C_\infty^2(\mathbb{R}^d) \subset \mathcal{D}(A)$. If $d = 1$, we have $C_\infty^2(\mathbb{R}) = \mathcal{D}(A)$ (cf. [284, Example 7.20, p. 102]). If $d > 2$, the inclusion is strict, cf. Günter [129, Chap. II.§14, pp. 82–83] or [128, Chap. II.§14, pp. 85–86] for a concrete example and Dautray–Lions [82, Remark 5, pp. 290–291] for an abstract argument. In general, $u \in \mathcal{D}(A)$ if, and only if, $u \in C_\infty(\mathbb{R}^d)$ and Δu exists in the sense of Schwartz' distributions (i.e. as generalized function) and is represented by a C_∞-function, cf. Itô [149, Sect. 3.§2, pp. 92–96].
d) (*Symmetric stable semigroups*) If $\alpha \in (0, 2)$ then $Au(x) = -(-\Delta)^{\alpha/2}u(x)$. The fractional power of the Laplacian is, at least for $u \in C_\infty^2(\mathbb{R}^d) \subset \mathcal{D}(A)$, given by

$$Au(x) = c_\alpha \int_{\mathbb{R}^d \setminus \{0\}} \left(u(x + y) - u(x) - \nabla u(x) \cdot y \chi(|y|) \right) \frac{dy}{|y|^{\alpha + d}} \tag{1.34}$$

where $c_\alpha = \alpha 2^{\alpha - 1} \pi^{-d/2} \Gamma\left(\frac{\alpha + d}{2}\right) / \Gamma\left(1 - \frac{\alpha}{2}\right)$ and for some truncation function $\chi \in B_b[0, \infty)$ such that $0 \leq 1 - \chi(s) \leq \kappa \min(s, 1)$ (for some $\kappa > 0$) and $s\chi(s)$ is bounded.

Since $\lim_{\epsilon \to 0} \int_{|y| > \epsilon} y \chi(|y|) \, dy = 0$, we can rewrite (1.34) under the integral without $\nabla u(x) \cdot y \chi(|y|)$ as a Cauchy principal value integral, if $\alpha \in [1, 2)$, and as a *bona fide* integral, if $\alpha \in (0, 1)$.

To get more details on the domain $\mathcal{D}(A)$ we use for $\alpha \in (0, 1)$ and a function $u \in C_\infty^2(\mathbb{R}^d)$ an alternative representation of the generator, see Sato [267, Example 32.7, p. 217]

$$Au(x) = c \int_{\mathbb{S}^d} \int_0^\infty \frac{u(x + r\gamma) - u(x)}{r^{1+\alpha}} \, dr \, \sigma_d(d\gamma) \qquad (1.35)$$

where σ_d is the uniform measure on the unit sphere $\mathbb{S}^d \subset \mathbb{R}^d$. For non-integer $\beta > 0$, let $C^\beta(\mathbb{R}^d)$ denote the Hölder space of $\lfloor\beta\rfloor$-times differentiable functions whose $\lfloor\beta\rfloor^{\text{th}}$ derivative is Hölder continuous with index $\beta - \lfloor\beta\rfloor$; as usual, $\|\cdot\|_{C^\beta}$ denotes the corresponding norm.

Splitting the inner integral in (1.35) yields for $u \in C^2_\infty(\mathbb{R}^d) \cap C^\beta(\mathbb{R}^d)$, $1 > \beta > \alpha$,

$$\|Au\|_\infty \leq c(\|u\|_{C^\beta} + \|u\|_\infty), \qquad (1.36)$$

and this shows that we have $C_\infty(\mathbb{R}^d) \cap C^{\alpha+\epsilon}(\mathbb{R}^d) \subset \mathcal{D}(A)$ for all $\epsilon > 0$.

For $\alpha \geq 1$ a similar argument yields the same statement. See also [16, Remark 5.3] for an extension to stable-like processes in the sense of Bass.

e) (*Convolution/Lévy semigroups*) The generator of a general Lévy semigroup is, for $u \in C^2_\infty(\mathbb{R}^d) \subset \mathcal{D}(A)$, of the form

$$Au(x) = l \cdot \nabla u(x) + \frac{1}{2}\text{div } Q\nabla u(x) + \int_{\mathbb{R}^d\backslash\{0\}} \left(u(x+y)-u(x)-\nabla u(x)\cdot y\,\chi(|y|)\right) \nu(dy)$$

$$(1.37)$$

where $l \in \mathbb{R}^d$, $Q \in \mathbb{R}^{d\times d}$ positive semidefinite and $\nu \in \mathcal{M}^+(\mathbb{R}^d)$ such that $\int_{\mathbb{R}^d\backslash\{0\}} \min(|y|^2, 1) \nu(dy) < \infty$; χ is a truncation function as in the previous example. For a proof we refer to Sato [267, Theorem 31.5, p. 208] or to Theorem 2.21 and Corollary 2.22 below.

f) (*Ornstein–Uhlenbeck semigroups*) Let A be the operator given by (1.37), and let $B \in \mathbb{R}^{d\times d}$. Then, the generator of the Ornstein–Uhlenbeck semigroup is, for every $u \in C^2_\infty(\mathbb{R}^d) \subset \mathcal{D}(A)$, of the form

$$Lu(x) = Au(x) + Bx \cdot \nabla u(x).$$

For a proof we refer to [268, Theorem 3.1].

The generator of the generalized Ornstein–Uhlenbeck semigroup, cf. Example 1.17(f), is defined for $u \in C^2_c(\mathbb{R})$ by

$$Au(x) = (\gamma_1 x + \gamma_2)u'(x) + \frac{1}{2}\left(x^2\sigma_{11}^2 + 2x\sigma_{12} + \sigma_{22}^2\right)u''(x) +$$

$$+ \iint_{\mathbb{R}^2\backslash\{0\}} \left[u(x + \gamma_1 x + \gamma_2) - u(x) - u'(x)(\gamma_1 x + \gamma_2)\mathbb{1}_{[0,1]}(|x|)\right] \nu(dy_1, dy_2)$$

where $\left(\begin{pmatrix} \gamma_1 \\ \gamma_2 \end{pmatrix}, \begin{pmatrix} \sigma_{11}^2 & \sigma_{12} \\ \sigma_{12} & \sigma_{22}^2 \end{pmatrix}, \nu(dy_1, dy_2)\right)$ is the Lévy triplet of the driving Lévy process $X_t \in \mathbb{R}^2$.

Moreover, $\{u \in C_\infty^2(\mathbb{R}) : xu'(x), x^2 u''(x) \in C_\infty(\mathbb{R})\} \subset \mathcal{D}(A)$ and the test functions $C_c^\infty(\mathbb{R}^d)$ are an operator core. For a proof we refer to Behme–Lindner [18, Theorem 3.1]. Using the technique of Theorem 3.8, in particular the remark following its statement, we can get a similar result with a different proof relying on the symbol, cf. Example 3.34(b).

g) (*Affine semigroups*) The generator of an affine semigroup exists if, and only if, $(T_t)_{t \geqslant 0}$ is *regular*, i.e. if $\frac{\partial^+}{\partial t} T_t e^{\xi \cdot x}|_{t=0}$ exists for all $(x, \xi) \in E \times (\mathbb{C}_-^m \times i\mathbb{R}^{d-m})$ and defines a function which is continuous at $\xi = 0$ for all $x \in E$; equivalently, $F(\xi) = \frac{\partial^+}{\partial t} \phi(t, \xi)|_{t=0}$ and $R(\xi) = \frac{\partial^+}{\partial t} \psi(t, \xi)|_{t=0}$ exist for all $\xi \in \mathbb{C}_-^m \times i\mathbb{R}^{d-m}$ and are continuous at $\xi = 0$. Regularity follows from the stochastic continuity of the corresponding affine process or from the condition for being Feller in Example 1.3, cf. Keller-Ressel–Schachermayer–Teichmann [176, Theorem 5.1] or [177, Theorem 3.10] for general state spaces. For $u \in C_c^2(E)$ the generator is given by

$$Au(x) = l \cdot \nabla u(x) + \frac{1}{2} \mathrm{div}\, Q \nabla u(x) + \int_{E \setminus \{0\}} \left(u(x+y) - u(x) - \nabla u(x) \cdot \chi(y) \right) \nu(dy)$$

$$+ \sum_{j=1}^m x_j \left[\frac{1}{2} \mathrm{div}\, Q^j \nabla u(x) + \int_{E \setminus \{0\}} \left(u(x+y) - u(x) - \nabla u(x) \cdot \chi^j(y) \right) \nu^j(dy) \right]$$

$$+ \sum_{j=1}^d x_j l^j \cdot \nabla u(x),$$

where $x = (x_1, \ldots, x_d) \in E$, $l, l^j \in \mathbb{R}^d$ for $j = 1, \ldots, d$, with $Q, Q^j \in \mathbb{R}^{d \times d}$ positive semidefinite matrices, $\nu, \nu^j \in \mathcal{M}^+(E)$ and χ, χ^j are truncation functions for $j = 1, \ldots, m$. These parameters are subject to further restrictions, cf. Duffie–Filipović–Schachermayer [91, Definition 2.6, p. 991].

The domain of the generator is strictly larger than $C_c^2(E)$ as it contains all functions $u \in C_\infty^2(E)$ which satisfy certain decay conditions at infinity, cf. Duffie–Filipović–Schachermayer [91, Sect. 8, p. 1026] for details.   □

By definition, the generator A is the strong (right-)derivative of T_t at $t = 0$. Using the semigroup property one can show the following analogue of the fundamental theorem of differential calculus.

Lemma 1.26. *Let $(A, \mathcal{D}(A))$ be the generator of the Feller semigroup $(T_t)_{t \geqslant 0}$. Then*

$$T_t u - u = A \int_0^t T_s u \, ds \quad \forall u \in C_\infty(\mathbb{R}^d)$$

$$= \int_0^t A T_s u \, ds \quad \forall u \in \mathcal{D}(A) \tag{1.38}$$

$$= \int_0^t T_s A u \, ds \quad \forall u \in \mathcal{D}(A).$$

A straightforward calculation using the formula (1.29) for the α-potential operator (e.g. [284, Theorem 7.13(f), pp. 97–98]) shows

Lemma 1.27. *Let* $(A, \mathcal{D}(A))$ *be a Feller generator. Then for each* $\alpha > 0$ *the operator* $\alpha - A$ *has a bounded inverse which is just the* α-*potential operator* U_α. *In particular,* $\mathcal{D}(A) = U_\alpha(C_\infty(E))$ *independently of* $\alpha > 0$.

In other words, the lemma shows that the equation

$$\alpha u - Au = f$$

has the solution $u = (\alpha - A)^{-1} f = U_\alpha f$. Therefore $(U_\alpha)_{\alpha > 0}$ is called the resolvent.

Because of the positivity of a Feller semigroup we find for any $u \in \mathcal{D}(A)$ which admits a global maximum $u_{\max} = u(x_0) = \sup_{y \in E} u(y)$

$$T_t u(x_0) - u(x_0) \leq T_t u_{\max} - u(x_0) \leq u_{\max} - u(x_0) = 0.$$

This proves the first part of the following lemma. The second part can be found in [284, Lemma 7.18, p. 101] and [100, Lemma 1.2.11, p. 16].

Lemma 1.28. *A Feller generator* $(A, \mathcal{D}(A))$ *satisfies the* **positive maximum principle**

$$u \in \mathcal{D}(A), \ u(x_0) = \sup_{y \in E} u(y) \geq 0 \implies Au(x_0) \leq 0. \tag{1.39}$$

Conversely, if the linear operator $(A, \mathcal{D})$, *where* $\mathcal{D} \subset C_\infty(E)$ *is a dense subspace, satisfies the positive maximum principle, then* $(A, \mathcal{D})$ *is* **dissipative**, *i.e.*

$$\|\lambda u - Au\|_\infty \geq \lambda \|u\|_\infty \quad \forall \lambda > 0. \tag{1.40}$$

In particular, $(A, \mathcal{D})$ *has a closed extension* $(A, \mathcal{D}(A))$, *and this extension satisfies again the positive maximum principle.*

For the Laplace operator (1.39) is quite familiar: At a (global) maximum the second derivative is negative.

Remark 1.29. The positive maximum principle can also be seen as the limiting case (for $p \to \infty$) of the notion of an $L^p(m)$-Dirichlet operator. Let $(A^{(p)}, \mathcal{D}(A^{(p)}))$ be the generator of a strongly continuous, sub-Markovian semigroup $(T_t^{(p)})_{t \geqslant 0}$ in $L^p(m)$ for some $p > 1$. Then $A^{(p)}$ is an $L^p(m)$-**Dirichlet operator**, i.e.

$$\int_E ((u(x) - 1)^+)^{p-1} A^{(p)} u(x)\, m(dx) \leqslant 0 \quad \forall u \in \mathcal{D}(A) \tag{1.41}$$

and this condition is necessary and sufficient for the Markov property of the semigroup, cf. [281, Theorem 2.2]. If $A^{(p)}$ generates for every $p > p_0$ a sub-Markovian semigroup and if there is a sufficiently rich set $\mathcal{D} \subset \mathcal{D}(A^{(p)})$ (for all $p \geqslant p_0$) such that $A^{(p)}(\mathcal{D})$ consists of lower semicontinuous and bounded functions, then (1.41) becomes as $p \to \infty$ the positive maximum principle (1.39), cf. [281, Theorem 2.7]. The notion of a Dirichlet operator in L^2 is due to Bouleau–Hirsch [49], for the spaces L^p it was introduced by Jacob [157, Sect. 4.6, pp. 364–382].   □

If $E \subset \mathbb{R}^d$, the positive maximum principle will have consequences for the structure of the generator, cf. Theorem 2.21 below. For the time being, we are more interested in the consequences the positive maximum principle imposes upon the semigroup: It allows to adapt the classical Hille–Yosida theorem, see e.g. Ethier-Kurtz [100, Theorem 1.2.12, p. 16], to the context of Feller semigroups, cf. [100, Theorem 4.2.2, p. 165].

Theorem 1.30 (Hille–Yosida–Ray). *Let $(A, \mathcal{D})$ be a linear operator on $C_\infty(E)$. $(A, \mathcal{D})$ is closable and the closure $(A, \mathcal{D}(A))$ is the generator of a Feller semigroup if, and only if,*

a) $\mathcal{D} \subset C_\infty(E)$ *is dense;*
b) $(A, \mathcal{D})$ *satisfies the positive maximum principle;*
c) $(\lambda - A)(\mathcal{D}) \subset C_\infty(E)$ *is dense for some (or all) $\lambda > 0$.*

Proof. The necessity of the conditions (a) and (c) follows from the Hille–Yosida theorem, while condition (b) is the first half of Lemma 1.28.

By the second part of Lemma 1.28, the condition (b) shows that $(A, \mathcal{D})$ is dissipative. Then the Hille–Yosida theorem ensures that the closure of $(A, \mathcal{D})$ generates a strongly continuous contraction semigroup on $C_\infty(E)$. Using once again condition (b), we see now that the associated resolvent, hence the semigroup, is positive, cf. [284, Lemma 7.18, p. 101].   □

Usually it is a problem to describe the domain $\mathcal{D}(A)$ of a Feller generator. The following result, due to Reuter and Dynkin, is often helpful if we want to determine the domain of the generator; our formulation follows Rogers–Williams [255, Lemma III.4.17, p. 237] and [284, Theorem 7.15, p. 100].

Lemma 1.31 (Dynkin; Reuter). *Let $(A, \mathcal{D}(A))$ be the infinitesimal generator of a Feller semigroup, and assume that $(G, \mathcal{D}(G))$, $\mathcal{D}(G) \subset C_\infty(E)$, extends $(A, \mathcal{D}(A))$, i.e. $\mathcal{D}(A) \subset \mathcal{D}(G)$ and $G|_{\mathcal{D}(A)} = A$. If for all $u \in \mathcal{D}(G)$*

$$Gu = u \implies u = 0, \tag{1.42}$$

then $(A, \mathcal{D}(A)) = (G, \mathcal{D}(G))$.

Since the positive maximum principle or dissipativity guarantee (1.42), Lemma 1.31 tells us that a Feller generator is maximally dissipative, i.e. that it has no proper dissipative extension.

Recall that the weak limit of a family $(u_t)_{t \geqslant 0} \subset C_\infty(E)$ is defined by

$$\text{weak-}\lim_{t \to 0} u_t = u \iff \forall \mu \in \mathcal{M}_b(E) \,:\, \lim_{t \to 0} \int u_t \, d\mu = \int u \, d\mu.$$

Definition 1.32. Let $(T_t)_{t \geqslant 0}$ be a Feller semigroup. The **pointwise (infinitesimal) generator** is a linear operator $(A_p, \mathcal{D}(A_p))$ defined by

$$\mathcal{D}(A_p) := \left\{ u \in C_\infty(E) \,\middle|\, \exists g \in C_\infty(E) \; \forall x \in E \,:\, g(x) = \lim_{t \to 0} \frac{T_t u(x) - u(x)}{t} \right\},$$

$$A_p u(x) := \lim_{t \to 0} \frac{T_t u(x) - u(x)}{t} \quad \forall u \in \mathcal{D}(A_p),\ x \in E. \tag{1.43}$$

The **weak (infinitesimal) generator** is a linear operator $(A_w, \mathcal{D}(A_w))$ defined by

$$\mathcal{D}(A_w) := \left\{ u \in C_\infty(E) \,\middle|\, \exists g \in C_\infty(E) \,:\, g = \text{weak-}\lim_{t \to 0} \frac{T_t u - u}{t} \right\},$$

$$A_w u := \text{weak-}\lim_{t \to 0} \frac{T_t u - u}{t} \quad \forall u \in \mathcal{D}(A_w). \tag{1.44}$$

Since strong convergence entails weak convergence and since weak convergence in $C_\infty(E)$ is actually bounded pointwise convergence, it is not hard to see that

$$\mathcal{D}(A) \subset \mathcal{D}(A_w) \subset \mathcal{D}(A_p) \quad \text{and} \quad A = A_w|_{\mathcal{D}(A)} = A_p|_{\mathcal{D}(A)}.$$

From the theory of operator semigroups we know that $(A, \mathcal{D}(A)) = (A_w, \mathcal{D}(A_w))$, cf. Pazy [236, Theorem 2.1.3, p. 43], and we even get $(A, \mathcal{D}(A)) = (A_p, \mathcal{D}(A_p))$ if we use Davies' proof of the "weak equals strong" theorem [78, Theorem 1.24, p. 17] and the fact that finite linear combinations of Dirac measures are vaguely (i.e. weak-*) dense in $\mathcal{M}_b(E)$.

Alternatively, we can use the positive maximum principle. Clearly $(A_p, \mathcal{D}(A_p))$ extends $(A, \mathcal{D}(A))$, and A_p satisfies the positive maximum principle. By (the analogue of) Lemma 1.28 we see that A_p is dissipative and from Lemma 1.31 we conclude that $\mathcal{D}(A) = \mathcal{D}(A_p)$. This proves the following result.

Theorem 1.33. *Let $(T_t)_{t \geq 0}$ be a Feller semigroup generated by $(A, \mathcal{D}(A))$. Then*

$$\mathcal{D}(A) = \left\{ u \in C_\infty(E) \,\middle|\, \exists g \in C_\infty(E) \; \forall\, x \in E \; : \; g(x) = \lim_{t \to 0} \frac{T_t u(x) - u(x)}{t} \right\}.$$
(1.45)

In particular, $(A, \mathcal{D}(A)) = (A_w, \mathcal{D}(A_w)) = (A_p, \mathcal{D}(A_p))$.

Operator Cores. Let $(A, \mathcal{D}(A))$ be a densely defined, closed linear operator and $\mathcal{D} \subset \mathcal{D}(A)$ be a dense subset. If $\mathcal{D}$ determines A in the sense that the closure of $(A, \mathcal{D})$ is $(A, \mathcal{D}(A))$, then $\mathcal{D}$ is called an **(operator) core**. In other words, $\mathcal{D}$ is an operator core if, and only if,

$$\forall u \in \mathcal{D}(A) \quad \exists (u_n)_{n \geq 1} \subset \mathcal{D} \, : \; \lim_{n \to \infty} \left(\|u - u_n\|_\infty + \|Au - Au_n\|_\infty \right) = 0. \quad (1.46)$$

Usually it is hard to determine operator cores, and the following abstract criterion often comes in handy.

Lemma 1.34. *Let $(T_t)_{t \geq 0}$ be a Feller semigroup, $(U_\alpha)_{\alpha > 0}$ the resolvent, $(A, \mathcal{D}(A))$ the generator and $\mathcal{D}_0 \subset \mathcal{D} \subset \mathcal{D}(A)$ dense subsets of $C_\infty(E)$. Then $\mathcal{D}$ is an operator core for $(A, \mathcal{D}(A))$ if one of the following conditions is satisfied.*

a) $T_t(\mathcal{D}_0) \subset \mathcal{D}$ *for all $t > 0$.*
b) $U_\alpha(\mathcal{D}_0) \subset \mathcal{D}$ *for some $\alpha > 0$.*

Proof. The first condition is a standard result from semigroup theory, see e.g. [100, Proposition 1.3.3, p. 17] or Davies [78, Theorem 1.9, p. 8]. The following simple proof for the second condition is taken from [9, proof of Theorem 4.4]. Fix any $u \in \mathcal{D}(A)$ and set $g := \alpha u - Au$. Since $\mathcal{D}_0$ is dense in $C_\infty(E)$, there exists a sequence $(g_n)_{n \geq 1} \subset \mathcal{D}_0$ converging to g. Then, for $u_n := U_\alpha g_n$,

$$(u_n, \, Au_n) = (U_\alpha g_n, \, \alpha u_n - g_n) \xrightarrow[n \to \infty]{\text{uniformly}} (u, \, \alpha u - g) = (u, \, Au).$$

By assumption, $u_n \in U_\alpha(\mathcal{D}_0) \subset \mathcal{D}$, and this shows that $\mathcal{D}$ is a core. □

The Full Generator. Sometimes it is useful to extend the notion of a generator even further. The starting point is the observation that $\frac{d}{dt} T_t = T_t A$ on $\mathcal{D}(A)$ or

$$T_t u(x) - u(x) = \int_0^t T_s Au(x) \, ds \quad \forall x \in E, \, u \in \mathcal{D}(A) \quad (1.47)$$

see Lemma 1.26. This motivates the following definition, cf. Ethier–Kurtz [100, pp. 23–24].

Definition 1.35. Let $(T_t)_{t \geq 0}$ be a Feller semigroup. The **full generator** is the set

$$\hat{A}_b := \left\{ (f,g) \in B_b(E) \times B_b(E) : T_t f - f = \int_0^t T_s g \, ds \right\}. \tag{1.48}$$

By (1.47), $\{(u, Au) : u \in \mathcal{D}(A)\} \subset \hat{A}_b$. Observe that the full generator need not be single-valued, i.e. for any $f \in B_b(E)$ there may be more than one $g \in B_b(E)$ such that $(f,g) \in \hat{A}_b$: For the shift semigroup $T_t f(x) := f(x+t)$ on $B_b(\mathbb{R})$ one has $(0,g) \in \hat{A}_b$ for each $g \in B_b(\mathbb{R})$ which is Lebesgue almost everywhere zero. The full generator is linear, dissipative and closed with respect to bounded pointwise limits bp-$\lim_{n \to \infty} (f_n, g_n) = (f,g)$. A thorough discussion of the full generator can be found in Ethier–Kurtz [100, Sect. 1.5, pp. 22–28]. The full generator is most useful in connection with the martingale problem. At this point we restrict ourselves to the following fact, cf. [100, Proposition 4.17, p. 162].

Theorem 1.36. *Let $(X_t)_{t \geq 0}$ be a Feller process (or a Markov process) with full generator $\hat{A}_b$. Then*

$$M_t := f(X_t) - f(X_0) - \int_0^t g(X_s) \, ds \quad \forall (f,g) \in \hat{A}_b \tag{1.49}$$

is a martingale with respect to the natural filtration $\mathscr{F}_t^X := \sigma(X_s : s \leq t)$.

Taking expectations in (1.49), we see that $\mathbb{E}^x M_t = 0$ for all $x \in E$; this is just (1.47), if $f = u \in \mathcal{D}(A)$ and $g = Au$.

Theorem 1.36 allows a stochastic characterization of the full generator $\hat{A}_b$.

Corollary 1.37. *Let $(X_t)_{t \geq 0}$ be a Feller process (or a strong Markov process) with full generator $\hat{A}_b$, denote by*

$$M_t^{[f,g]} = f(X_t) - f(X_0) - \int_0^t g(X_s) \, ds, \quad t \geq 0,$$

and write $\mathscr{F}_t^X = \sigma(X_s : s \leq t)$ for the natural filtration of the process $(X_t)_{t \geq 0}$. Then

$$\hat{A}_b = \left\{ (f,g) \in B_b(E) \times B_b(E) : \left(M_t^{[f,g]}, \mathscr{F}_t^X \right)_{t \geq 0} \text{ is a martingale} \right\}.$$

Sometimes it is important to consider unbounded measurable functions f, g. While it is, in general, not clear how to define $T_t f = \mathbb{E}^x f(X_t)$ for an *unbounded* function f, the expression $f(X_t)$ is well-defined, and the stochastic version of $\hat{A}_b$ can be extended to this situation. We set

$$\hat{A} = \left\{ (f,g) \in B(E) \times B(E) : \left(M_t^{[f,g]}, \mathscr{F}_t^X \right)_{t \geq 0} \text{ is a local martingale} \right\} \tag{1.50}$$

and, by a stopping argument and the strong Markov property of $(X_t)_{t \geq 0}$, we see that $\hat{A}_b = \hat{A} \cap (B_b(E) \times B_b(E))$.

Note that the full generator $\hat{A}$ need not be single-valued, i.e. there might be two (or more) functions $g_1 \neq g_2$ such that $(f, g_1), (f, g_2) \in \hat{A}$. Therefore we avoid the notion of **extended generator** which is sometimes found in the literature, e.g. Davis [80, (14.15), p. 32] or Meyn–Tweedie [226]

$$\hat{D}(A) := \{f \in B(E) : \exists! \, g \in B(E), \, (f, g) \in \hat{A}\}. \tag{1.51}$$

Dynkin's Characteristic Operator. The following extension is due to Dynkin, see [97, Chap. V.§§3–4, pp. 140–149], our presentation follows [284, Sect. 7.5, pp. 103–109]. Let $(X_t)_{t \geq 0}$ be a Feller process, denote by $\mathscr{F}_t^X = \sigma(X_s : s \leq t)$ the natural filtration, and by

$$\tau_r^x := \inf\{t > 0 : X_t \in \overline{\mathbb{B}}^c(x, r)\}, \quad r \geq 0, \, x \in E \tag{1.52}$$

the first hitting time of the open set $E \setminus \overline{\mathbb{B}}(x, r)$ (this is always a stopping time for $\mathscr{F}_{t+}^X$). Note that

$$\tau_0^x = \inf\{t > 0 : X_t \neq x\}, \quad x \in E.$$

Using the strong Markov property of a Feller process one can show, cf. [284, Theorem A.26, p. 350], that

$$\mathbb{P}^x(\tau_0^x \geq t) = e^{-\lambda(x)t} \quad \text{for some} \quad \lambda(x) \in [0, \infty], \, x \in E.$$

This allows us to characterize points in the state space:

$$x \in E \text{ is called } \begin{cases} \text{an } \textbf{exponential holding point,} & \text{if } 0 < \lambda(x) < \infty, \\ \text{an } \textbf{instantaneous point,} & \text{if } \lambda(x) = \infty, \\ \text{an } \textbf{absorbing point} \text{ or a } \textbf{trap,} & \text{if } \lambda(x) = 0. \end{cases} \tag{1.53}$$

If x is not absorbing, then there is some $r > 0$ such that $\mathbb{E}^x \tau_r^x < \infty$, and the following definition makes sense.

Definition 1.38. Let $(X_t)_{t \geq 0}$ be a Feller process and denote by τ_r^x the first hitting time of the set $\overline{\mathbb{B}}^c(x, r)$. Dynkin's **characteristic operator** is the linear operator defined by

$$\mathfrak{A}u(x) := \begin{cases} \lim_{r \to 0} \dfrac{\mathbb{E}^x u(X_{\tau_r^x}) - u(x)}{\mathbb{E}^x \tau_r^x}, & \text{if } x \text{ is not absorbing,} \\ 0, & \text{if } x \text{ is absorbing,} \end{cases} \tag{1.54}$$

on the set $\mathcal{D}(\mathfrak{A})$ consisting of all $u \in B_b(E)$ such that the limit in (1.54) exists for each non-absorbing point $x \in E$.

From (1.49) and the optional stopping theorem for martingales we easily derive **Dynkin's formula**

$$\mathbb{E}^x u(X_\sigma) - u(x) = \mathbb{E}^x \int_0^\sigma Au(X_s)\,ds, \quad u \in \mathcal{D}(A) \qquad (1.55)$$

where σ is a stopping time such that $\mathbb{E}^x \sigma < \infty$. If we use $\sigma = \tau_r^x$, this formula allows us to show that the characteristic operator $(\mathfrak{A}, \mathcal{D}(\mathfrak{A}))$ extends the generator $(A, \mathcal{D}(A))$, see Dynkin [97, Chap. V.§3, Theorem 5.5, pp. 142–143] or [284, Theorem 7.26, p. 107].

Theorem 1.39. *Let $(X_t)_{t \geq 0}$ be a Feller process with generator $(A, \mathcal{D}(A))$ and characteristic operator $(\mathfrak{A}, \mathcal{D}(\mathfrak{A}))$. Then $\mathfrak{A}$ is an extension of A and $\mathfrak{A}|_{\mathcal{D}} = A$ where $\mathcal{D} = \{u \in \mathcal{D}(\mathfrak{A}) \cap C_\infty(E) : \mathfrak{A}u \in C_\infty(E)\}$.*

As an application of the characteristic operator we can characterize the structure of the generators of Feller processes in $\mathbb{R}^d$ with continuous sample paths. A linear operator $L : \mathcal{D}(L) \subset B_b(\mathbb{R}^d) \to B_b(\mathbb{R}^d)$ is called **local**, if $Lu(x) = Lw(x)$ whenever $u, w \in \mathcal{D}(L)$ coincide in some neighbourhood of the point x, i.e. $u|_{B(x,\epsilon)} = w|_{B(x,\epsilon)}$.

Theorem 1.40. *Let $(X_t)_{t \geq 0}$ be a Feller process with values in $\mathbb{R}^d$ and continuous sample paths. Then the generator $(A, \mathcal{D}(A))$ is a local operator.*

If the test functions $C_c^\infty(\mathbb{R}^d) \subset \mathcal{D}(A)$ are in the domain of the generator we can use a result due to Peetre [237] to see that local operators are differential operators.

Theorem 1.41 (Peetre). *Let $L : C_c^\infty(\mathbb{R}^d) \to C_c^k(\mathbb{R}^d)$ be a linear operator where $k \geq 0$ is fixed. If $\operatorname{supp} Lu \subset \operatorname{supp} u$, then $Lu = \sum_{\alpha \in \mathbb{N}_0^d} a_\alpha(\cdot) \frac{\partial^{|\alpha|}}{\partial x^\alpha} u$ with finitely many, uniquely determined distributions $a_\alpha \in \mathcal{D}'(\mathbb{R}^d)$ (i.e. the topological dual of $C_c^\infty(\mathbb{R}^d)$) which are locally represented by functions of class $C^k(\mathbb{R}^d)$.*

If we apply this to Feller generators, we get

Corollary 1.42. *Let $(X_t)_{t \geq 0}$ be a Feller process with continuous sample paths and generator $(A, \mathcal{D}(A))$. If $C_c^\infty(\mathbb{R}^d) \subset \mathcal{D}(A)$, then A is a second-order differential operator.*

Proof. By Theorem 1.41, A is a differential operator. Since A has to satisfy the positive maximum principle, A is at most a *second order differential operator*. This follows from the fact that we can find test functions $\phi \in C_c^\infty(\mathbb{R}^d)$ such that x_0 is a global maximum while $\partial_j \partial_k \partial_l \phi(x_0)$ has arbitrary sign and arbitrary modulus. $\qquad\square$

Recall the so-called **Dynkin–Kinney criterion** which guarantees the continuity of the trajectories of a stochastic process:

$$\forall \epsilon > 0,\ r > 0 \ : \ \limsup_{h \to 0} \sup_{t \leq h\ |x| \leq r} \frac{1}{h}\, \mathbb{P}^x(r > |X_t - x| > \epsilon) = 0, \qquad (1.56)$$

see Dynkin [96, Kapitel 6.§5, Satz 6.6, p. 139].[3] In fact, the Dynkin–Kinney criterion (1.56) is equivalent to the locality of the generator, cf. [284, Theorem 7.30, p. 108].

Corollary 1.43. *Let* $(X_t)_{t \geq 0}$ *be a Feller process such that the test functions* $C_c^\infty(\mathbb{R}^d)$ *are in the domain of the generator* $(A, \mathcal{D}(A))$. *Then* $(A, C_c^\infty(\mathbb{R}^d))$ *is a local operator if, and only if, the Dynkin–Kinney criterion* (1.56) *holds.*

1.5 Feller Semigroups and L^p-Spaces

A Feller semigroup $(T_t)_{t \geq 0}$ is, *a priori*, defined on the bounded measurable functions $B_b(E)$ or the continuous functions vanishing at infinity $C_\infty(E)$. We will briefly discuss some standard situations which allow to extend $T_t|_{C_c(E)}$ onto a space of integrable functions. Throughout this section we assume that $(E, \mathcal{B}(E), m)$ is a measure space such that the m is a positive Radon measure with full topological support, i.e. for any open set $U \subset E$ we have $m(U) > 0$.

We assume that the operators T_t are **m-symmetric** in the following sense

$$\int_E T_t u(x) \cdot w(x)\, m(dx) = \int_E u(x) \cdot T_t w(x)\, m(dx) \quad \forall u, w \in C_c(E). \qquad (1.57)$$

If $|w| \leq 1$, we have $|T_t w| \leq T_t |w| \leq T_t 1 \leq 1$, and so

$$\|T_t u\|_{L^1(m)} = \sup_{\substack{w \in C_c(E) \\ |w| \leq 1}} \left| \int_E T_t u \cdot w\, dm \right| \leq \sup_{\substack{w \in C_c(E) \\ |w| \leq 1}} \int_E |u| \cdot |T_t w|\, dm \leq \|u\|_{L^1(m)}.$$

Since $C_c(E)$ is dense in $L^1(m)$ this shows that $T_t|_{C_c(E)}$ has an extension $T_t^{(1)}$ such that $T_t^{(1)} : L^1(m) \to L^1(m)$ is a contraction operator. It is easy to see that $(T_t^{(1)})_{t \geq 0}$ is a strongly continuous sub-Markovian contraction semigroup on $L^1(m)$.

To proceed, we need a version of the Riesz convexity theorem which we take from Butzer–Berens [57, Sect. 3.3.2, pp. 187–191].

---

[3]In [96] the criterion reads

$$\forall \epsilon > 0,\ r > 0\ :\ \lim_{h \to 0} \sup_{t \leq h} \sup_{|x| \leq r} \frac{1}{h} \mathbb{P}^x(|X_t - x| > \epsilon) = 0.$$

A careful check of the proof reveals that (1.56) is sufficient. Alternatively, if we already have a càdlàg modification, we can use the simplified argument in [271, Theorem 2]: Just observe in that proof the following identity $\{|X_t - X_s| > \epsilon,\ \sup_{u \leq T} |X_u| \leq r\} = \{2r \geq |X_t - X_s| > \epsilon,\ \sup_{u \leq T} |X_u| \leq r\}$.

Theorem 1.44 (M. Riesz). *Let $(E, \mathscr{B}(E), \mu)$ and $(F, \mathscr{B}(F), \nu)$ be two σ-finite measure spaces, $1 \leqslant p \leqslant q \leqslant \infty$ and assume that*

$$T : L^p(\mu) + L^q(\mu) \to L^p(\nu) + L^q(\nu)$$

is a bounded linear operator. Then $T : L^r(\mu) \to L^r(\nu)$ is bounded for any $r \in [p, q]$, and we have the following estimate for the operator norm

$$\|T\|_{L^r(\mu) \to L^r(\nu)} \leqslant \|T\|_{L^p(\mu) \to L^p(\nu)}^{\theta} \cdot \|T\|_{L^q(\mu) \to L^q(\nu)}^{1-\theta} \quad if \quad \frac{1}{r} = \frac{1-\theta}{p} + \frac{\theta}{q}, \ \theta \in [0, 1].$$

Using $p = 1$ and the L^1-semigroup $(T_t^{(1)})_{t \geqslant 0}$ as the left end-point and $q = \infty$ and the Feller semigroup $(T_t)_{t \geqslant 0}$ as the right end-point, we see that $T_t|_{C_c(E)}$ extends to all intermediate spaces $L^p(m)$, $1 < p < \infty$ in such a way that the extensions yield strongly continuous, sub-Markovian contraction semigroups $(T_t^{(p)})_{t \geqslant 0}$ in $L^p(m)$.

The symmetry assumption (1.57) is quite restrictive and we can relax it in the following way. Let $(T_t)_{t \geqslant 0}$ be a Feller semigroup and denote by T_t^* the formal adjoint of T_t with respect to the space $L^2(m)$, i.e. the linear operator defined by

$$\int_E T_t u(x) \cdot w(x) \, m(dx) = \int_E u(x) \cdot (T_t^* w)(dx) \quad \forall u, w \in C_c(E). \tag{1.58}$$

Note that $T_t^* w \in \mathcal{M}_b(E)$ since the bounded Radon measures $\mathcal{M}_b(E)$ are the topological dual of $C_\infty(E)$. If we know that

$$T_t^{\circledast} := T_t^*|_{L^1(m)} \text{ maps } L^1(m) \text{ into itself,}$$

then the calculations can be modified under the assumption that $T_t^{\circledast}$ is sub-Markovian, i.e. $0 \leqslant T_t^{\circledast} u \leqslant 1$ for all $u \in L^1(m)$ such that $0 \leqslant u \leqslant 1$.

Lemma 1.45. *Let $(T_t)_{t \geqslant 0}$ be a Feller semigroup and assume that the operators T_t are m-symmetric or that the $L^2(m)$-adjoints $T_t^{\circledast} := T_t^*|_{L^1(m)}$ are sub-Markovian. Then $(T_t)_{t \geqslant 0}$ has for every $1 \leqslant p < \infty$ an extension $(T_t^{(p)})_{t \geqslant 0}$ to a strongly continuous, positivity preserving, sub-Markovian contraction semigroup on $L^p(m)$.*

If the domain $\mathcal{D}(A)$ of the Feller generator A contains a subset $\mathcal{D}$ which is dense both in $C_\infty(E)$ and $L^1(m)$, one can show that the $L^p(m)$-generators $(A^{(p)}, \mathcal{D}(A^{(p)}))$ coincide on this set with $A|_{\mathcal{D}}$. A proof for the m-symmetric case and $p = 2$ is given in Proposition 3.15.

Remark 1.46. There are good reasons to consider semigroups in an L^p-setting and not only in L^2. In general, L^p-theories lead to better regularity and embedding results than the corresponding L^2-theory; moreover, one has much better control on capacities. Therefore, L^p-semigroups have been studied by Fukushima [114]; building on earlier work of Malliavin, see [215, Part II] for a survey, (r, p)-capacities were studied by Fukushima and Kaneko [116] in order to get a better grip on exceptional (capacity-zero) sets. These are also discussed in [165, 166]. For further regularity results using Paley–Littlewood theory, we refer to Stein [307]. If we happen to know that $(T_t)_{t \geqslant 0}$ is an analytic semigroup on all spaces L^p, standard results from semigroup theory tell us that $T_t(L^p) \subset \bigcap_{k \geqslant 1} \mathcal{D}((-A)^k)$ and, should we have Sobolev embeddings, then $\bigcap_{k \geqslant 1} \mathcal{D}((-A)^k)$ can be embedded into spaces of continuous and differentiable functions, see [167, Sect. 2]. A concrete application to gradient perturbations is given in [349, Theorem 1.1 and its proof]. Finally, many results on functional inequalities are set in L^p spaces, see e.g. Wang [340, Chap. 5].

Chapter 2
Feller Generators and Symbols

In this chapter we are going to study the structure of Feller generators. Throughout we assume that $E = \mathbb{R}^d$ or $E = [0, \infty)$. Our exposition will heavily depend on methods from Fourier analysis. Let us briefly review the most important definitions and formulae. The **Fourier transform** of a (signed or even complex-valued) measure $\mu \in \mathcal{M}_b(\mathbb{R}^d)$ is defined as

$$\mathcal{F}\mu(\xi) = \hat{\mu}(\xi) = (2\pi)^{-d} \int_{\mathbb{R}^d} e^{-ix\cdot\xi}\, \mu(dx), \tag{2.1}$$

and the **inverse Fourier transform** is

$$\mathcal{F}^{-1}\mu(x) = \breve{\mu}(x) = \int_{\mathbb{R}^d} e^{ix\cdot\xi}\, \mu(d\xi). \tag{2.2}$$

The (inverse) Fourier transform of a function $u \in L^1(dx)$ is defined as the (inverse) Fourier transform of the measure $\mu(dx) = u(x)\, dx$. On the Schwartz space $\mathcal{S}(\mathbb{R}^d)$ of rapidly decreasing C^∞-functions, $\mathcal{F}$ and $\mathcal{F}^{-1}$ are indeed inverse operations. If $X \sim \mu$ is a random variable with law μ, then the **characteristic function** is the inverse Fourier transform of μ: $\mathbb{E}\, e^{i\xi\cdot X} = \mathcal{F}^{-1}\mu(\xi)$. The (inverse) Fourier transform trivializes the convolution product

$$\mathcal{F}[\mu * \nu](\xi) = (2\pi)^d\, \mathcal{F}\mu(\xi)\, \mathcal{F}\nu(\xi), \quad \mu, \nu \in \mathcal{M}_b(\mathbb{R}^d), \tag{2.3}$$

and satisfies the following **Plancherel's identity**

$$\int \mathcal{F}u(\xi)\, \mu(d\xi) = \int u(x)\, \mathcal{F}\mu(x)\, dx, \tag{2.4}$$

B. Böttcher et al., *Lévy Matters III*, Lecture Notes in Mathematics 2099, DOI 10.1007/978-3-319-02684-8_2,
© Springer International Publishing Switzerland 2013

for all $u \in L^1(dx)$ and $\mu \in \mathcal{M}_b(\mathbb{R}^d)$. In particular, $\|\mathcal{F}u\|_{L^2} = (2\pi)^{-d/2}\|u\|_{L^2}$ for all $u \in L^2(dx) \cap L^1(dx)$ and $\|\mathcal{F}u\|_\infty \leqslant (2\pi)^{-d}\|u\|_{L^1}$ for $u \in L^1(dx)$. Finally,

$$P(\xi)\mathcal{F}[u](\xi) = \mathcal{F}[P(-i\nabla)u](\xi) \quad \text{and} \quad P(i\nabla)\mathcal{F}[u](\xi) = \mathcal{F}[Pu](\xi) \qquad (2.5)$$

for all polynomials $P : \mathbb{R}^d \to \mathbb{C}$ and $u \in \mathcal{S}(\mathbb{R}^d)$.

2.1 Orientation: Convolution Semigroups and Lévy Processes

Assume that $(T_t)_{t\geqslant0}$ is a Feller semigroup on $\mathbb{R}^d$ or $[0,\infty)$ which is **invariant under translations**, i.e. it satisfies

$$\tau_{-z}(T_t u) = T_t(\tau_{-z}u) \quad \text{where} \quad \tau_{-z}u := u(\cdot + z). \qquad (2.6)$$

By a classical result from functional analysis, cf. Rudin [259, Theorem 6.33, p. 173], a continuous linear operator $L : C_c^\infty(\mathbb{R}^d) \to C(\mathbb{R}^d)$ is translation invariant if, and only if, it is a convolution operator, i.e. $Lu = \lambda * u$ for some Schwartz distribution $\lambda \in \mathcal{D}'(\mathbb{R}^d)$. Since T_t is positivity preserving and sub-Markovian, λ is a positive distribution of order zero, i.e. a positive Radon measure, cf. Hörmander [142, Theorem 2.1.7, p. 38]. Thus,

$$T_t u(x) = \int_{\mathbb{R}^d} u(y + x)\,\mu_t(dy) = \tilde{\mu}_t * u(x) \quad \forall u \in C_\infty(\mathbb{R}^d) \qquad (2.7)$$

$(\tilde{\mu}_t(B) = \mu_t(-B))$ denotes the measure obtained by reflection at the origin), and by the sub-Markov property of T_t we see that $\tilde{\mu}_t(\mathbb{R}^d) \leqslant 1$ is a sub-probability measure. If we compare $\tilde{\mu}_t$ with the kernel (1.17) obtained from the Riesz representation theorem, we see that $p_t(x, dy) = \tilde{\mu}_t(x - dy) = \mu_t(dy - x)$. The formula (2.7) also shows that strong continuity of $(T_t)_{t\geqslant0}$ is the same as vague continuity of the family of measures $(\mu_t)_{t\geqslant0}$, and that the semigroup property of $(T_t)_{t\geqslant0}$ is the same as the convolution property for the measures $(\mu_t)_{t\geqslant0}$

$$T_0 = \text{id}, \quad T_{t+s} = T_t T_s \iff \mu_0 = \delta_0, \quad \mu_{t+s} = \mu_t * \mu_s; \qquad (2.8)$$

therefore, $(\mu_t)_{t\geqslant0}$ is often called a **(vaguely continuous) convolution semigroup** of measures.

Lemma 2.1. *There is a one-to-one correspondence between translation invariant Feller semigroups $(T_t)_{t\geqslant0}$ and vaguely continuous convolution semigroups of sub-probability measures $(\mu_t)_{t\geqslant0}$.*

Since the Fourier transform trivializes the convolution product, it is the method of choice in the analysis of translation invariant Feller semigroups. From (2.8) we see that $\tilde{\mu}_t$ is an **infinitely divisible measure**, i.e.

$$\mu_t = \mu_{t/n}^{*n} \quad \text{and, under the inverse Fourier transform,} \quad \breve{\mu}_t = \left(\breve{\mu}_{t/n}\right)^n.$$

For rational $t = q$ we get $\breve{\mu}_q = \exp(q \log \breve{\mu}_1)$—the logarithm of the complex function $\breve{\mu}_1$ is uniquely defined, e.g. Sato [267, Lemma 7.6, p. 33], Bingham–Goldie–Teugels [29, second paragraph on p. 338] or Dieudonné [86, Chap. IX, Appendix 2, pp. 252–253]—and because of the vague continuity of the family $(\mu_t)_{t \geqslant 0}$ we get the following theorem which is known from the theory of limit distributions. A probabilistic proof is given in Sato [267, Theorem 8.1, pp. 37–38], analytic proofs can be found in [73, 162] as well as in Corollary 2.22 in the next section.

Theorem 2.2. *Let $(T_t)_{t \geqslant 0}$ be a translation invariant Feller semigroup. Then*

$$\widehat{T_t u}(\xi) = e^{-t\psi(\xi)} \hat{u}(\xi) \quad \forall \xi \in \mathbb{R}^d, \ u \in C_c^\infty(\mathbb{R}^d) \tag{2.9}$$

where the exponent $\psi : \mathbb{R}^d \to \mathbb{C}$ is of the form

$$\psi(\xi) = \psi(0) - il \cdot \xi + \frac{1}{2}\xi \cdot Q\xi + \int_{\mathbb{R}^d \setminus \{0\}} \left(1 - e^{iy \cdot \xi} + i\xi \cdot y \chi(|y|)\right) \nu(dy) \tag{2.10}$$

with $\psi(0) \geqslant 0$, a vector $l \in \mathbb{R}^d$, a symmetric positive semi-definite $d \times d$-matrix $Q \in \mathbb{R}^{d \times d}$, a measure $\nu \in \mathcal{M}^+(\mathbb{R}^d \setminus \{0\})$ such that $\int_{\mathbb{R}^d \setminus \{0\}} \min(|y|^2, 1) \, \nu(dy) < \infty$, and some arbitrary function $\chi \in B_b[0, \infty)$ such that $0 \leqslant 1 - \chi(s) \leqslant \kappa \min(s, 1)$ (for some $\kappa > 0$) and $s\chi(s)$ stays bounded.

Conversely, (2.9) defines, for any ψ of the form (2.10), a translation invariant Feller semigroup.

Up to the choice of χ, the exponent $\psi - \psi(0)$ is uniquely determined by (l, Q, ν) and vice versa. The choice of χ only influences the parameter l.

Common choices for the function χ are $\chi(s) = \mathbb{1}_{[0,1]}(s)$ or $(1 + s^2)^{-1}$ or $s^{-1} \sin s$. Theorem 2.2 says, in particular, that we can embed every infinitely divisible probability measure μ into a vaguely continuous convolution semigroup $(\mu_t)_{t \geqslant 0}$ if we set $\mu_1 := \mu$.

Definition 2.3. Let $(T_t)_{t \geqslant 0}$, ψ, (l, Q, ν) and χ be as in Theorem 2.2. Then

ψ	is called the **characteristic exponent**,
$\psi(0)$	is the **killing rate**,
(l, Q, ν)	is the **Lévy triplet**,
l	is the **drift coefficient**,
Q	is the **covariance matrix**,
ν	is the **jump** or **Lévy measure**,
χ	is the **truncation function**,
(2.10)	is called the **Lévy–Khintchine formula**.

Observe that $\operatorname{Re} \psi \geqslant 0$ is the characteristic exponent of the symmetrized Lévy process $X_{t/2} - \tilde{X}_{t/2}$ where $(\tilde{X}_t)_{t \geqslant 0}$ is an independent copy of $(X_t)_{t \geqslant 0}$. This follows from $e^{-t \operatorname{Re}\psi(\xi)} = |e^{-\frac{1}{2}\psi(\xi)}|^2 = \mathbb{E}\,e^{i\xi\cdot(X_{t/2}-\tilde{X}_{t/2})}$. By Taylor's theorem, the integrand in (2.10) is bounded by $c \min(|y|^2, 1)(1+|\xi|^2)$, so

$$|\psi(\xi)| \leqslant c_\psi (1+|\xi|^2) \quad \forall \xi \in \mathbb{R}^d. \tag{2.11}$$

A different proof will be given in Sect. 2.2 where we identify the characteristic exponents as the continuous negative definite functions in the sense of Schoenberg.

Example 2.4. We will determine the characteristic exponents of the semigroups and processes from Example 1.3 and Example 1.17, respectively; cf. Example 1.25 for the generators. The Lévy triplet (l, Q, ν) is given relative to the truncation function $\mathbb{1}_{[0,1]}$. Throughout we will always assume that $\psi(0) = 0$.

a) (*Shift semigroup—deterministic drift*) $\psi(\xi) = -il \cdot \xi$ and $(l, Q, \nu) = (l, 0, 0)$ where $l \in \mathbb{R}^d$ is the speed.
b) (*Poisson semigroup—Poisson process*) $\psi(\xi) = \lambda(1 - e^{i\xi\cdot l})$; here $l \in \mathbb{R}^d$ is the jump size and $\lambda > 0$ the intensity of the jumps. We have $(l, Q, \nu) = (0, 0, \lambda\delta_l)$ if $|l| > 1$, otherwise $(l, Q, \nu) = (\lambda l, 0, \lambda\delta_l)$.
b') (*Compound Poisson processes*) Assume that μ is a finite measure with total mass $\|\mu\|$. Then

$$\psi(\xi) := \|\mu\| \int_{\mathbb{R}^d\setminus\{0\}} \left(1 - e^{i\xi\cdot l}\right) \frac{\mu(dl)}{\|\mu\|} \tag{2.12}$$

is the characteristic exponent of a d-dimensional **compound Poisson process** and $(l, Q, \nu) = (\int_{\overline{\mathbb{B}(0,1)}} l\,\mu(dl), 0, \mu)$. The formula (2.12) has an interesting probabilistic interpretation: A compound Poisson process is a mixture of Poisson processes with different jump sizes. The jump size is randomly distributed according to the law $F(dy) = \mu(dy)/\|\mu\|$, and jumps occur with the overall intensity $\|\mu\|$.

c) (*Heat/Brownian semigroup—Brownian motion*) $\psi(\xi) = \frac{1}{2}\xi \cdot Q\xi$ where $Q \in \mathbb{R}^{d\times d}$ is positive semi-definite and describes the, possibly degenerate, covariance structure of a Brownian motion, that is $B_t = \sqrt{Q}W_t$ where $(W_t)_{t\geqslant 0}$ is a standard Wiener process.
d) (*Symmetric stable semigroups—symmetric stable processes*) $\psi(\xi) = |\xi|^\alpha$ and $(l, Q, \nu) = (0, 0, c_\alpha|y|^{-d-\alpha}\,dy)$, $c_\alpha = \alpha 2^{\alpha-1}\pi^{-d/2}\Gamma\left(\frac{\alpha+d}{2}\right)/\Gamma\left(1-\frac{\alpha}{2}\right)$, where $0 < \alpha < 2$ is the stability parameter.
e) (*Convolution semigroups—Lévy processes*) This corresponds to general characteristic exponents ψ and Lévy triplets (l, Q, ν).
f) (*Ornstein–Uhlenbeck semigroups—Ornstein–Uhlenbeck processes*) These are only Lévy processes in the trivial case $B = 0$.
g) (*One-sided stable semigroups—stable subordinators*) $\psi(\xi) = (-i\xi)^\alpha$ where $\alpha \in (0, 1)$ and $f(\lambda) = \lambda^\alpha$ is the Laplace exponent of the process: This means $\mathbb{E}\,e^{-\lambda X_t} = e^{-t\lambda^\alpha}$. A simple change of variables in the formula for the Gamma function $\Gamma(\alpha)$ reveals that $\lambda^\alpha = \gamma_\alpha \int_0^\infty (1 - e^{-\lambda s})\,s^{-\alpha-1}\,ds$ where $\gamma_\alpha = \alpha/\Gamma(1-\alpha)$. Thus, $(l, Q, \nu) = (\gamma_\alpha/(1-\alpha), 0, \gamma_\alpha s^{-\alpha-1}\mathbb{1}_{(0,\infty)}(s)\,ds)$.

Since a one-sided process $(X_t)_{t \geq 0}$ takes values in the half-line $[0, \infty)$, it is common to use the Laplace transform $\mathbb{E}\, e^{-\lambda X_t}$ instead of the characteristic function $\mathbb{E}\, e^{i\xi \cdot X_t}$. Since the Laplace transform converges, our previous discussion on Lévy processes remains valid—up to the transformation $\xi \rightsquigarrow i\lambda$—and we do not need to introduce a special notation for subordinators. The Laplace exponents are exactly the **Bernstein functions**. We refer to the monograph [293] for a detailed discussion. $\qquad\qquad\square$

Let us now turn to the stochastic processes associated with a translation invariant Feller semigroup.

Definition 2.5. A **Lévy process** is a Feller process $(X_t)_{t \geq 0}$ with translation invariant, conservative semigroup $(T_t)_{t \geq 0}$. If the semigroup is not conservative, $(X_t)_{t \geq 0}$ is a **Lévy process with (exponential) killing**.

Note that μ_t is a probability measure if, and only if,

$$1 = \mu_t(\mathbb{R}^d) = \breve{\mu}_t(0) = e^{-t\psi(0)}.$$

Thus, conservativeness is the same as to say that $\psi(0) = 0$. On the other hand, if $\psi(0) > 0$, we have $\mu_t(\mathbb{R}^d) = \mathbb{P}(X_t \in \mathbb{R}^d) = e^{-t\psi(0)} < 1$ which means that the process $(X_t)_{t \geq 0}$ "dies (i.e. leaves the state space)" according to an exponential distribution with rate $\psi(0)$.

At first sight, the Definition 2.5 of a Lévy process looks unfamiliar, but the next result will fix this.

Theorem 2.6. *A stochastic process $(X_t)_{t \geq 0}$ with values in $\mathbb{R}^d$ is a Lévy process if, and only if, it has (a modification satisfying)*

a) *stationary increments:*

$$\mathbb{P}^x(X_t - X_s \in B) = \mathbb{P}^0(X_{t-s} \in B) \quad \text{for} \quad x \in \mathbb{R}^d,\ B \in \mathscr{B}(\mathbb{R}^d),\ 0 \leq s < t;$$

b) *independent increments:* $(X_{t_j} - X_{t_{j-1}})_{j=1}^n$ *are independent for* $0 \leq t_0 < \cdots < t_n$;
c) *càdlàg paths:* $t \mapsto X_t(\omega)$ *is almost surely right-continuous with finite left limits.*

Note that for Lévy processes it is usually assumed that the process starts at $x = 0$, i.e. $X_0 = 0$. In this setting condition 2.6(a) simplifies to

$$X_t - X_s \sim X_{t-s} \quad \text{for} \quad 0 \leq s < t.$$

Proof of Theorem 2.6. Assume that $(X_t)_{t \geq 0}$ is a Lévy process in the sense of Definition 2.5, and equip it with the filtration $\mathscr{F}_t := \mathscr{F}_{t+}^X$ (or a completion thereof). By the Markov property, we find for all $x, \xi \in \mathbb{R}^d$ and $0 \leq s < t$

$$\mathbb{E}^x\left(e^{i\xi \cdot (X_t - X_s)} \mid \mathscr{F}_s\right) = \mathbb{E}^{X_s}\, e^{i\xi \cdot (X_{t-s} - X_0)} = \mathbb{E}^y\, e^{i\xi \cdot (X_{t-s} - y)}\Big|_{y = X_s} = \mathbb{E}^0\, e^{i\xi \cdot X_{t-s}},$$

where we used translation invariance in the last step. This proves that $X_t - X_s$ is distributed like X_{t-s}, $X_0 = 0$ a.s. and that $X_t - X_s$ is independent of $\mathscr{F}_s$. Since all increments occurring up to time s are $\mathscr{F}_s$ measurable, we have shown the properties (a) and (b), while (c) is always satisfied for (a suitable version of) a Feller process, cf. Theorem 1.19.

Conversely, assume that $(X_t)_{t \geq 0}$ is a process on $(\Omega, \mathscr{F}, \mathbb{P})$ satisfying (a)–(c). We have to show the Markov property and that the associated operator semigroup is a Feller semigroup. First we note that Property (b) is equivalent to the independence of $X_{t+s} - X_s$ and $\mathscr{F}_s := \mathscr{F}_{s+}^X$ since the natural filtration $\mathscr{F}_s^X$ is also generated by all increments in $[0, s]$. Thus,

$$\mathbb{E}\left(u(X_{t+s}) \mid \mathscr{F}_s\right) = \mathbb{E}\left(u(X_{t+s} - X_s + X_s) \mid \mathscr{F}_s\right) = \mathbb{E}\,u(X_t + y)\big|_{y = X_s}$$

(see [284, Theorem 6.2, p. 63] for more details) and this shows that

$$T_t u(x) := \mathbb{E}^x u(X_t) := \mathbb{E}\,u(X_t + x) = \int_{\mathbb{R}^d} u(y + x)\,\mathbb{P}(X_t \in dy), \quad u \in B_b(\mathbb{R}^d),$$

defines a sub-Markovian semigroup. Being a convolution operator, it is not hard to see that T_t maps $C_\infty(\mathbb{R}^d)$ to $C_\infty(\mathbb{R}^d)$. The strong continuity now follows from the càdlàg property of the sample paths: Let $\phi \in C_c(\mathbb{R}^d)$ with $\mathbb{1}_{B(0,\delta)} \leq \phi \leq \mathbb{1}_{B(0,2\delta)}$. Then

$$|T_t u(x) - u(x)| \leq \left| \int (u(X_t + x) - u(x))\phi(X_t)\,d\mathbb{P} \right| + 2\|u\|_\infty\,\mathbb{E}(1 - \phi(X_t))$$

$$\leq \int_{|X_t| \leq 2\delta} |u(X_t + x) - u(x)|\,d\mathbb{P} + 2\|u\|_\infty\,\mathbb{E}(1 - \phi(X_t)).$$

The assertion follows using the uniform continuity of u (for the first expression), dominated convergence and $\lim_{t \to 0+} \mathbb{E}(1 - \phi(X_t)) = 0$ (for the second). $\square$

We can use the Fourier transform to determine the generator of a Lévy process. Since $\mathrm{Re}\,\psi \geq 0$, we see with Theorem 2.2 that $|\widehat{T_t u}| = |e^{-t\psi}\hat{u}| \leq |\hat{u}|$, i.e. $\widehat{T_t u}$ is contained in $L^1(d\xi)$ if, say, $u \in C_c^\infty(\mathbb{R}^d)$. Thus, for all $u \in C_c^\infty(\mathbb{R}^d)$

$$\frac{T_t u(x) - u(x)}{t} = \frac{\mathcal{F}^{-1}\big[\widehat{T_t u} - \hat{u}\big](x)}{t} = \mathcal{F}^{-1}\left[\frac{e^{-t\psi} - 1}{t}\,\hat{u}\right](x) \xrightarrow[t \to 0]{} \mathcal{F}^{-1}[-\psi\,\hat{u}](x)$$

where we used the elementary estimate $|e^{-\zeta} - 1| \leq |\zeta|$ for all $\zeta \in \mathbb{C}$ with $\mathrm{Re}\,\zeta \geq 0$ and the polynomial bound (2.11) for ψ.

Theorem 2.7. *Let $(T_t)_{t \geq 0}$ be a translation invariant Feller semigroup with generator $(A, \mathcal{D}(A))$ and $(X_t)_{t \geq 0}$ the corresponding Lévy process with killing and characteristic exponent ψ. Then $C_c^\infty(\mathbb{R}^d) \subset A$ and $A|_{C_c^\infty(\mathbb{R}^d)}$ is of the form*

$$Au(x) = \mathcal{F}^{-1}[-\psi\,\hat{u}](x) = -\int e^{ix\cdot\xi}\psi(\xi)\hat{u}(\xi)\,d\xi, \quad u \in C_c^\infty(\mathbb{R}^d). \quad (2.13)$$

Conversely, every operator of the form (2.13) *with* ψ *given by* (2.10)*, generates a translation invariant Feller semigroup and a Lévy process with killing.*

Remark 2.8. Let $m(\xi)$ be a polynomially bounded, real or complex function on $\mathbb{R}^d$. An operator of the form

$$m(D)u := \mathcal{F}^{-1}[m\,\mathcal{F}u] \quad \forall u \in C_c^\infty(\mathbb{R}^d)$$

is a **Fourier multiplier operator**. The function m is called the **Fourier multiplier** or **symbol**.

Theorem 2.7 shows that the generator of a translation invariant Feller semigroup $(T_t)_{t\geq 0}$ (or a Lévy process $(X_t)_{t\geq 0}$) is a Fourier multiplier operator with the symbol $-\psi$. Similarly, we know from Theorem 2.2 that T_t is a Fourier multiplier operator with the symbol $e^{-t\psi(\xi)}$, and it is only a short calculation to see that the resolvent U_α (cf. Definition 1.21) is a Fourier multiplier operator with the symbol $(\alpha + \psi(\xi))^{-1}$.

Using the rules for the Fourier transform one sees readily that the composition of two Fourier multiplier operators with the symbols $m(\xi)$ and $n(\xi)$ is again a Fourier multiplier operator with the symbol $m(\xi)n(\xi)$, and that any two Fourier multiplier operators commute. If we choose $m(\xi)$ as a polynomial, the corresponding Fourier multiplier operator is a differential operator with constant coefficients. In general, Fourier multiplier operators are pseudo-differential operators with *constant "coefficients"*, see Definition 2.25. Good sources for Fourier multiplier operators are Duoandikoetxea [94, Chap. 8], Edwards [98, Chap. 16], Grafakos [126, Chap. 5] and Larsen [201].

A discussion on general L^p–L^q Fourier multipliers in connection with characteristic exponents of Lévy processes (which are also known as negative definite functions, cf. Sect. 2.2 below) can be found in [105, Sect. 2.3, pp. 46–52] and [165, Sect. 3.1, pp. 1244–1245]. Bañuelos–Bogdan [11] prove the L^p-boundedness and provide sharp continuity constants for certain Fourier multipliers with a negative definite ψ; these results are remarkable as they do not use the Hörmander–Mikhlin multiplier theorem due to the missing smoothness of ψ.

Abels–Husseini [1] give necessary and sufficient conditions on ψ which guarantee that the operator $-\psi(D)$ is locally hypoelliptic.[1]

In order to introduce *variable "coefficients"*, we have to change the multiplier $m(\xi) \rightsquigarrow m(x,\xi)$, but then the nice composition rule, commutativity and many other properties break down.   □

If we replace in (2.13) ψ by the Lévy-Khintchine representation (2.10) and invert the Fourier transform, we get

---

[1] That is if $f, u \in W^{2,s}(\mathbb{R}^d)$ for some $s \in [-\infty, \infty)$ and $\psi(D)u = f \in C^\infty(U)$ for some open set $U \in \mathbb{R}^d$, then $u \in C^\infty(U)$.

$$Au(x) = Lu(x) + Su(x)$$

with

$$Lu(x) = -\gamma u(x) + l \cdot \nabla u(x) + \frac{1}{2}\operatorname{div} Q \nabla u(x)$$

and

$$Su(x) = \int_{\mathbb{R}^d \setminus \{0\}} \left(u(x+y) - u(x) - \nabla u(x) \cdot y\chi(y)\right) \nu(dy);$$

(l, Q, ν) is the Lévy triplet of ψ, $\gamma = \psi(0) \geqslant 0$, and χ is any truncation function as in Theorem 2.2. Set

$$\|u\|_{(2)} := \sum_{0 \leqslant |\alpha| \leqslant 2} \|\nabla^\alpha u\|_\infty, \quad u \in C_b^2(\mathbb{R}^d).$$

Then it is obvious that

$$\|Lu\|_\infty \leqslant c(\gamma + |l| + |Q|)\|u\|_{(2)}.$$

In order to get a similar estimate for the operator S we observe that

$$\left|u(x+y) - u(x) - \nabla u(x) \cdot y\chi(|y|)\right|$$

$$= \left|\int_0^1 \int_0^\theta y^\top D^2 u(x+ty)y\, dt\, d\theta + (1 - \chi(|y|))y \cdot \nabla u(x)\right|$$

$$\leqslant c|y|^2 \left(\||D^2 u|\|_\infty + \||\nabla u|\|_\infty\right) \qquad\qquad (|y| \leqslant 1)$$

($D^2 u$ denotes the Hessian) and

$$\left|u(x+y) - u(x) - \nabla u(x) \cdot y\chi(|y|)\right| \leqslant c'\left(\|u\|_\infty + \||\nabla u|\|_\infty\right) \qquad (|y| > 1)$$

(mind the properties of the truncation function χ at $s = 0$ and $s = \infty$). Thus we have

$$\|Su\|_\infty \leqslant c'' \int_{\mathbb{R}^d \setminus \{0\}} \min(|y|^2, 1)\nu(dy)\|u\|_{(2)},$$

(we may assume that $\nu \neq 0$, otherwise the estimate would be trivial), and we get

$$\|Au\|_\infty \leqslant C \left(\gamma + |l| + |Q| + \int_{\mathbb{R}^d \setminus \{0\}} \min(|y|^2, 1)\, \nu(dy)\right)\|u\|_{(2)} \qquad (2.14)$$

for some absolute constant C.

By the bounded linear transform (B.L.T.) theorem (cf. Reed–Simon [248, Theorem I.7, p. 9]) we see that the estimate (2.14) allows us to extend $(A, C_c^\infty(\mathbb{R}^d))$ uniquely onto $C_\infty^2(\mathbb{R}^d)$ such that the estimate remains valid on the larger space. This means that the graph norm $\|u\|_\infty + \|Au\|_\infty$ of A is bounded by $\|u\|_{(2)}$. Since the generator $(A, \mathcal{D}(A))$ is a closed operator, we conclude that

$$\overline{C_c^\infty(\mathbb{R}^d)}^{\|\cdot\|_{(2)}} = C_\infty^2(\mathbb{R}^d) \subset \mathcal{D}(A),$$

see also [277, Lemma 2.5].

Corollary 2.9. *Let* $(T_t)_{t \geqslant 0}$, $(A, \mathcal{D}(A))$ *and* $(X_t)_{t \geqslant 0}$ *be as in Theorem 2.7. Then* $C_\infty^2(\mathbb{R}^d) \subset \mathcal{D}(A)$ *and* $A|_{C_\infty^2(\mathbb{R}^d)}$ *is an **integro-differential operator***

$$Au(x) = -\gamma u(x) + l \cdot \nabla u(x) + \frac{1}{2}\mathrm{div}\, Q\, \nabla u(x)$$

$$+ \int_{\mathbb{R}^d \setminus \{0\}} \big(u(x+y) - u(x) - \nabla u(x) \cdot y\chi(y)\big)\nu(dy) \tag{2.15}$$

where (l, Q, ν) *is the Lévy triplet of* ψ, $\gamma = \psi(0) \geqslant 0$ *and* $\chi \in B_b[0, \infty)$ *is a truncation function such that* $0 \leqslant 1 - \chi(s) \leqslant \kappa \min(s, 1)$ *(for some* $\kappa > 0$*) and* $s\chi(s)$ *stays bounded.*

Moreover, the estimate $\|Au\|_\infty \leqslant c_{l,Q,\nu}\|u\|_{(2)}$ *holds for all* $u \in C_\infty^2(\mathbb{R}^d)$.

Corollary 2.10. *Let* $(A, \mathcal{D}(A))$ *be as in Theorem 2.7. Then* $C_c^\infty(\mathbb{R}^d)$ *is an operator core.*

Proof. Denote by ψ be the symbol of the generator and write U_α for the α-resolvent operator, $\alpha > 0$. For $u \in C_c^\infty(\mathbb{R}^d)$ we know that $\hat{u} \in S(\mathbb{R}^d)$; since $\mathrm{Re}\,\psi \geqslant 0$,

$$|\xi|^n |(\alpha + \psi)^{-1}\hat{u}(\xi)| \leqslant \frac{1}{\alpha}|\xi|^n|\hat{u}(\xi)| \in L^1(\mathbb{R}^d, d\xi) \quad \forall n \geqslant 0$$

and the Riemann–Lebesgue lemma proves that

$$\nabla^\beta U_\alpha u(x) = \int_{\mathbb{R}^d} e^{ix\cdot\xi}(i\xi)^\beta(\alpha + \psi(\xi))^{-1}\hat{u}(\xi)\,d\xi$$

is in $C_\infty(\mathbb{R}^d)$ for all $\beta \in \mathbb{N}_0^d$. Thus, $U_\alpha C_c^\infty(\mathbb{R}^d) \subset C_\infty^\infty(\mathbb{R}^d)$, and Lemma 1.34(b) shows that $C_\infty^\infty(\mathbb{R}^d)$ is a core for $(A, \mathcal{D}(A))$.

In order to show that $C_c^\infty(\mathbb{R}^d)$ is a core for $(A, \mathcal{D}(A))$, we take any sequence $(\phi_n)_{n \geqslant 1} \subset C_c^\infty(\mathbb{R}^d)$ satisfying $\mathbb{1}_{B(0,n)} \leqslant \phi_n \leqslant 1$ and $\|\phi_n\|_{(2)} \leqslant 2$ for every $n \geqslant 1$. Pick any $u \in C_\infty^\infty(\mathbb{R}^d)$. Then $u_n := \phi_n u \in C_c^\infty(\mathbb{R}^d)$, and it is not hard to see that $\lim_{n\to\infty}\|u - u_n\|_{(2)} = 0$. From (2.14) we get

$$\|u - u_n\|_\infty + \|Au - Au_n\|_\infty \leqslant c\|u - u_n\|_{(2)} \xrightarrow[n\to\infty]{} 0,$$

and this proves that $C_c^\infty(\mathbb{R}^d)$ is also an operator core. $\square$

Corollary 2.11. *Let* $(A, \mathcal{D}(A))$ *be as in Theorem 2.7. Then* A *has a unique extension to* $C_b^2(\mathbb{R}^d)$, *again denoted by* A, *satisfying*

$$Au(x) = \lim_{n\to\infty} A(u\phi_n)(x)$$

for $u \in C_b^2(\mathbb{R}^d)$ *and any sequence* $(\phi_n)_{n\geqslant 1} \subset C_c^\infty(\mathbb{R}^d)$ *with* $\mathbb{1}_{B(0,n)} \leqslant \phi_n \leqslant 1$. *Moreover, the extension is again represented by the formula* (2.15) *and*

$$\|Au\|_\infty \leqslant c\|u\|_{(2)} \quad \forall u \in C_b^2(\mathbb{R}^d).$$

Proof. Note that the formula (2.15) for Au remains valid if $u \in C_b^2(\mathbb{R}^d)$. Let us, for the moment, write $\tilde{A}$ if we mean this extension. By dominated convergence we get $\tilde{A}u(x) = \lim_{n\to\infty} A(u\phi_n)(x)$ for any $u \in C_b^2(\mathbb{R}^d)$ and any sequence $(\phi_n)_{n\geqslant 1}$ as described in the statement of the corollary. In this sense, the extension $\tilde{A}$ is unique. If we pick the ϕ_n such that $\|\phi_n\|_{(2)} \leqslant 1$, we get from (2.14)

$$\|\tilde{A}u\|_\infty = \Big\| \lim_{n\to\infty} A(u\phi_n)\Big\|_\infty \leqslant \varlimsup_{n\to\infty} \|A(u\phi_n)\|_\infty \leqslant c \varlimsup_{n\to\infty} \|u\phi_n\|_{(2)} \leqslant c'\|u\|_{(2)}.$$

$\square$

The characteristic exponent ψ and the Lévy triplet (l, Q, ν) play a key role for the further study of Lévy processes. For sample path properties, the survey papers by Fristedt [112] and Taylor [318] are still unsurpassed. An up-to-date account on Hausdorff dimensions and other fractal properties can be found in Xiao [353]. Sato [267] is the best source for general properties of a Lévy process, while Bertoin [27] and Kyprianou [198] concentrate on potential theory and fluctuation theory. For a quick introduction (without proofs) from a stochastic analysis point of view, Protter [243, Chap. I.§§3–4] is a good choice; his exposition is based on Bretagnolle [53]; a full treatment is given in Applebaum [5]. The presentations in Berg–Forst [24] and Jacob [157] are closest to the point of view in this section.

Let us close with the Lévy–Itô or semimartingale representation of a Lévy process. This shows that the Lévy characteristics are essentially the semimartingale characteristics in the sense of Jacod–Shiryaev [170, Chap. II.§2a, pp. 75–81 and II.§4c, pp. 106–111]. Following Protter [243, Theorem I.42, p. 31] we have

Theorem 2.12. *Let* $(X_t)_{t\geqslant 0}$ *be a Lévy process with characteristic exponent* ψ, $\psi(0) = 0$ *and Lévy triplet* (l, Q, ν) *w.r.t. the truncation function* $\chi = \mathbb{1}_{[0,1]}$. *Let* $(W_t)_{t\geqslant 0}$ *be a standard Brownian motion on* $\mathbb{R}^d$ *and* $N_t(B) = \sum_{0<s\leqslant t} \mathbb{1}_B(\Delta X_s)$, $B \in \mathscr{B}(\mathbb{R}^d \setminus \{0\})$, *be the counting measure of the jumps* $\Delta X_s = X_s - X_{s-}$ *of* $(X_t)_{t\geqslant 0}$ *of size* B *occurring in* $[0, t]$; *for each* B, $(N_t(\cdot, B))_{t\geqslant 0}$ *is a Poisson process with intensity* $\nu(B) \in [0, \infty]$.

*Then $(X_t)_{t \geqslant 0}$ has the following **Lévy–Itô decomposition***

$$X_t = lt + \sqrt{Q}W_t + \int_{0<|y|\leqslant 1} y\,(N_t(\cdot,dy) - t\nu(dy)) + \sum_{0<s\leqslant t} \Delta X_s\, \mathbb{1}_{\{|\Delta X_s|>1\}}.$$

$$(2.16)$$

The four components appearing in the decomposition are stochastically independent Lévy processes.

An *elementary* proof of Theorem 2.12 can be found in Breiman [52, Chap. 14.8, pp. 312–315]. The four processes appearing in (2.16) yield various decompositions of the process $(X_t)_{t \geqslant 0}$,

$$X_t = X_t^{\mathrm{C}} + X_t^{\mathrm{D}} = X_t^{\mathrm{MG}} + X_t^{\mathrm{FV}}$$

where

- $X_t^{\mathrm{C}} = lt + \sqrt{Q}W_t$ is the continuous part,
- $X_t^{\mathrm{D}} = \int_{0<|y|\leqslant 1} y\,(N_t(\cdot,dy) - t\nu(dy)) + \sum_{0<s\leqslant t} \Delta X_s\, \mathbb{1}_{\{|\Delta X_s|>1\}}$ is the pure jump part,
- $X_t^{\mathrm{MG}} = \sqrt{Q}W_t + \int_{0<|y|\leqslant 1} y\,(N_t(\cdot,dy) - t\nu(dy))$ is the L^2-martingale part,
- $X_t^{\mathrm{FV}} = lt + \sum_{0<s\leqslant t} \Delta X_s\, \mathbb{1}_{\{|\Delta X_s|>1\}}$ is the finite variation part of $(X_t)_{t \geqslant 0}$.

2.2   Positive and Negative Definite Functions

A function $\phi : \mathbb{R}^d \to \mathbb{C}$ is said to be **positive definite** (in the sense of Bochner), if

$$\sum_{j,k=1}^{n} \phi(\xi_j - \xi_k)\lambda_j \bar{\lambda}_k \geqslant 0 \quad \forall n \geqslant 1,\ \xi_1,\ldots,\xi_n \in \mathbb{R}^d,\ \lambda_1,\ldots,\lambda_n \in \mathbb{C}. \quad (2.17)$$

Fourier transforms of positive measures and characteristic functions of random variables are typical examples. In fact, these are already *all* possibilities. Easily accessible proofs of the following classical theorem by Bochner can be found in Bogachev [35, Theorem 7.13.1, p. 121], Reed–Simon [248, Theorem IX.9, p. 330] or Sasvári [266, Theorem 1.7.3, pp. 42–43].

Theorem 2.13. *A function $\phi : \mathbb{R}^d \to \mathbb{C}$ is continuous and positive definite if, and only if, it is the inverse Fourier transform of a measure $\mu \in \mathcal{M}_b^+(\mathbb{R}^d)$*

$$\phi(\xi) = \breve{\mu}(\xi) = \int e^{ix\cdot\xi}\,\mu(dx), \quad \xi \in \mathbb{R}^d.$$

The measure μ is uniquely determined by ϕ, and vice versa.

If $(\mu_t)_{t \geq 0}$ is a vaguely continuous convolution semigroup, we have seen in Sect. 2.1, see the discussion preceding Theorem 2.2, that $\breve{\mu}_t = \breve{\mu}_1^t = \exp(t \log \breve{\mu}_1)$ for all $t > 0$ and that the exponent $\psi := -\log \breve{\mu}_1$ characterizes the measures $(\mu_t)_{t \geq 0}$. The characteristic exponent ψ has an interesting analytic property: $-\psi$ is conditionally positive definite in the following sense. Pick $\xi_1, \ldots, \xi_n \in \mathbb{R}^d$ and $\lambda_1, \ldots, \lambda_n \in \mathbb{C}$ satisfying $\sum_{j=1}^{n} \lambda_j = 0$; then

$$\sum_{j,k=1}^{n} \left(1 - e^{-t\psi(\xi_j - \xi_k)}\right) \lambda_j \bar{\lambda}_k = -\sum_{j,k=1}^{n} e^{-t\psi(\xi_j - \xi_k)} \lambda_j \bar{\lambda}_k \leq 0$$

since $e^{-t\psi}$ is positive definite as it is the Fourier transform of a probability measure. Dividing by $t > 0$ and letting $t \to 0$ yields

$$\sum_{j,k=1}^{n} -\psi(\xi_j - \xi_k) \lambda_j \bar{\lambda}_k \geq 0 \qquad \begin{array}{l} \forall n \geq 1, \ \xi_1, \ldots, \xi_n \in \mathbb{R}^d, \\[2mm] \lambda_1, \ldots, \lambda_n \in \mathbb{C}, \ \sum_{j=1}^{n} \lambda_j = 0. \end{array} \tag{2.18}$$

Definition 2.14. A function $\psi : \mathbb{R}^d \to \mathbb{C}$ is **negative definite** (in the sense of Schoenberg) if $\psi(0) \geq 0$, $\psi(\xi) = \overline{\psi(-\xi)}$, and if it satisfies (2.18).

Observe that (2.18) resembles (2.17), up to the *condition* that $\sum_{j=1}^{n} \lambda_j = 0$. Therefore, some authors call $-\psi$ **conditionally positive definite**, if ψ is negative definite in the sense of Definition 2.14.

The following characterisation of (continuous) negative definite functions is compiled from Berg–Forst [24, §7, pp. 39–48, and Theorem 10.8, p. 75].

Theorem 2.15. *For a function* $\psi : \mathbb{R}^d \to \mathbb{C}$ *the following assertions are equivalent.*

a) ψ *is negative definite.*
b) $-\psi$ *is conditionally positive definite.*
c) $\psi(0) \geq 0$ *and* $e^{-t\psi}$ *is positive definite for every* $t > 0$.
d) *The matrix* $\left(\psi(\xi_j) + \overline{\psi(\xi_k)} - \psi(\xi_j - \xi_k)\right) \in \mathbb{C}^{n \times n}$ *is positive hermitian for all* $n \geq 1$ *and* $\xi_1, \ldots, \xi_n \in \mathbb{R}^d$, *i.e.*

$$\sum_{j,k=1}^{n} \left(\psi(\xi_j) + \overline{\psi(\xi_k)} - \psi(\xi_j - \xi_k)\right) \lambda_j \bar{\lambda}_k \geq 0 \qquad \begin{array}{l} \forall n \geq 1, \ \xi_1, \ldots, \xi_n \in \mathbb{R}^d, \\[2mm] \lambda_1, \ldots, \lambda_n \in \mathbb{C}. \end{array} \tag{2.19}$$

e) $\psi(0) \geq 0$ *and* $\frac{1}{t}\left(e^{-t\psi(0)} - e^{-t\psi}\right)$ *is negative definite for every* $t > 0$.
   Moreover, if ψ *is continuous, the assertions (a)–(e) are also equivalent to*
f) $\psi(0) \geq 0$ *and* ψ *is given by a Lévy–Khintchine representation*

$$\psi(\xi) = \psi(0) - il \cdot \xi + \frac{1}{2}\xi \cdot Q\xi + \int_{\mathbb{R}^d \setminus \{0\}} \left(1 - e^{iy \cdot \xi} + i\xi \cdot y\chi(|y|)\right)\nu(dy)$$

(2.20)

with the Lévy triplet (l, Q, ν) comprising $l \in \mathbb{R}^d$, $Q \in \mathbb{R}^{d \times d}$ symmetric and positive semidefinite, $\nu \in \mathcal{M}^+(\mathbb{R}^d \setminus \{0\})$ with $\int_{\mathbb{R}^d \setminus \{0\}} \min(|y|^2, 1)\,\nu(dy) < \infty$, and any truncation function $\chi \in B_b[0, \infty)$ such that, for some $\kappa > 0$, we have $0 \leqslant 1 - \chi(s) \leqslant \kappa \min(s, 1)$ and $s\chi(s)$ stays bounded.

g) *$\psi(0) \geqslant 0$ and $\psi - \psi(0)$ is the characteristic exponent of an infinitely divisible probability distribution μ_1 (or a vaguely continuous convolution semigroup $(\mu_t)_{t \geqslant 0}$ or a translation invariant Feller semigroup $(T_t)_{t \geqslant 0}$).*

Using Taylor's formula, it is not hard to see that the integrand appearing in the Lévy–Khintchine formula is for fixed $\xi \in \mathbb{R}^d$ of order $O((y \cdot \xi)^2)$ as $|y| \to 0$ and is bounded for $|y| \to \infty$. Thus, with a constant c_χ depending only on the truncation function,

$$\left|1 - e^{iy \cdot \xi} + i\xi \cdot y\chi(|y|)\right| \leqslant c_\chi \min(|y|^2, 1)(1 + |\xi|^2).$$

(2.21)

A continuous negative definite function is uniquely determined by $\psi(0)$ and the Lévy triplet (l, Q, ν), and it is possible to calculate the Lévy triplet from ψ, cf. Berg–Forst [24, Proposition 18.2, p. 172 and Theorem 18.19, p. 182].

Lemma 2.16. *Let $\psi : \mathbb{R}^d \to \mathbb{C}$ be a continuous negative definite function, and denote by $(T_t)_{t \geqslant 0}$ the corresponding operator semigroup. Then*

$$\int u\,d\nu = \lim_{t \to 0} \frac{1}{t} T_t u(0) \quad \forall u \in C_c(\mathbb{R}^d \setminus \{0\}),$$

$$\frac{1}{2}\xi \cdot Q\xi = \lim_{n \to \infty} \frac{\psi(n\xi)}{n^2}.$$

(2.22)

The formula for the Lévy measure can be restated as $\nu = \lim_{t \to 0} \frac{1}{t}\mu_t|_{\mathcal{B}(\mathbb{R}^d \setminus \{0\})}$ where $(\mu_t)_{t \geqslant 0}$ is the convolution semigroup associated with ψ and with the limit taken in the vague topology w.r.t. $C_\infty(\mathbb{R}^d \setminus \{0\})$—to wit: We have to test the measures against functions which are continuous functions with compact support in $\mathbb{R}^d \setminus \{0\}$.

(Continuous) negative definite functions enjoy a number of important properties which can be easily derived from the equivalent characterizations listed in Theorem 2.15.

Proposition 2.17. *Let $\psi : \mathbb{R}^d \to \mathbb{C}$ be a negative definite function.*

a) $\operatorname{Re} \psi(\xi) \geqslant \psi(0) \geqslant 0$ *for all $\xi \in \mathbb{R}^d$.*

b) $\overline{\psi}$ *and $\operatorname{Re} \psi$ are negative definite.*

c) $\sqrt{|\psi|}$ *is subadditive: $\sqrt{|\psi(\xi + \eta)|} \leqslant \sqrt{|\psi(\xi)|} + \sqrt{|\psi(\eta)|}$ for all $\xi, \eta \in \mathbb{R}^d$.*

d) ψ *is polynomially bounded: $|\psi(\xi)| \leqslant 2 \sup_{|\eta| \leqslant 1} |\psi(\eta)|(1 + |\xi|^2)$ for all $\xi \in \mathbb{R}^d$.*

e) *The family of all negative definite functions is a convex cone which is closed under pointwise convergence.*

f) *The family of all continuous negative definite functions is a convex cone which is closed under locally uniform convergence. If $\psi_n \to \psi$ and if (l_n, Q_n, v_n) is the Lévy triplet of ψ_n, then the Lévy triplet (l, Q, v) of ψ is given by*

$$\lim_{n\to\infty} \int_{\mathbb{R}^d\setminus\{0\}} u(y) \frac{|y|^2}{1+|y|^2} v_n(dy) = \int_{\mathbb{R}^d\setminus\{0\}} u(y) \frac{|y|^2}{1+|y|^2} v(dy) \quad \forall u \in C_b(\mathbb{R}^d)$$

$$\lim_{\epsilon\to 0} \overline{\lim_{n\to\infty}} \left| \xi \cdot Q_n\xi + \int_{0<|y|\leq\epsilon} (\xi \cdot y)^2 v_n(dy) - \xi \cdot Q\xi \right| = 0 \quad \text{and} \quad \lim_{n\to\infty} l_n = l.$$

Proof. Most of the properties are proved in Berg–Forst [24, §7, pp. 39–48]. The properties (a), (b), (e) and the cone property in (f) follow directly from the definition of negative definiteness. The convergence of the Lévy triplets can be found in Sato [267, Theorem 8.7, pp. 41–42] or Cuppens [77, Theorem 4.3.5, pp. 88–89].

For (c) we use Theorem 2.15(d) and the fact that the matrix

$$\begin{pmatrix} \psi(\xi) + \overline{\psi(\xi)} - \psi(0) & \psi(\xi) + \overline{\psi(\eta)} - \psi(\xi - \eta) \\ \psi(\eta) + \overline{\psi(\xi)} - \psi(\eta - \xi) & \psi(\eta) + \overline{\psi(\eta)} - \psi(0) \end{pmatrix}$$

has a positive determinant. Finally, (d) follows from subadditivity: Fix $\xi \in \mathbb{R}^d$ and denote by $n_\xi := \lfloor|\xi|\rfloor$ the integer part and by $f_\xi := |\xi| - \lfloor|\xi|\rfloor$ the fractional part of $|\xi|$. Then

$$\sqrt{|\psi(\xi)|} \leq n_\xi \sqrt{\left|\psi\left(\tfrac{\xi}{|\xi|}\right)\right|} + \sqrt{\left|\psi\left(f_\xi \tfrac{\xi}{|\xi|}\right)\right|} \leq (1 + |\xi|) \sup_{|\eta|\leq 1} \sqrt{|\psi(\eta)|}. \qquad \square$$

The following estimates for the derivatives of a continuous negative definite function are due to Hoh [137, Proposition 2.1] for real-valued ψ, see also Jacob [157, Theorem 3.7.13, p. 154].

Lemma 2.18 (Hoh). *Let $\psi : \mathbb{R}^d \to \mathbb{C}$ be the characteristic exponent of a Lévy process (with killing) with Lévy triplet (l, Q, v). If the Lévy measure v has bounded support, then ψ is arbitrarily often differentiable, and we have for $\alpha \in \mathbb{N}_0^d$ the estimates*

$$|\nabla^\alpha \psi(\xi)| \leq c_{|\alpha|} \begin{cases} |\psi(\xi)|, & \text{if } \alpha = 0, \\ (|l| + \operatorname{Re} \psi(\xi))^{\frac{1}{2}}, & \text{if } |\alpha| = 1, \\ 1, & \text{if } |\alpha| \geq 2. \end{cases} \qquad (2.23)$$

The constant $c_{|\alpha|}$ depends only on l, Q and the moments of v.

Proof. Without loss of generality we may assume that ν is supported in $\overline{\mathbb{B}}(0,1)$ and that the truncation function $\chi(s)$ in the Lévy–Khintchine formula (2.20) is $\mathbb{1}_{[0,1]}(s)$. For $\alpha = 0$, there is nothing to show. If $|\alpha| = 1$ we get

$$\frac{\partial \psi(\xi)}{\partial \xi_j} = -i l \cdot e_j + Q \xi \cdot e_j + i \int_{\mathbb{R}^d \setminus \{0\}} y_j \left(1 - e^{i y \cdot \xi}\right) \nu(dy)$$

and, by the elementary estimate $(a + b + c)^2 \leqslant 3(a^2 + b^2 + c^2)$ and the Cauchy–Schwarz inequality,

$$\left| \frac{\partial \psi(\xi)}{\partial \xi_j} \right|^2 \leqslant 3 \left[|l|^2 + |Q^{\frac{1}{2}} e_j|^2 |Q^{\frac{1}{2}} \xi|^2 + \int_{\mathbb{R}^d \setminus \{0\}} |y|^2 \nu(dy) \int_{\mathbb{R}^d \setminus \{0\}} \left|1 - e^{i y \cdot \xi}\right|^2 \nu(dy) \right]$$

$$= 3 \left[c_l |l| + c_Q |Q^{\frac{1}{2}} \xi|^2 + 2 c_\nu \int_{\mathbb{R}^d \setminus \{0\}} (1 - \cos(y \cdot \xi)) \nu(dy) \right]$$

$$\leqslant c_1^2 \left(|l| + \operatorname{Re} \psi(\xi) \right)$$

with a constant c_1 depending only on (l, Q, ν). Similarly,

$$\frac{\partial^2 \psi(\xi)}{\partial \xi_j \partial \xi_k} = Q e_k \cdot e_j + \int_{\mathbb{R}^d \setminus \{0\}} y_j y_k \, e^{i y \cdot \xi} \, \nu(dy)$$

and

$$\left| \frac{\partial^2 \psi(\xi)}{\partial \xi_j \partial \xi_k} \right| \leqslant \max_{1 \leqslant j \leqslant d} |Q e_j| + \int_{\mathbb{R}^d \setminus \{0\}} |y|^2 \nu(dy) = c_2.$$

A similar estimate holds for all higher derivatives. □

Remark 2.19. The concepts of positive and negative definiteness can be extended to abelian semigroups with involution. Up to now we have been considering the semigroup $(\mathbb{R}^d, +)$ with the involution $\xi \mapsto \xi^* := -\xi$. If we use, instead, $([0, \infty), +)$ with the trivial involution $\xi^* := \xi$, most of the material of this section remains valid and we arrive naturally at the notion of **completely monotone functions** (which are the positive definite functions on $[0, \infty)$) and **Bernstein functions** (which are the negative definite functions on $[0, \infty)$). These functions correspond to the Laplace transforms and characteristic (Laplace) exponents, respectively, of increasing Lévy processes on $[0, \infty)$ (subordinators). A short discussion is given in [293, Chap. 4], for a systematic treatment we refer to Berg–Christensen–Ressel [25]. □

2.3 Generator and Symbol of a Feller Semigroup on $\mathbb{R}^d$

In this section we consider Feller semigroups $(T_t)_{t \geq 0}$ on $(C_\infty(\mathbb{R}^d), \|\cdot\|_\infty)$. Recall from Definition 1.2 that a Feller semigroup is a strongly continuous, positivity preserving and sub-Markovian semigroup $T_t : C_\infty(\mathbb{R}^d) \to C_\infty(\mathbb{R}^d)$. The generator $(A, \mathcal{D}(A))$ enjoys the **positive maximum principle** (Lemma 1.28),

$$u \in \mathcal{D}(A), \ u(x_0) = \sup_{y \in E} u(y) \geq 0 \implies Au(x_0) \leq 0, \qquad (2.24)$$

which will play a major role in this section: It allows to characterize the structure of the generators. *Locally*, a Feller generator looks like the generator of a Lévy process (with killing)—i.e. the process corresponding to a translation invariant Feller semigroup—in the sense that it is given by a Lévy–Khintchine type representation (2.15) with an x-dependent Lévy triplet $(l, Q, \nu) = (l(x), Q(x), N(x, \cdot))$.

We begin with a result from the theory of distributions. The Schwartz distributions form the topological dual $\mathcal{D}'(\mathbb{R}^d)$ of the test functions $C_c^\infty(\mathbb{R}^d)$. A distribution $\Lambda \in \mathcal{D}'(\mathbb{R}^d)$ is of **order** $k \geq 0$, if for every $r > 0$ there is a constant c_r such that

$$|\Lambda[u]| \leq c_r \|u\|_{(k)} \quad \forall u \in C_c^\infty(\overline{\mathbb{B}}(0, r))$$

($\|u\|_{(k)} = \sum_{0 \leq |\alpha| \leq k} \|\nabla^\alpha u\|_\infty$). We say that $\Lambda \in \mathcal{D}'(\mathbb{R}^d)$ is **supported in** $\{x\}$ if

$$\operatorname{supp}(\Lambda) \subset \{x\} : \iff \Lambda[u] = 0 \quad \forall u \in C_c^\infty(\mathbb{R}^d \setminus \{x\}).$$

The following fundamental results can be found, e.g. in Hörmander [142, Theorem 2.1.7, p. 38 and Theorem 2.3.4, pp. 46–47] or Duistermaat–Kolk [92, Theorem 3.18, p. 42 and Theorem 8.10, p. 77].

Theorem 2.20. *Let* $\Lambda : C_c^\infty(\mathbb{R}^d) \to \mathbb{R}$ *be a linear functional.*

a) *If* Λ *is positive, i.e.* $\Lambda[u] \geq 0$ *for all* $u \geq 0$, *then* Λ *is a distribution of order zero, and* Λ *is represented by a positive Radon measure* $\mu \in \mathcal{M}^+(\mathbb{R}^d)$:

$$\Lambda[u] = \int_{\mathbb{R}^d} u(y) \, \mu(dy) \quad \forall u \in C_c^\infty(\mathbb{R}^d).$$

b) *If* $\Lambda \in \mathcal{D}'(\mathbb{R}^d)$ *is a distribution of order* $k \geq 0$ *with* $\operatorname{supp}(\Lambda) \subset \{x\}$, *then there exist coefficients* $a_\alpha \in \mathbb{R}$, $\alpha \in \mathbb{N}_0^d$, $|\alpha| \leq k$, *such that*

$$\Lambda[u] = \sum_{0 \leq |\alpha| \leq k} a_\alpha \nabla^\alpha u(x) \quad \forall u \in C^\infty(\mathbb{R}^d).$$

Let $(A, \mathcal{D}(A))$ be a Feller generator such that $C_c^\infty(\mathbb{R}^d) \subset \mathcal{D}(A)$. We are going to show that the positive maximum principle implies that $u \mapsto Au(x), u \in C_c^\infty(\mathbb{R}^d)$, is for every $x \in \mathbb{R}^d$ a distribution of order at most 2. Therefore, Theorem 2.20 will lead to an explicit formula for a Feller generator. As a special case, we obtain a new proof for the Lévy–Khinchine formula (2.10). The following theorem is due to Courrège [74], see also [38], and von Waldenfels [333, 334, 335].

Theorem 2.21 (Courrège; v. Waldenfels). *Let $(A, \mathcal{D}(A))$ be a Feller generator. If $C_c^\infty(\mathbb{R}^d) \subset \mathcal{D}(A)$, then A is of the following form*

$$Au(x) = -c(x)u(x) + l(x) \cdot \nabla u(x) + \frac{1}{2} \operatorname{div} Q(x) \nabla u(x)$$
$$+ \int_{\mathbb{R}^d \setminus \{0\}} \big(u(x+y) - u(x) - \nabla u(x) \cdot y \chi(|y|) \big) \, N(x, dy) \tag{2.25}$$

where $c(x) \geqslant 0$, $(l(x), Q(x), N(x, \cdot))$ is, for fixed $x \in \mathbb{R}^d$, a Lévy triplet: $l(x) \in \mathbb{R}^d$, $Q(x) \in \mathbb{R}^{d \times d}$ symmetric, positive semidefinite, and $N(x, \cdot) \in \mathcal{M}^+(\mathbb{R}^d \setminus \{0\})$ satisfying $\int_{\mathbb{R}^d \setminus \{0\}} \min(|y|^2, 1) \, N(x, dy) < \infty$, and $\chi \in B_b[0, \infty)$ is a truncation function such that $0 \leqslant 1 - \chi(s) \leqslant \kappa \min(s, 1)$ (for some $\kappa > 0$) and $s \chi(s)$ stays bounded.

In (2.25) the differential operator $-c(x)u(x) + l(x) \cdot \nabla u(x) + \frac{1}{2} \operatorname{div} Q(x) \nabla u(x)$ is often called the **local part**[2] of A, while the integral part is the **non-local part**.

The following proof was shown to one of us by F. Hirsch (private communication, Evry, 1996), a careful presentation can be found in Hoh [138, Sect. 2.2, pp. 17–27]. The idea to use distributions can be traced back to Herz [134].

Proof (of Theorem 2.21). Throughout this proof we will abbreviate *positive maximum principle* by PMP. We fix $x \in \mathbb{R}^d$ and observe that $A_x[u] := Au(x)$, $u \in C_c^\infty(\mathbb{R}^d)$, defines a linear functional on $C_c^\infty(\mathbb{R}^d)$.

Step 1. Since the zero of a positive function is a negative minimum, we infer from the PMP for A that

$$\phi \in C_c^\infty(\mathbb{R}^d), \ \phi \geqslant 0, \ \phi(x) = 0 \implies A_x[\phi] \geqslant 0.^3$$

Let $u \in C_c^\infty(\mathbb{R}^d)$, $u \geqslant 0$, and consider $\phi(y) = |x - y|^2 u(y)$; this is a positive test function with $\phi(x) = 0$. Thus, $u \mapsto |\cdot -x|^2 A_x[u] := A_x[|\cdot -x|^2 u]$ is a positive linear functional. By Theorem 2.20(a), $|\cdot -x|^2 A_x$ is a positive distribution, and we may identify it with a Radon measure $\nu(x, \cdot) \in \mathcal{M}^+(\mathbb{R}^d)$. Notice that the compact sets $K \subset \mathbb{R}^d \setminus \{x\}$ are contained in annuli of the form $\mathbb{B}(x, R) \setminus \mathbb{B}(x, r)$ for suitable radii $0 < r < R$. Thus, the Borel measure

---

[2]Recall that an operator L is **local**, if $\operatorname{supp} Lu \subset \operatorname{supp} u$ or, equivalently, if $Lu(x) = Lw(x)$ whenever u and w coincide in some open neighbourhood of x.

[3]This property is, in fact, a weaker version of the PMP which is often referred to as **présque positif** (Courrège [74, §1.2 (P), pp. 2-03/04]) or **fast positiv** (von Waldenfels [335]).

$$\mu(x, dy) := \frac{1}{|x - y|^2} \, v(x, dy)$$

is finite on compact sets in $\mathbb{R}^d \setminus \{x\}$, hence a Radon measure on $\mathbb{R}^d \setminus \{x\}$. By our construction,

$$\mu(x, \cdot) = A_x|_{C_c^\infty(\mathbb{R}^d \setminus \{x\})} \quad \text{as distributions in } \mathcal{D}'(\mathbb{R}^d \setminus \{x\}),$$

and we find

$$\int_{0 < |x-y| \leqslant 1} |x - y|^2 \, \mu(x, dy) = \int_{0 < |x-y| \leqslant 1} v(x, dy) < \infty. \qquad (2.26)$$

Step 2. Pick $u, v \in C_c^\infty(\mathbb{R}^d)$ with $0 \leqslant u, v \leqslant 1$, $\mathrm{supp}(u) \subset \overline{\mathbb{B}}(x, 1)$, $u(x) = 1$ and $\mathrm{supp}(v) \subset \overline{\mathbb{B}}^c(x, 1)$. Then x is a positive maximum for the function $\|v\|_\infty u + v$,

$$\sup_{y \in \mathbb{R}^d} \left(\|v\|_\infty u(y) + v(y) \right) = \|v\|_\infty = \|v\|_\infty u(x) + v(x),$$

and by the PMP, $A_x[\|v\|_\infty u + v] \leqslant 0$ or $A_x[v] \leqslant -\|v\|_\infty A_x[u]$. Now choose a sequence $(v_n)_{n \geqslant 1} \subset C_c^\infty(\mathbb{R}^d)$ such that $v_n \uparrow \mathbb{1}_{\overline{\mathbb{B}}^c(x,1)}$ as $n \to \infty$. Monotone convergence shows

$$\int_{|x-y|>1} \mu(x, dy) = \lim_{n \to \infty} \int_{|x-y|>1} v_n(y) \, \mu(x, dy) \leqslant - \lim_{n \to \infty} \|v_n\|_\infty A_x[u] = - A_x[u] < \infty.$$

$$(2.27)$$

Step 3. Let $\mu(x, \cdot)$ be as in Steps 1 and 2, assume that χ is a truncation function (as specified in the theorem), and define for $u \in C_c^\infty(\mathbb{R}^d)$

$$S_x[u] := \int_{|x-y|>0} \left(u(y) - u(x) - \nabla u(x) \cdot (y - x) \chi(|y - x|) \right) \mu(x, dy). \qquad (2.28)$$

Taylor's theorem and the definition of a truncation function imply

$$\left| u(y) - u(x) - \nabla u(x) \cdot (y - x) \chi(|y - x|) \right| \leqslant C_\chi \|u\|_{(2)} \min(|x - y|^2, 1)$$

for some constant C_χ depending only on the truncation function. This shows, because of (2.26) and (2.27), that S_x is a distribution of order 2.

Step 4. Set $L_x := A_x - S_x$. From the definition of S_x it is clear that L_x is a distribution with $\mathrm{supp}(L_x) \subset \{x\}$. Moreover, L_x satisfies the following weaker version of the positive maximum principle

$$u \in C_c^\infty(\mathbb{R}^d), \ u \geqslant 0, \ u(x) = 0 \implies L_x[u] \geqslant 0. \tag{2.29}$$

To see this, pick $\phi \in C^\infty(\mathbb{R}^d)$ with $0 \leqslant \phi \leqslant 1$, $\phi|_{\overline{B}(x,1)} = 0$ and $\phi|_{\overline{B}^c(x,2)} = 1$, and set $\phi_n(y) := \phi(ny)$. Since $\lim_{n\to\infty} \phi_n = \mathbb{1}_{\mathbb{R}^d \setminus \{x\}}$, we find for $u \in C_c^\infty(\mathbb{R}^d)$ as in (2.29) and any sufficiently large $n \geqslant 1$

$$
\begin{aligned}
L_x[u] = L_x[(1 - \phi_n)u] &= A_x[(1 - \phi_n)u] - S_x[(1 - \phi_n)u] \\
&= A_x[(1 - \phi_n)u] - \int_{\mathbb{R}^d \setminus \{x\}} (1 - \phi_n(y))u(y)\,\mu(x, dy) \\
&\geqslant - \int_{\mathbb{R}^d \setminus \{x\}} (1 - \phi_n(y))u(y)\,\mu(x, dy)
\end{aligned}
$$

where we used that $\operatorname{supp}(L_x) \subset \{x\}$, $u(x) = 0$, $\nabla u(x) = 0$ (x is a minimum), and the positive maximum principle for A_x. Letting $n \to \infty$ we conclude with monotone convergence that $L_x[u] \geqslant \lim_{n\to\infty} - \int_{\mathbb{R}^d \setminus \{x\}} (1 - \phi_n(y))u(y)\,\mu(x, dy) = 0$.

Step 5. As in the first two steps we conclude from (2.29) that L_x is a distribution of order 2: Using a Taylor expansion we see

$$u(y) - u(x) - \nabla u(x) \cdot (y - x) \leqslant C \|u\|_{(2)} |y - x|^2$$

from which we get

$$C \|u\|_{(2)} |y - x|^2 + u(x) + \nabla u(x) \cdot (y - x) - u(y) \geqslant 0,$$

and by (2.29),

$$L_x[u] \leqslant u(x) L_x[1] + \nabla u(x) \cdot L_x[(\cdot - x)] + C \|u\|_{(2)} L_x[|\cdot - x|^2] \leqslant C' \|u\|_{(2)}.$$

Step 6. Since $\operatorname{supp}(L_x) \subset \{x\}$ is a distribution of order 2, we get from Theorem 2.20(b) that

$$L_x[u] = \frac{1}{2} \sum_{j,k=1}^d q_{jk}(x) \frac{\partial^2 u(x)}{\partial x_j \partial x_k} + \sum_{j=1}^d l_j(x) \frac{\partial u(x)}{\partial x_j} - c(x)u(x). \tag{2.30}$$

Let $(u_n)_{n \geqslant 1} \subset C_c^\infty(\mathbb{R}^d)$ with $0 \leqslant u_n \leqslant 1$, $u_n|_{B(x,1)} = 1$ and $u_n \uparrow 1$. From the PMP we see $A_x[u_n] \leqslant 0$ for all $n \geqslant 1$, and since $\nabla u_n(x) = 0$, dominated convergence gives $S_x[u_n] = - \int_{\mathbb{R}^d \setminus \{x\}} (1 - u_n(y))\,\mu(x, dy) \longrightarrow 0$ as $n \to \infty$. Thus,

$$\varlimsup_{n\to\infty} L_x[u_n] = \varlimsup_{n\to\infty} A_x[u_n] - \lim_{n\to\infty} S_x[u_n] \leqslant 0;$$

from (2.30) we conclude that $c(x) \geqslant 0$.

If we set $v_n(y) := ((x - y) \cdot \xi)^2 u_n(y)$ for any $\xi \in \mathbb{R}^d$, we have $v_n(x) = 0$, and from Step 4 we infer that $L_x[v_n] \geqslant 0$, hence $\sum_{j,k=1}^d q_{jk}(x)\xi_j\xi_k \geqslant 0$.

Step 7. Finally, with $N(x, dy) := \mu(x, dy{-}x)$ we get (2.25) from (2.28) and (2.30).

$\square$

If $(T_t)_{t\geqslant 0}$ is translation invariant, this gives a new proof of the Lévy–Khintchine formula.

Corollary 2.22. *Let $(T_t)_{t\geqslant 0}$ be a translation invariant Feller semigroup with generator $(A, \mathcal{D}(A))$ and denote by $(\mu_t)_{t\geqslant 0}$ the corresponding vaguely continuous convolution semigroup. Then $C_c^\infty(\mathbb{R}^d) \subset \mathcal{D}(A)$ and the characteristic exponent ψ of the inverse Fourier transform $\widecheck{\mu}_t = \exp(-t\psi)$ has the Lévy–Khintchine representation (2.10).*

Proof. As $(\mu_t)_{t\geqslant 0}$ is a vaguely continuous convolution semigroup, $\widecheck{\mu}_t = \exp(-t\psi)$ is, for every $t > 0$, a positive definite function with a continuous exponent ψ. By Schoenberg's theorem, Theorem 2.15(c), ψ is a continuous negative definite function, and as such it is polynomially bounded: $|\psi(\xi)| \leqslant c_\psi(1 + |\xi|^2)$, see Proposition 2.17(d). This allows to show that $C_c^\infty(\mathbb{R}^d) \subset \mathcal{D}(A)$: We have for $u \in C_c^\infty(\mathbb{R}^d)$

$$\frac{T_t u(x) - u(x)}{t} = \mathcal{F}^{-1}\left[\frac{e^{-t\psi}\hat{u} - \hat{u}}{t}\right](x) \xrightarrow[t\to 0]{} \mathcal{F}^{-1}[-\psi\hat{u}](x) \quad \forall x \in \mathbb{R}^d$$

and the right-hand side is, by the Riemann–Lebesgue lemma, a C_∞-function since $\hat{u} \in \mathcal{S}(\mathbb{R}^d)$ and $\psi\hat{u} \in L^1(d\xi)$. Thus, Theorem 1.33 shows that $C_c^\infty(\mathbb{R}^d)$ is in the domain $\mathcal{D}(A)$ of the generator.

As $Au = \mathcal{F}^{-1}[-\psi\hat{u}]$, we can deduce the formula for ψ from Corollary 2.23. $\square$

Using Fourier inversion, we deduce immediately the following formula from the representation (2.25).

Corollary 2.23. *Let $(A, \mathcal{D}(A))$ be a Feller generator. If $C_c^\infty(\mathbb{R}^d) \subset \mathcal{D}(A)$, then A is of the following form*

$$Au(x) := -q(x, D)u(x) := -\int_{\mathbb{R}^d} e^{ix\cdot\xi} q(x, \xi)\, \hat{u}(\xi)\, d\xi \quad \forall u \in C_c^\infty(\mathbb{R}^d) \quad (2.31)$$

where, for every $x \in \mathbb{R}^d$, the function $q(x, \cdot)$ is a continuous negative definite function with x-dependent Lévy triplet $(l(x), Q(x), N(x, dy))$ relative to some truncation function χ, i.e.

$$q(x, \xi) = q(x, 0) - i\,l(x) \cdot \xi + \frac{1}{2}\xi \cdot Q(x)\xi$$
$$+ \int_{\mathbb{R}^d\setminus\{0\}} \left(1 - e^{iy\cdot\xi} + i\xi \cdot y\chi(|y|)\right) N(x, dy). \quad (2.32)$$

Example 2.24 (Ornstein–Uhlenbeck Semigroups). (Compare Example 1.17(f)) Let $(Z_t)_{t\geq 0}$ be a Lévy process with characteristic exponent ψ and $B \in \mathbb{R}^{d\times d}$. The stochastic process $X_t^x := e^{tB}x + \int_0^t e^{(t-s)B}\,dZ_s, x \in \mathbb{R}^d$, is an Ornstein–Uhlenbeck process and its generator has the symbol

$$q(x,\xi) = -ixB \cdot \xi + \psi(\xi). \qquad \square$$

An operator of the form (2.31) is known in analysis as a *pseudo-differential operator* and the x-dependent Fourier multiplier $-q(x,\xi)$ is referred to as the *symbol* of the operator. There is a well-developed theory of pseudo-differential operators, see e.g. Hörmander [143], Kumano-go [194], Taylor [317] or Ruzhansky–Turunen [261], to mention just a few standard treatises, but the symbol classes considered there require more smoothness of $(x,\xi) \mapsto q(x,\xi)$ than we usually can admit for Feller generators: In the most general case $\xi \mapsto q(x,\xi)$ is continuous and negative definite.

An interesting approach is due to Barndorff-Nielsen–Levendorskiĭ [13] who start with the characteristic exponent of a Lévy process which depends on a certain set of parameters, for example a normal inverse Gaussian Lévy process. Then they make the parameters x-dependent and construct in this way Feller processes which are sensitive with respect to their current position; see also [42]. Note that their construction requires C^∞-smoothness in x. Nevertheless, there is a well developed symbolic calculus for *non-classical, rough* pseudo-differential operators which was initiated by Jacob, see his survey [155] and the monographs [157, 158, 159], and further developed by Hoh [137, 138]. Since $|q(x,\xi)| \leq c_x(1 + |\xi|^2) < \infty$, the following definition makes sense.

Definition 2.25. Let $q : \mathbb{R}^d \times \mathbb{R}^d \to \mathbb{C}$, $(x,\xi) \mapsto q(x,\xi)$, be a function which is, for every $x \in \mathbb{R}^d$, continuous and negative definite in the variable ξ, i.e. given by (2.32). Then

$$-q(x,D)u(x) := -\int_{\mathbb{R}^d} e^{ix\cdot\xi}q(x,\xi)\hat{u}(\xi)\,d\xi, \quad u \in C_c^\infty(\mathbb{R}^d)$$

is a **pseudo-differential operator (with negative definite symbol)** and $q(x,\xi)$ is called the **symbol** of the operator.

By Corollary 2.23, every Feller generator $(A, \mathcal{D}(A))$ with $C_c^\infty(\mathbb{R}^d) \subset \mathcal{D}(A)$ is a pseudo-differential operator with negative definite symbol. Since the symbol depends on the state space variable x, we call $-q(x,D)$ also an operator with variable **"coefficients"** (where the Lévy triplet assumes the role of the coefficients). In abuse of notation we also call $q(x,\xi)$ the **symbol of a Feller process**. The generators with **constant coefficients** correspond to the translation invariant Feller semigroups or Lévy processes and the pseudo-differential operator becomes a Fourier multiplier operator, cf. Theorem 2.7 and Remark 2.8:

$$-\psi(D)u(x) = -\int_{\mathbb{R}^d} e^{ix\cdot\xi}\psi(\xi)\hat{u}(\xi)\,d\xi, \quad u \in C_c^\infty(\mathbb{R}^d).$$

In the translation invariant case there is a one-to-one correspondence between symbols ψ, Lévy triplets (l, Q, ν) and translation invariant Feller semigroups/Lévy processes (with killing), cf. Theorems 2.2 and 2.7, but this is no longer the case for general Feller semigroups and processes. Although every Feller semigroup and Feller process whose generator satisfies $C_c^\infty(\mathbb{R}^d) \subset \mathcal{D}(A)$ is a pseudo-differential operator with negative definite symbol, the converse question is a difficult problem: *When does a given x-dependent Lévy triplet $(l(x), Q(x), N(x, \cdot))$ or a given negative definite symbol $q(x, \cdot)$ give rise to a Feller process?* This question was for the first time asked by Jacob [152]. The (positive) answer is by far not trivial, and a thorough discussion of all known solution strategies is given in Chap. 3 below. At this point we will only provide two counterexamples.

Example 2.26. a) Consider the continuous negative definite symbol $q : \mathbb{R} \times \mathbb{R} \to \mathbb{C}$ given by

$$q(x, \xi) = -i\,\xi\,2\,\mathrm{sgn}(x)\sqrt{|x|}$$

which gives rise to a deterministic process (a pure drift process with state space dependent drift) with sample path starting in $X_0 = x$ given by

$$X_t = \begin{cases} (t + \sqrt{x})^2, & x > 0, \\ 0, & x = 0, \\ -(t + \sqrt{-x})^2, & x < 0. \end{cases}$$

Then $T_t u(x) := \mathbb{E}^x(u(X_t))$ defines a semigroup on $B_b(\mathbb{R})$. But for any function $u \in C_\infty(\mathbb{R})$ with $u(t^2) \neq u(-t^2)$, the map $x \mapsto T_t u(x)$ is not continuous at 0. Thus, it is not a Feller semigroup. Note, however, that $\lim_{|x| \to \infty} T_t u(x) = 0$ for all $u \in C_\infty(\mathbb{R})$.

b) (cf. van Casteren [331, Remark 2.12, p. 156]) Analogously, the symbol

$$q(x, \xi) = i\,\xi\,x^3$$

gives rise to the deterministic process

$$X_t = \frac{x}{\sqrt{1 + 2tx^2}}.$$

Its semigroup satisfies $T_t : C_\infty(\mathbb{R}) \to C(\mathbb{R})$, but for suitable $u \in C_\infty(\mathbb{R})$ we have $\lim_{|x| \to \infty} T_t u(x) = u(1/\sqrt{2t}) \neq 0$.   □

Set $e_\xi(x) := e^{ix\xi}$, fix $\sigma \in C_c^\infty(\mathbb{R}^d)$ with $\mathbb{1}_{B(0,1)} \leq \sigma \leq 1$ and $\sigma_x(y) := \sigma(y - x)$.

Proposition 2.27. *Let $(A, \mathcal{D}(A))$ be a Feller generator, $C_c^\infty(\mathbb{R}^d) \subset \mathcal{D}(A)$, and denote by $q(x, \xi)$ its symbol.*

a) $\sup_{|x| \leq r} |Au(x)| \leq c_r \|u\|_{(2)}$ *for all $r > 0$ and $u \in C_c^\infty(\overline{\mathbb{B}}(0, r))$;*
b) $q(x, 0)$ *is locally bounded;*

c) $-q(x, \xi) = e_{-\xi}(x) A e_\xi(x)$ *where* $e_\xi(x) = e^{ix \cdot \xi}$;

d) $|q(x, \xi)| \leqslant \gamma(x)(1 + |\xi|^2)$ *for some locally bounded function* $\gamma : \mathbb{R}^d \to [0, \infty)$;

e) *Let* $(l(x), Q(x), N(x, \cdot))$ *be the Lévy triplet of* $q(x, \cdot) - q(x, 0)$ *and denote by* χ *a truncation function (cf. Theorem 2.21). Then the Lévy kernel* $N(x, dy)$ *is uniquely determined by* $Au(x) = \int_{\mathbb{R}^d \backslash \{0\}} u(x + y) \, N(x, dy)$ *for* $x \notin \mathrm{supp}(u)$ *and* $u \in C_c^\infty(\mathbb{R}^d)$. *Moreover, denoting by* $l_j(x)$ *and* $q_{jk}(x)$ *the entries of* $l(x)$ *and* $Q(x)$, *respectively,*

$$q(x, 0) = \int_{\mathbb{R}^d \backslash \{0\}} (\sigma(y) - 1) \, N(x, dy) - A[\sigma_x](x),$$

$$l_j(x) = A[(\cdot - x)_j \sigma_x](x) - \int_{\mathbb{R}^d \backslash \{0\}} y_j (\sigma(y) - \chi(|y|)) \, N(x, dy),$$

$$q_{jk}(x) = A[(\cdot - x)_j (\cdot - x)_k \sigma_x](x) - \int_{\mathbb{R}^d \backslash \{0\}} y_j y_k \sigma(y) \, N(x, dy),$$

for all $x \in \mathbb{R}^d$ *and* $j, k = 1, \dots, d$.

f) *For all* $u \in C_c^\infty(\mathbb{R}^d)$ *the map* $x \mapsto \int_{\mathbb{R}^d \backslash \{0\}} u(y) \frac{|y|^2}{1 + |y|^2} N(x, dy)$ *is measurable; in particular,* $x \mapsto N(x, B)$ *is measurable for all Borel sets* $B \in \mathscr{B}(\mathbb{R}^d \backslash \{0\})$.

Proof. a) Let $r > 0$. Consider the family of linear operators

$$u \mapsto A_x[u] := Au(x) \text{ for } x \in \overline{\mathbb{B}}(0, r).$$

We have seen in the proof of Theorem 2.21 that $A_x : C_c^\infty(\overline{\mathbb{B}}(0, r)) \to \mathbb{R}$ satisfies $|A_x[u]| \leqslant c_{r,x} \|u\|_{(2)}$ which means that we can understand $(A_x)_{x \in \overline{\mathbb{B}}(0,r)}$ as a family of linear operators from the Banach space $(C_c^2(\overline{\mathbb{B}}(0, r)), \| \cdot \|_{(2)})$ to the Banach space $(\mathbb{R}, | \cdot |)$. Therefore, the Banach–Steinhaus theorem proves (a).

b) Fix $r > 0$ and pick some $\theta_r \in C_c^\infty(\mathbb{R}^d)$ such that $0 \leqslant \theta_r \leqslant 1$ and $\theta_r|_{\mathbb{B}(0,r)} = 1$. Then we find

$$A\theta_r(x) = -c(x)\theta_r(x) + \int_{\mathbb{R}^d \backslash \{0\}} (\theta_r(x + y) - 1) \, N(x, dy) \leqslant -c(x) \quad \forall x \in \mathbb{B}(0, r).$$

Since $c \geqslant 0$ and since $A\theta_r \in C_\infty$, we get (b).

c) Since we may plug in e_ξ into the formula (2.25), we get (c) from a direct (but formal) computation. Theorem 2.37(a) justifies this formal calculation since the representation (2.25) remains valid on $C_b^2(\mathbb{R}^d)$.

d) Fix $r > 0$ and $\theta_r \in C_c^\infty(\mathbb{R}^d)$ as in (b). A calculation similar to the one needed for (c) shows for $x \in \mathbb{B}(0, r)$

$$-q(x, \xi) = e_{-\xi} A[\theta_r e_\xi] + e_{-\xi} A[(1 - \theta_r) e_\xi]$$

$$= e_{-\xi} A[\theta_r e_\xi] + \int_{\mathbb{R}^d \backslash \{0\}} e^{i\xi \cdot y} (1 - \theta_r(x + y)) \, N(x, dy),$$

and we get for all $x \in \mathbb{B}(0, r)$

$$|q(x, \xi)| \leq |A[\theta_r e_\xi](x)| + |c(x)\theta_r(x) - A\theta_r(x)| \leq |A[\theta_r e_\xi](x)| + |A\theta_r(x)| + c(x).$$

Now we can use parts (a) and (b) and the continuity of $A\theta_r$ to conclude that $(x, \xi) \mapsto q(x, \xi)$ is bounded for (x, ξ) from compact sets. The claimed estimate follows from the estimate for $\|\theta_r e_\xi\|_{(2)}$ which produces a polynomial of order 2 in the parameter ξ.

e) This follows from a direct calculation using (2.25).

f) The following argument is taken from Hoh [138, Lemma 2.19, p. 26]. For every $u \in C_c^\infty(\mathbb{R}^d \setminus \{0\})$ we set $v(y) := u(y)\frac{|y|^2}{1+|y|^2}$. Since $\nabla^\alpha v(0) = 0$ for all multiindices $0 \leq |\alpha| \leq 2$, we have

$$\int_{\mathbb{R}^d \setminus \{0\}} u(y)\frac{|y|^2}{1 + |y|^2} N(x, dy) = A[v(\cdot - x)](x) = -\int_{\mathbb{R}^d} q(x, \xi)\hat{v}(\xi) \, d\xi.$$

The last expression depends measurably on x; the measurability of $x \mapsto N(x, B)$ follows now from standard approximation and Dynkin system (monotone class) arguments.

$\square$

Continuity of the Symbol. Every symbol $q(x, \xi)$ is continuous in ξ, but the continuity in x is not obvious. The following result is from [45, Remark (A4.b), p. 78].

Lemma 2.28. *Let $(A, \mathcal{D}(A))$ be a Feller generator with $C_c^\infty(\mathbb{R}^d) \subset \mathcal{D}(A)$ and symbol $q(x, \xi)$. If the corresponding Feller process has uniformly bounded jumps, i.e. if $N(x, dy)$ satisfies $\sup_{x \in \mathbb{R}^d} N(x, \mathbb{R}^d \setminus \mathbb{B}(0, r)) = 0$ for some $r > 0$, then $x \mapsto q(x, \xi)$ is continuous.*

This follows from the fact that $A : C_c^\infty(\mathbb{R}^d) \to C_\infty(\mathbb{R}^d)$ and inserting the function $u \in C_c^\infty(\mathbb{R}^d)$ with $\mathbb{1}_{\mathbb{B}(0,3r)} \leq u \leq 1$ into (2.25). The next example, again from [45, pp. 78–79], shows what can go wrong if there are unbounded jumps.

Example 2.29. Consider a Markov process $(X_t)_{t \geq 0}$ on $[0, \infty]$ with the following transition function:

$$\begin{cases} \mathbb{P}^x(X_t = x) = e^{-t}\cosh t, \quad \mathbb{P}^x(X_t = x^{-1}) = e^{-t}\sinh t, & \text{if } x \in [0, \infty), \\ \mathbb{P}^\infty(X_t = \infty) = 1, \quad\quad\quad \mathbb{P}^\infty(X_t = 0) = 0, & \text{if } x = \infty, \end{cases}$$

where we use the convention $1/0 = \infty$ and $1/\infty = 0$. The transition semigroup is given by

$$\begin{cases} T_t u(x) = e^{-t}\left(u(x)\cosh t + u(x^{-1})\sinh t\right), & \text{if } x \in [0, \infty), \\ T_t u(\infty) = u(\infty), & \text{if } x = \infty, \end{cases}$$

and the infinitesimal generator is

$$Au(x) = -\mathbb{1}_{\{0\}}(x)u(x) + \int_{(0,\infty)} (u(x+y) - u(x))\,\mathbb{1}_{\{x\neq1\}}(x)\delta_{\frac{1}{x}-x}(dy).$$

This is a Feller generator on $\{u \in C_\infty[0,\infty] : u(0) = u(\infty) = 0\}$ with the symbol

$$q(x,\xi) = \mathbb{1}_{\{0\}}(x) + \left(\exp\left[i\xi\left(\tfrac{1}{x} - x\right)\right] - 1\right)\mathbb{1}_{(0,\infty)\backslash\{1\}}(x);$$

but $q(x,\xi)$ is clearly discontinuous at $x = 0$. $\qquad\qquad\qquad\qquad\square$

If $x \mapsto q(x,0)$ is continuous, in particular if $q(x,0) = 0$, then $x \mapsto q(x,\xi)$ is always continuous, cf. [274, Theorem 4.4, p. 249].

Theorem 2.30. *Let $(A, \mathcal{D}(A))$ be a Feller generator with $C_c^\infty(\mathbb{R}^d) \subset \mathcal{D}(A)$ and symbol $q(x,\xi)$ with Lévy triplet $(l(x), Q(x), N(x,\cdot))$. Then the following assertions are equivalent.*

a) *$x \mapsto q(x,0)$ is continuous;*
b) *$x \mapsto q(x,\xi)$ is continuous for all $\xi \in \mathbb{R}^d$;*
c) *$\lim_{|\xi|\to 0} \sup_{x\in K} |q(x,\xi) - q(x,0)| = 0$ for all compact sets $K \subset \mathbb{R}^d$;*
d) *$\lim_{r\to\infty} \sup_{x\in K} N(x, \mathbb{R}^d \setminus \mathbb{B}(0,r)) = 0$ for all compact sets $K \subset \mathbb{R}^d$.*

Boundedness of the Coefficients. The Lévy triplet $(l(x), Q(x), N(x,\cdot))$ is the analogue of the "coefficients" for the operator $A = -q(x, D)$. In fact, if $N \equiv 0$, then $l(x)$ and $Q(x)$ are indeed the coefficients of the local part of A. We say that a pseudo-differential operator with negative definite symbol has **bounded coefficients**, if

$$\sup_{x\in\mathbb{R}^d} |q(x,0)| + \sup_{x\in\mathbb{R}^d} |l(x)| + \sup_{x\in\mathbb{R}^d} |Q(x)|$$

$$+ \sup_{x\in\mathbb{R}^d} \int_{\mathbb{R}^d\backslash\{0\}} \min(|y|^2, 1)N(x, dy) < \infty \qquad (2.33)$$

where $|\cdot|$ denotes any vector resp. matrix norm. The following result is from [285, Lemma 6.2].

Theorem 2.31. *Let $(A, \mathcal{D}(A))$ be a Feller generator, $C_c^\infty(\mathbb{R}^d) \subset \mathcal{D}(A)$ with the symbol $q(x,\xi)$ having the Lévy triplet $(l(x), Q(x), N(x,\cdot))$. Then the following assertions are equivalent for any compact set $K \subset \mathbb{R}^d$ or for $K = \mathbb{R}^d$.*

a) *$\sup_{x\in K} |q(x,\xi)| \leqslant c_{q,K}(1 + |\xi|^2) \quad \forall \xi \in \mathbb{R}^d$;*

b) *$\sup_{x\in\mathbb{R}^d} |q(x,0)| + \sup_{x\in K} |l(x)| + \sup_{x\in K} |Q(x)| + \sup_{x\in K} \int_{\mathbb{R}^d\backslash\{0\}} \min(|y|^2, 1)N(x, dy) < \infty$;*

c) *$\sup_{|\eta|\leqslant 1} \sup_{x\in K} |q(x,\eta)| < \infty$.*

If one, hence all, of the above conditions hold, then there exists a constant such that $c_{q,K} \leqslant 2 \sup_{|\eta| \leqslant 1} \sup_{x \in K} |q(x, \eta)|$. *In particular, A has bounded coefficients if, and only if,*

$$|q(x, \xi)| \leqslant 2 \sup_{\substack{|\eta| \leqslant 1 \\ x \in \mathbb{R}^d}} |q(x, \eta)|(1 + |\xi|^2) < \infty \quad \forall x \in \mathbb{R}^d. \tag{2.34}$$

Conservativeness. Recall that a Feller semigroup $(T_t)_{t \geqslant 0}$ is *conservative* if we have $T_t 1 = 1$ for all $t > 0$. For the corresponding stochastic process conservativeness means a.s. infinite life-time. For a Lévy process this is the same as $\psi(0) = 0$, i.e. there is no killing. For a general Feller process a new phenomenon comes in: It is possible that the coefficients grow so fast that they cause an explosion. This situation is known from the theory of stochastic differential equations (see [285] and Sect. 3.5 for the connection of SDEs and symbols) where non-explosion is ensured by the linear growth condition and from the theory of elliptic diffusions, cf. Stroock–Varadhan [312, Chap. 10].

As in the case of Lévy processes, a conservative Feller process necessarily satisfies $q(x, 0) = 0$. The rationale behind this are the identities (1.38)

$$T_t u - u = \int_0^t T_s A u \, ds$$

where we formally insert $u = 1$ and observe that $A1 \approx -q(x, 0)$. With a few precautions, this programme can be made rigorous. The following result is taken from [275, Corollary 3.4] [4] or [274, Lemma 5.1]

Lemma 2.32. *Let* $(A, \mathcal{D}(A))$ *be a Feller generator with* $C_c^\infty(\mathbb{R}^d) \subset \mathcal{D}(A)$ *such that the symbol* $q(x, \xi)$ *is continuous as a function of* x. *If the Feller semigroup* $(T_t)_{t \geqslant 0}$ *is conservative, then* $q(x, 0) \equiv 0$.

If the symbol has bounded coefficients, cf. Theorem 2.31, we get a partial converse.

Theorem 2.33. *Let* $(A, \mathcal{D}(A))$ *be a Feller generator with* $C_c^\infty(\mathbb{R}^d) \subset \mathcal{D}(A)$ *such that the symbol* $q(x, \xi)$ *has bounded coefficients. If* $q(x, 0) = 0$, *then* $(T_t)_{t \geqslant 0}$ *is conservative and* $x \mapsto q(x, \xi)$ *is continuous.*

The influence of the growth of the coefficients can also be observed for Feller processes. The following sufficient criterion holds for all dimensions. Recall that for a recurrent elliptic diffusion, typically in dimension $d = 1$ and 2, the coefficients can grow arbitrarily fast without causing explosion in finite time, cf. [312, 10.3.3, p. 260]. If $d \geqslant 3$ we see that the coefficients should not grow faster than $|x|^2$. Having in mind that the symbol of a nondegenerate elliptic diffusion is

---

[4]The additional assumptions on the symbol in [275] just ensure the existence of a Feller process. Since we *assume* in Lemma 2.32 that $(A, \mathcal{D}(A))$ generates a Feller process, we can dispense of them.

$$q(x,\xi) = c(x) - il(x) \cdot \xi + \frac{1}{2}\sum_{j,k} q_{jk}(x)\xi_j\xi_k$$

—and this symbol also grows like $|\xi|^2$—the following result from [345, Theorem 2.1, Remark 2.1 (2)] is a natural counterpart to the results from Stroock–Varadhan [312]. Earlier versions of such results are due to Hoh–Jacob [140, Corollary 2.2], Jacob [156, Proposition 2.1] and [274, Theorem 5.5].

Theorem 2.34. *Let $(A, \mathcal{D}(A))$ be a Feller generator such that $C_c^\infty(\mathbb{R}^d) \subset \mathcal{D}(A)$ and denote by $q(x,\xi)$ the symbol. If for any $x \in \mathbb{R}^d$, $q(x,0) = 0$ and*

$$\lim_{r\to\infty} \sup_{|y-x|\leqslant 2r} \sup_{|\eta|\leqslant 1/r} |q(y,\eta)| < \infty \quad \forall x \in \mathbb{R}^d \tag{2.35}$$

then the Feller semigroup $(T_t)_{t\geqslant 0}$ is conservative, and the associated Feller process $(X_t)_{t\geqslant 0}$ has infinite life-time.

For a smaller class of symbols Hoh [138, Proposition 9.1] has the following non-explosion result which actually follows from Theorem 2.34. Hoh's result is stated in the context of the martingale problem (cf. Sect. 3.5) and applies to all pseudo-differential operators with negative definite symbol $q(x,\xi)$ such that the martingale problem for $(-q(x,D), C_c^\infty(\mathbb{R}^d))$ has a solution. We state it for Feller processes.

Theorem 2.35. *Let $(A, \mathcal{D}(A))$ be a Feller generator such that $C_c^\infty(\mathbb{R}^d) \subset \mathcal{D}(A)$ and with real-valued symbol $q(x,\xi)$. If $q(x,0) = 0$ and if there is a fixed continuous, real-valued negative definite function $\psi : \mathbb{R}^d \to \mathbb{R}$ such that*

$$q(x,\xi) \leqslant \frac{c\psi(\xi)}{\sup_{|\eta|\leqslant 1/|x|} \psi(\eta)} \quad \forall |x| \geqslant 1 \tag{2.36}$$

then the Feller semigroup $(T_t)_{t\geqslant 0}$ is conservative and the associated Feller process $(X_t)_{t\geqslant 0}$ has infinite life-time.

Related conservativeness results for jump processes (often defined on general metric spaces) generated by symmetric Dirichlet forms are discussed by Uemura and co-authors [219, 220, 330]. These criteria are based on the speed and scale measures of the processes. For a Feller process, the role of the scale measure is taken by the jump kernel $N(x, dy)$, while the speed measure is Lebesgue measure dx.

2.4 The Symbol of a Stochastic Process

For a Lévy process $(Y_t)_{t\geqslant 0}$ the characteristic exponent ψ and the symbol of the infinitesimal generator coincide, cf. Theorem 2.7. This observation allows us to calculate the symbol stochastically:

$$-\psi(\xi) = \frac{d}{dt}e^{-t\psi(\xi)}\Big|_{t=0} = \lim_{t\to 0}\frac{\mathbb{E}\,e^{i\xi\cdot Y_t}-1}{t} = \lim_{t\to 0}\frac{\mathbb{E}^x\,e^{i\xi\cdot(Y_t-x)}-1}{t}. \quad (2.37)$$

We are going to show that this formula remains valid for Feller processes $(X_t)_{t\geqslant 0}$ which are generated by pseudo-differential operators. Following Jacob [156] we write $e_\xi(x) := e^{ix\cdot\xi}$ and

$$\lambda_t(x,\xi) := \mathbb{E}^x\,e^{i\xi\cdot(X_t-x)} = e_{-\xi}(x)T_t e_\xi(x), \quad x,\xi \in \mathbb{R}^d. \quad (2.38)$$

The function $\lambda_t : \mathbb{R}^d \times \mathbb{R}^d \to \mathbb{C}$ is the symbol of the transition operator T_t in the sense of Definition 2.25

$$T_t u(x) = \lambda_t(x,D)u(x) = \int_{\mathbb{R}^d} e^{ix\cdot\xi}\lambda_t(x,\xi)\,\hat{u}(\xi)\,d\xi \quad \forall u \in C_c^\infty(\mathbb{R}^d). \quad (2.39)$$

Under the assumption that the uniformly continuous C^2-functions are in the domain, $C_u^2(\mathbb{R}^d) \subset \mathcal{D}(A)$, Jacob established the analogue of (2.37) for Feller processes. This was subsequently generalized in [275, Theorem 3.1]; the starting point in [275] is a pseudo-differential operator with negative definite symbol, and the assumptions stated in that paper ensure that this operator is indeed a Feller generator. With some modifications of the original argument, we can extract the following result.

Theorem 2.36. *Let $(X_t)_{t\geqslant 0}$ be a Feller process with generator $(A, \mathcal{D}(A))$ such that $C_c^\infty(\mathbb{R}^d) \subset \mathcal{D}(A)$. If the symbol $q(x,\xi)$ is continuous as a function of x, and if $q(x,D)$ has bounded coefficients, see (2.33), then*

$$-q(x,\xi) = \frac{d}{dt}\lambda_t(x,\xi)\Big|_{t=0} = \lim_{t\to 0}\frac{\mathbb{E}^x\,e^{i\xi\cdot(X_t-x)}-1}{t}.$$

It is enough to assume in the above theorem that $x \mapsto q(x,0)$ is continuous, cf. Theorem 2.30.

Proof of Theorem 2.36. Let $\phi \in C_c^\infty(\mathbb{R}^d)$ be such that $\mathbb{1}_{B(0,1)} \leqslant \phi \leqslant \mathbb{1}_{B(0,2)}$ and set $\phi_k(x) := \phi(x/k)$ for $k \in \mathbb{N}$. Since $\phi_k e_\xi \in C_c^\infty(\mathbb{R}^d) \subset \mathcal{D}(A)$, we use (1.38) to get

$$T_t[\phi_k e_\xi](x) - \phi_k(x)e_\xi(x) = \int_0^t AT_s[\phi_k e_\xi](x)\,ds.$$

Observe that $\widehat{\phi_k e_\xi}(\eta) = \hat{\phi}_k(\eta-\xi) = k^d\,\hat{\phi}(k(\eta-\xi))$. Thus,

$$A[\phi_k e_\xi](x) = -\int e^{ix\cdot\eta}q(x,\eta)\widehat{\phi_k e_\xi}(\eta)\,d\eta$$

$$= -\int e^{ix\cdot\eta}q(x,\eta)k^d\,\hat\phi(k(\eta-\xi))\,d\eta$$

$$= -\int e^{ix\cdot(k^{-1}\eta+\xi)}q(x,k^{-1}\eta+\xi)\hat\phi(\eta)\,d\eta.$$

Since $q(x,\xi)$ has bounded coefficients, we can use Theorem 2.31(a) with $K = \mathbb{R}^d$ and the estimate $(1+|k^{-1}\eta+\xi|^2) \leqslant (1+2|\eta|^2+2|\xi|^2) \leqslant 2(1+|\xi|^2)(1+|\eta|^2)$ for all $k \geqslant 1$, to infer

$$\|A[\phi_k e_\xi]\|_\infty \leqslant c_q\int|\hat\phi(\eta)|(1+|k^{-1}\eta+\xi|^2)\,d\eta$$

$$\leqslant 2c_q(1+|\xi|^2)\int|\hat\phi(\eta)|(1+|\eta|^2)\,d\eta.$$

By dominated convergence we get

$$\text{bp-}\lim_{k\to\infty}A[\phi_k e_\xi](x) = -e_\xi(x)q(x,\xi) \quad\text{and}$$

$$\text{bp-}\lim_{k\to\infty}T_s A[\phi_k e_\xi](x) = -T_s[e_\xi q(\cdot,\xi)](x).$$

Therefore,

$$\lim_{t\to0}\frac{e_{-\xi}(x)T_t e_\xi(x)-1}{t} = -\lim_{t\to0}\frac{1}{t}\int_0^t e_{-\xi}(x)T_s[e_\xi q(\cdot,\xi)](x)\,ds = -q(x,\xi)$$

$$\tag{2.40}$$

where we used that $q(\cdot,\xi) \in C_b(\mathbb{R}^d)$ in combination with Lemma 1.8. $\square$

The proof of Theorem 2.36 actually shows that

$$T_t e_\xi - e_\xi = \int_0^t T_s \tilde A e_\xi\,ds$$

where $\tilde A e_\xi = \text{bp-}\lim_{n\to\infty}A(\phi_n e_\xi)$. By standard arguments we see that this limit does not depend on the choice of the sequence $(\phi_n)_{n\geqslant1}$, see Corollary 2.11 for the constant-coefficient analogue. This shows that $(e_\xi, \tilde A e_\xi) \in \hat A_b$ where $\hat A_b$ is the full generator as in Definition 1.35.

It is possible to avoid the assumption that $q(x,D)$ has bounded coefficients. This was first shown in [285, Sect. 2]; here we combine this approach with the presentation in [290, Sect. 4].

Theorem 2.37. Let $(X_t)_{t\geqslant0}$ be a Feller process with transition semigroup $(T_t)_{t\geqslant0}$ and generator $(A,\mathcal{D}(A))$. Assume that $C_c^\infty(\mathbb{R}^d) \subset \mathcal{D}(A)$ so that $A|_{C_c^\infty(\mathbb{R}^d)}$ is a pseudo-differential operator with negative definite symbol $q(x,\xi)$;

let $(l(x), Q(x), N(x, dy))$ denote the Lévy triplet of $q(x, \xi) - q(x, 0)$ as in Theorem 2.21.

a) *A has a unique extension to $C_b^2(\mathbb{R}^d)$, again denoted by A, satisfying*

$$Au(x) = \lim_{n \to \infty} A(u\phi_n)(x)$$

for any sequence $(\phi_n)_{n \geqslant 1} \subset C_c^\infty(\mathbb{R}^d)$ with $\mathbb{1}_{B(0,n)} \leqslant \phi_n \leqslant 1$. This extension has again the representation (2.25).

b) *There exists an absolute constant $C < \infty$ such that*

$$|Au(x)| \leqslant C\gamma(x) \|u\|_{(2)} \quad \forall x \in \mathbb{R}^d, \ u \in C_b^2(\mathbb{R}^d) \quad where$$

$$\gamma(x) = q(x, 0) + |l(x)| + |Q(x)| + \int_{\mathbb{R}^d \setminus \{0\}} \min(|y|^2, 1) \, N(x, dy). \tag{2.41}$$

c) $C_c^2(\mathbb{R}^d) \subset \mathcal{D}(A)$.
d) $A : C_b^2(\mathbb{R}^d) \to B(\mathbb{R}^d)$ *and* $x \mapsto Au(x)$ *is locally bounded.*
e) $A : C_\infty^2(\mathbb{R}^d) \to C(\mathbb{R}^d)$.
f) $\{(u, Au) : u \in C_c^2(\mathbb{R}^d)\} \subset \hat{A}$.
   If $q(x, D)$ has bounded coefficients, then
g) $A : C_b^2(\mathbb{R}^d) \to B_b(\mathbb{R}^d)$.
h) $A : C_\infty^2(\mathbb{R}^d) \to C_\infty(\mathbb{R}^d)$ *and* $C_\infty^2(\mathbb{R}^d) \subset \mathcal{D}(A)$.
i) $\{(u, Au) : u \in C_b^2(\mathbb{R}^d)\} \subset \hat{A}_b$.
j) $\{u \in \mathcal{S}'(\mathbb{R}^d) : \hat{u}, \widehat{\Delta u} \in L^1(d\xi)\}$

$$\subset \{u \in \mathcal{S}'(\mathbb{R}^d) : (1 + \|q(\cdot, \xi)\|_\infty)\hat{u} \in L^1(d\xi)\} \subset \mathcal{D}(A).^5$$

Proof (Sketch). Since the Feller generator $(A, C_c^\infty(\mathbb{R}^d))$ has the integro-differential representation (2.25), the arguments are very similar to those used for generators with bounded coefficients, cf. Theorem 2.7 and its Corollaries 2.9 and 2.11.

a) This follows as the corresponding assertion from Corollary 2.11.
b) Write $A = L(x, D) + S(x, D)$ and use the argument leading to the estimate (2.14).
c) Let $u \in C_c^2(\mathbb{R}^d)$, set $K := \operatorname{supp}(u)$ and denote by $j_\epsilon \in C_c^\infty(\overline{\mathbb{B}}(0, \epsilon))$ the Friedrichs mollifier, see e.g. [157, p. 31]. Then $\operatorname{supp}(j_\epsilon * u) \subset K + \overline{\mathbb{B}}(0, \epsilon)$, $j_\epsilon * u \in C_c^\infty(\mathbb{R}^d)$, and $\lim_{\epsilon \to 0} \|j_\epsilon * u - u\|_{(2)} = 0$. As all supports are contained in the fixed compact set $K + \overline{\mathbb{B}}(0, 1)$, we can use the estimate (2.41) to infer that $(A[j_{1/n} * u])_{n \geqslant 1}$ is a Cauchy sequence if we consider uniform convergence on the set $K + \overline{\mathbb{B}}(0, 1)$. Since $A(j_{1/n} * u) \in C_\infty(\mathbb{R}^d)$, we conclude that this sequence

---

[5] $\mathcal{S}'(\mathbb{R}^d)$ denotes the space of tempered distributions (generalized functions), i.e. the topological dual of the space of rapidly decreasing smooth functions $\mathcal{S}(\mathbb{R}^d)$, see e.g. Rudin [259, pp. 173–185].

is a Cauchy sequence in $(C_\infty(\mathbb{R}^d), \|\cdot\|_\infty)$. Using the closedness of $(A, \mathcal{D}(A))$ we get that $u \in \mathcal{D}(A)$ and $C_c^2(\mathbb{R}^d) \subset \mathcal{D}(A)$. See Schnurr [294, Theorem 3.8, p. 65] and [295, Theorem 3.7] for details.

d) Since the extension of A is defined as a limit of measurable functions, it is clearly measurable. The local boundedness follows from (2.41) since the coefficient is locally bounded as a function of x, cf. Proposition 2.27(d) and Theorem 2.31.

e) Note that $A : C_c^\infty(\mathbb{R}^d) \to C(\mathbb{R}^d)$ and that $C_c^\infty(\mathbb{R}^d)$ is dense in $C_\infty^2(\mathbb{R}^d)$ with respect to the norm $\|\cdot\|_{(2)}$. Thus, the estimate (2.41) guarantees that for any $u \in C_\infty^2(\mathbb{R}^d)$ and any sequence $(u_n)_{n\geq 1} \subset C_c^\infty(\mathbb{R}^d)$ with $\lim_{n\to\infty} \|u - u_n\|_{(2)}$ the sequence $(Au_n)_{n\geq 1} \subset C(\mathbb{R}^d)$ converges locally uniformly to $Au \in C(\mathbb{R}^d)$.

f) Set $\tau := \tau_r^x := \inf\{t > 0 : |X_t - x| > r\}$ and pick some $\phi_r \in C_c^\infty(\mathbb{R}^d)$ such that $\mathbb{1}_{B(x,2r)} \leq \phi_r \leq \mathbb{1}_{B(x,4r)}$. By Theorem 1.36 and Part (c)

$$M_t^{[u]} = u(X_t) - u(x) - \int_{[0,t)} Au(X_s)\, ds$$

is a martingale for all $u \in C_c^2(\mathbb{R}^d) \subset \mathcal{D}(A)$. If $f \in C_b^2(\mathbb{R}^d)$, then $f\phi_r$ is a function in $C_c^2(\mathbb{R}^d)$; thus $M_{t\wedge\tau}^{[f]} = \lim_{R\to\infty} M_{t\wedge\tau}^{[f\phi_R]}$, and optional stopping shows that $(M_t^{[f]})_{t\geq 0}$ is a local martingale. Therefore, $\{(f, Af) : f \in C_b^2(\mathbb{R}^d)\} \subset \hat{A}$.

The assertions (g)–(i) follow analogously, just observe that the constant appearing on the right-hand side of the estimate (2.41) is bounded if $q(x, D)$ has bounded coefficients, cf. Theorem 2.31. For (h) we can argue as in Corollary 2.9 using the closedness of $(A, \mathcal{D}(A))$ and the fact that the closure of $C_c^\infty(\mathbb{R}^d)$ with respect to the norm $\|\cdot\|_{(2)}$ is just $C_\infty^2(\mathbb{R}^d)$. Finally, the first inclusion in (j) follows from

$$(1 + \sup_{y \in \mathbb{R}^d} |q(y, \xi)|)|\hat{u}(\xi)| \leq c_q(1 + |\xi|^2)|\hat{u}(\xi)| = c_q(|\hat{u}(\xi)| + |\widehat{Au}(\xi)|),$$

while the proof of the second inclusion is essentially due to the closedness of $(A, \mathcal{D}(A))$, cf. [277, Theorem 2.7, Remark 2.8]. Notice that, by the Riemann–Lebesgue lemma, $\hat{u} \in L^1(d\xi)$ implies $u \in C_\infty(\mathbb{R}^d)$. $\qquad\square$

Remark 2.38. The spaces appearing in Theorem 2.37(j) are closely connected with the Hörmander scale $B_{p,k}(\mathbb{R}^d)$, cf. Hörmander [141, Sect. 10.1, pp. 3–16], and its anisotropic counterpart $B_{\psi,p}^s(\mathbb{R}^d)$ introduced by Jacob [157, Sect. 3.10, pp. 207–222]. $\qquad\square$

If we combine Theorem 2.37(i) with the technique developed in [285, Sect. 2] we can improve the probabilistic formula to calculate the symbol of a Feller process from [285, Theorem 2.6]. Recall that, for any stopping time τ, we write $X_t^\tau = X_{t\wedge\tau}$ for the stopped process.

Corollary 2.39. *Let $(X_t)_{t\geq 0}$ be a Feller process with transition semigroup $(T_t)_{t\geq 0}$ and generator $(A, \mathcal{D}(A))$. Assume that $C_c^\infty(\mathbb{R}^d) \subset \mathcal{D}(A)$ so that $A|_{C_c^\infty(\mathbb{R}^d)}$ is a pseudo-differential operator with negative definite symbol $q(x, \xi)$.*

Let $\tau := \tau_r^x = \inf\{s > 0 : |X_s - x| > r\}$ be the first exit time from the ball $\overline{\mathbb{B}}(x, r)$ of the process started at $X_0 = x$. If $x \mapsto q(x, \xi)$ is continuous, then

$$- q(x, \xi) = \lim_{t \to 0} \frac{\mathbb{E}^x(e^{i(X_t^\tau - x) \cdot \xi}) - 1}{t}. \tag{2.42}$$

Proof. Let τ be the first exit time from $\overline{\mathbb{B}}(x, r)$. By Theorem 2.37(f) $(e_\xi, Ae_\xi) \in \hat{A}$, i.e.

$$e_\xi(X_t^\tau) - e_\xi(x) - \int_{[0, \tau \wedge t)} Ae_\xi(X_s)\, ds, \quad t \geqslant 0,$$

is a martingale under $\mathbb{P}^x$. Using Proposition 2.27(c) and taking expectations we get

$$\lim_{t \downarrow 0} \frac{\mathbb{E}^x(e^{i(X_t^\tau - x) \cdot \xi}) - 1}{t} = -e_{-\xi}(x) \lim_{t \downarrow 0} \mathbb{E}^x \left(\frac{1}{t} \int_{[0, t)} e_\xi(X_s^\tau) q(X_s^\tau, \xi) \mathbb{1}_{[\![0, \tau[\![}(s, \cdot)\, ds \right)$$

$$= -e_{-\xi}(x) e_\xi(x) q(x, \xi) = -q(x, \xi)$$

($[\![0, \tau[\![$ denotes the stochastic interval $\{(s, \omega) : 0 \leqslant s < \tau(\omega)\}$) where we used the fact that $s \mapsto q(X_s, \xi)$ is continuous if $s \downarrow 0$. $\qquad\square$

Remark 2.40. a) Note that the limit (2.42) does not depend on the stopping time τ_r^x. At first glance this is surprising since the right-hand side of (2.42) should be the symbol associated with the stopped process $X^\tau = (X_t^\tau)_{t \geqslant 0}$, but the latter is *not a Markov process*: The stopping time $\tau = \tau_r^x$ depends on the initial position $X_0 = x$. Thus, the notion of a generator of X^τ is not defined.

b) One could use (2.42) to get yet another proof of the Courrège–v. Waldenfels theorem, Theorem 2.21. Let us briefly sketch the argument. Using the positive maximum principle we show that $|Au(x)| \leqslant c_A(x)\|u\|_{(2)}$. A very short direct proof is contained in the monograph Hoh [138, Lemma 2.9, pp.18–19].[6] Since $\xi \mapsto \mathbb{E}^x(e^{i(X_t^\tau - x) \cdot \xi})$ is a characteristic function, it is continuous and positive definite, and the very definition of a negative definite function, Definition 2.14, shows that $\xi \mapsto 1 - \mathbb{E}^x(e^{i(X_t^\tau - x) \cdot \xi})$ is continuous and negative definite (cf. also Berg–Forst [24, Corollary 7.7, p. 41]) and so is $\xi \mapsto t^{-1}(1 - \mathbb{E}^x(e^{i(X_t^\tau - x) \cdot \xi}))$. The estimate $|Au(x)| \leqslant c_A(x)\|u\|_{(2)}$ together with the identity

$$T_t e_\xi - T_t e_\eta - e_\xi + e_\eta = \int_0^t T_s A(e_\xi - e_\eta)\, ds$$

can then be used to deduce that the limit (2.42) is actually locally uniform in ξ, i.e. $\xi \mapsto q(x, \xi)$ is continuous for every fixed $x \in \mathbb{R}^d$. Then the "usual" Lévy–Khintchine formula implies Theorem 2.21. $\qquad\square$

---

[6]This follows also from our proof of Theorem 2.21, cf. Proposition 2.27(a), but in the present situation this would yield a circular conclusion.

2.5 The Semimartingale Connection

Remark 2.40 shows that we may use the limit (2.42) not only in connection with Feller processes but with any Markov process—provided that the limit exists. The following definition is a slight generalization of [285, Definition 2.1], see also Schnurr [294, Definition 4.1].

Definition 2.41. Let $(X_t)_{t \geq 0}$ be an $\mathbb{R}^d$-valued Markov process. For any starting point x denote by $\tau = \tau_r^x := \inf\{t > 0 : |X_t - x| > r\}$ the first exit time from the ball $\overline{\mathbb{B}}(x, r)$. The function $q : \mathbb{R}^d \times \mathbb{R}^d \to \mathbb{C}$ given by

$$q(x, \xi) := -\lim_{t \downarrow 0} \frac{\mathbb{E}^x (e^{i(X_\tau^\tau - x) \cdot \xi}) - 1}{t} \qquad (2.43)$$

is called the **symbol of the process**, if the limit exists for every $x, \xi \in \mathbb{R}^d$ independently of the choice of $r > 0$.

From the discussion in Remark 2.40(b) it is clear that $\xi \mapsto q(x, \xi)$ is negative definite; to ensure continuity and a Lévy–Khintchine representation, we need further assumptions, e.g. $C_c^\infty(\mathbb{R}^d) \subset \mathcal{D}(A)$ or similar "richness" conditions, cf. Definition 2.47 below. A thorough discussion along these lines can be found in the papers by Çinlar et al. [71] and Çinlar–Jacod [70] (from the point of view of the general theory of stochastic processes) and in Schnurr [294, 295] (from the point of view of pseudo-differential operators).

Example 2.42. Let us mention a few processes which do admit symbols in the sense of Definition 2.41

a) Lévy processes, cf. Theorem 2.2 and (2.37).
b) Feller processes, cf. Theorem 2.36 and Corollary 2.39.
c) Temporally homogeneous jump–diffusion semimartingales which are Markov processes (so-called Itô processes), cf. Definition 2.43 and Theorem 2.45 below.
d) Certain classes of Dirichlet processes, cf. [286]. Consider the pure-jump quadratic form defined for $u, w \in C_c^\infty(\mathbb{R}^d)$,

$$\mathcal{E}(u, w) = \frac{1}{2} \iint_{h \neq 0} (u(x + h) - u(x))(w(x + h) - w(x)) \, j(x, x + h) \, dx \, dh,$$

with a symmetric jump density $j(x, y) = j(y, x)$ satisfying

$$\sup_{x \in \mathbb{R}^d} \int_{\mathbb{R}^d \setminus \{0\}} \min(|h|^2, 1) \, j(x, x + h) \, dh < \infty,$$

$$\sup_{x \in \mathbb{R}^d} \int_{0 < |h| \leq 1} |h| \, (j(x, x + h) - j(x, x - h)) \, dh < \infty.$$

This form can be extended to a Dirichlet form $(\mathcal{E}, \mathcal{D}(\mathcal{E}))$ and the infinitesimal generator is a pseudo-differential operator with the negative definite symbol

$$q(x, \xi) = \int_{\mathbb{R}^d \setminus \{0\}} (1 - \cos h \cdot \xi) j(x, x + h) \, dh$$

$$- \frac{i}{2} \int_{\mathbb{R}^d \setminus \{0\}} \sin h \cdot \xi \, (j(x, x + h) - j(x, x - h)) \, dh.$$

e) Solutions of Lévy driven SDEs, cf. [285]. Consider the stochastic differential equation $dX_t^x = \Phi(X_t^x) \, dY_t$, $X_0 = x$ where $(Y_t)_{t \geq 0}$ is a Lévy process with characteristic exponent ψ. If the matrix-valued function $\Phi : \mathbb{R}^n \to \mathbb{R}^{n \times d}$ is such that for every initial value $x \in \mathbb{R}^n$ the solution $(X_t^x)_{t \geq 0}$ is unique and defines a Markov process in $\mathbb{R}^n$, then the process $(X_t^x)_{t \geq 0}$ admits a symbol and the symbol is of the form $q(x, \xi) = \psi(\Phi(x)^\top \xi)$ for $x, \xi \in \mathbb{R}^n$. This follows from some obvious modifications of the arguments in [285]. □

We will see that there is a close connection between (certain subclasses of) semimartingales and those processes for which the limit (2.43) exists. Our standard reference for semimartingales is the monograph [170] by Jacod–Shiryaev. Recall that a (d-dimensional) **semimartingale** has, relative to a truncation function[7] χ, a unique canonical decomposition

$$X_t = X_0 + X_t^C + \int_0^t y \chi(|y|)(\mu^X(\cdot, ds, dy) - v(\cdot, ds, dy)) + \sum_{s \leq t} (1 - \chi(|\Delta X_s|)) \Delta X_s + B_t$$

where X^C is the continuous martingale part, B is a previsible process with paths of finite variation (on compact time intervals) and the jump measure

$$\mu^X(\omega, ds, dy) = \sum_{s : \Delta X_s \neq 0} \delta_{(s, \Delta X_s(\omega))}(ds, dy)$$

with compensator (i.e. dual predictable projection) $v(\omega, ds, dy)$. The triplet (B, C, v) with the (predictable) quadratic variation $C = [X^C, X^C]$ of X^C is called the **semimartingale characteristics**.

Definition 2.43. A **homogeneous diffusion with jumps** is a semimartingale $(X_t)_{t \geq 0}$ whose semimartingale characteristics are of the form

$$B_t = \int_0^t l(X_s) \, ds, \quad C_t = \int_0^t Q(X_s) \, ds, \quad v(\cdot, ds, dy) = N(X_s, dy) \, ds, \quad (2.44)$$

---

[7] That is $\chi \in B_b[0, \infty)$ such that $0 \leq 1 - \chi(s) \leq \kappa \min(s, 1)$ (for some $\kappa > 0$) and $s\chi(s)$ stays bounded.

where $l : \mathbb{R}^d \to \mathbb{R}^d$ and $Q : \mathbb{R}^d \to \mathbb{R}^{d \times d}$ are measurable functions such that $Q(x)$ is a positive semidefinite $d \times d$ matrix, and $N(x, dy)$ is a Borel transition kernel such that $N(x, \{0\}) = 0$. The coefficients $(l(x), Q(x), N(x, dy))$ are called the **differential characteristics** of the semimartingale.

An **Itô process** is a homogeneous diffusion with jumps which is also a strong Markov process.

For an example of a (deterministic!) non-Markovian diffusion with jumps we refer to Schnurr [296, Example 2.1]. Using the fact that semimartingales are stable under pre-stopping[8] in conjunction with Theorem 2.37(f), one can show that a Feller process with $C_c^\infty(\mathbb{R}^d) \subset \mathcal{D}(A)$ is a homogeneous diffusion with jumps, cf. [277, Theorem 3.5] and Schnurr [295, Theorem 3.9].

Theorem 2.44. *Let $(X_t)_{t \geq 0}$ be a Feller process with generator $(A, \mathcal{D}(A))$. Assume that $C_c^\infty(\mathbb{R}^d) \subset \mathcal{D}(A)$ so that $A|_{C_c^\infty(\mathbb{R}^d)}$ is a pseudo-differential operator with negative definite symbol $q(x, \xi)$. Then $(X_t)_{t \geq 0}$ is an Itô process and its differential characteristics coincide with the Lévy triplet $(l(x), Q(x), N(x, dy))$ of the symbol $q(x, \xi) - q(x, 0)$ (relative to the truncation function χ as in Definition 2.43).*

Almost the same proof covers Itô processes, cf. Schnurr [294, Theorem 4.4].

Theorem 2.45. *Assume that $(X_t)_{t \geq 0}$ is an Itô process with differential characteristics $(l(x), Q(x), N(x, dy))$ (relative to the truncation function χ). If $s \mapsto l(X_s)$, $s \mapsto Q(X_s)$ and $s \mapsto \int_{\mathbb{R}^d \setminus \{0\}} \min(|y|^2, 1) \, N(X_s, dy)$ are a.s. right-continuous at $s = 0$, then the limit (2.43) exists and defines the symbol $q(x, \xi)$ of the process. Moreover, $\xi \mapsto q(x, \xi)$ enjoys for every fixed $x \in \mathbb{R}^d$ a Lévy–Khintchine representation with Lévy triplet $(l(x), Q(x), N(x, dy))$.*

Remark 2.46. Theorems 2.44 and 2.45 show that Feller and Itô processes are semimartingales and that the differential characteristics and the x-dependent Lévy triplet of the negative definite symbol coincide. This means that $l(x)$, $Q(x)$ and $N(x, dy)$ correspond to the drift part, diffusion part and the jumps of the process $(X_t)_{t \geq 0}$. The measure $N(x, B)$ describes the intensity of jumps originating from x and of size $B \subset \mathbb{R}^d \setminus \{0\}$; therefore, $N(x, dy)$ is often called **jump(ing) kernel**. It is closely connected to the Lévy systems which were introduced by Ikeda–Watanabe [145] and further developed by Benveniste–Jacod [22]: A **Lévy system** is a pair (M, A) consisting of a kernel $M(x, dy)$ on $\mathbb{R}^d \times \mathcal{B}(\mathbb{R}^d)$ and a continuous additive functional A such that

$$\mathbb{E}^x \left(\sum_{0 < s \leq t} f(X_{s-}, X_s) \mathbb{1}_{\{X_{s-} \neq X_s\}} \right) = \mathbb{E}^x \left(\int_0^t \int_{\mathbb{R}^d} f(X_{s-}, y) \, M(X_{s-}, dy) \, dA_s \right)$$

holds for all positive $f \in B_b(\mathbb{R}^d \times \mathbb{R}^d)$. For a Feller process, cf. Ikeda–Watanabe [145] we have $M(x, dy) = N(x, dy - x)$ and $A_t = t$. Playing with different

---

[8]That is for every stopping time τ set $X_t^{\tau-} := X_t \cdot \mathbb{1}_{\{t < \tau\}} + X_{\tau-} \cdot \mathbb{1}_{\{t \geq \tau\}}$.

functions $f(x, y)$ and using some stopping arguments, Ikeda–Watanabe obtained formulae of the type

$$\mathbb{E}^x \left(e^{-\lambda \tau_D} \mathbb{1}_F(X_{\tau_D-}) \mathbb{1}_E(X_{\tau_D}) \right) = \mathbb{E}^x \left(\int_0^{\tau_D} \int_F e^{-\lambda t} M(y, E) \delta_{X_t}(dy) dt \right)$$

for all $x \in \mathbb{R}^d$ and $\lambda > 0$ (the case $\lambda = 0$ requires $\mathbb{E}^x \tau_D < \infty$) or, in case we have $\mathbb{P}^x(X_{\tau_D} \in \partial D) = 0$,

$$\mathbb{P}^x \left(X_{\tau_D} \in E \mid X_{\tau_D-} = y \right) = \frac{M(y, E)}{M(y, \mathbb{R}^d \setminus \bar{D})} \quad \forall y \in D, \ x \in \mathbb{R}^d.$$

Here τ_D is the first exit time from the set D, moreover $F \subset D$ and $\mathrm{dist}(D, E) > 0$.
□

Theorem 2.45 is also a special case of the more complicated result by Çinlar et al. [71, Theorem (7.16), pp. 211–211]. Let $\psi : \mathbb{R}^d \to \mathbb{C}$ be a continuous negative definite function given by the Lévy–Khintchine representation (2.10) with Lévy triplet (l, Q, ν) and truncation function χ. We write $\psi(D)$ for the pseudo-differential operator with symbol ψ and A_ψ for the unique extension of $-\psi(D)$ to $C_b^2(\mathbb{R}^d)$, cf. Corollary 2.11.

Definition 2.47. A family $\mathcal{F} \subset B_b(\mathbb{R}^d)$ of functions is called a **full class** if

$$\forall r > 0, 1 \leqslant j \leqslant d \quad \exists f_1, \dots, f_d \in \mathcal{F}, \ g \in C^2(\mathbb{R}^d) \text{ such that:}$$

$$\forall x \in \overline{\mathbb{B}}(0, r) : \ x_j = g(f_1(x), \dots, f_d(x)).$$

The family $\mathcal{F}$ is a **complete class** if it contains a countable subset $\mathcal{F}_0 \subset \mathcal{F} \cap C_b^2(\mathbb{R}^d)$ such that for every continuous negative definite function ψ each of the countable sets

$$\{A_\psi f(x) : f \in \mathcal{F}_0\} \subset \mathbb{R}, \quad x \in \mathbb{R}^d,$$

uniquely determines the Lévy characteristics (l, Q, ν) of ψ (relative to the truncation function χ).

Typical examples of full and complete classes are $C_c^\infty(\mathbb{R}^d)$ or $\{e_\xi : \xi \in \mathbb{R}^d\}$ and any function space containing one of these families.

Theorem 2.48. *Let $(X_t)_{t \geqslant 0}$ be a Markov process which is also a semimartingale and denote by $(A, \mathcal{D}(A))$ its generator and by $\hat{A}$ the extended generator.*

a) *$(X_t)_{t \geqslant 0}$ is an Itô process if, and only if, the family*

$$\{u \in B_b(\mathbb{R}^d) : \exists w \in B(\mathbb{R}^d) \text{ such that } (u, w) \in \hat{A}\}$$

is a full and complete class.

b) *If $(X_t)_{t \geqslant 0}$ is an Itô process, then $A|_{C_b^2(\mathbb{R}^d)}$ is of the form (2.25). In this case, the Lévy triplet and the differential characteristics coincide.*

In fact, there is a variant of Theorem 2.48 which holds for homogeneous Hunt semimartingales, but since every Hunt semimartingale is a random time-changed Itô process, cf. Çinlar–Jacod [70, Theorem (3.35), p. 207] or [71, Proposition (7.13), p. 211], we do not include the statement here, but refer to Çinlar et al. [71, Theorem (7.14), p. 211].

We close this section with a condition that guarantees that a C_b-Feller semigroup, cf. Definition 1.6, is a Feller semigroup, i.e. maps $C_\infty(\mathbb{R}^d)$ into itself. This is a partial converse to Theorem 1.9. Note that, mutatis mutandis, the result extends to Itô processes which have the C_b-Feller property in the sense that the associated transition semigroup maps $C_b(\mathbb{R}^d)$ into itself; we follow [291, Appendix, Proposition 4.4].

Theorem 2.49. *Let $(X_t)_{t \geqslant 0}$ be an Itô process (or a C_b-Feller process) and write $(T_t)_{t \geqslant 0}$ for the transition semigroup and $(A, \mathcal{D}(A))$ for the infinitesimal generator. Denote by $q(x, \xi)$ the symbol of the process and assume that $C_c^\infty(\mathbb{R}^d) \subset \mathcal{D}(A)$. If the symbol satisfies*

$$\varlimsup_{r \to \infty} \sup_{|x| \leqslant r} \sup_{|\xi| \leqslant 1/r} |q(x, \xi)| = 0, \tag{2.45}$$

then $T_t : C_\infty(\mathbb{R}^d) \to C_\infty(\mathbb{R}^d)$.

If $(X_t)_{t \geqslant 0}$ is a C_b-Feller process, (2.45) ensures that $(X_t)_{t \geqslant 0}$ is also a Feller process.

Chapter 3
Construction of Feller Processes

So far we have studied *necessary* conditions for an operator $(A, \mathcal{D}(A))$ to be the generator of a Feller process and we will now discuss *sufficient* conditions. Unless otherwise mentioned, we will assume throughout this chapter that $E = \mathbb{R}^d$. We have seen in Corollary 2.23 and Theorem 2.21 that a Feller generator A is a pseudo-differential operator with negative definite symbol if $C_c^\infty(\mathbb{R}^d) \subset \mathcal{D}(A)$, i.e.

$$Au(x) = -\int_{\mathbb{R}^d} e^{ix\cdot\xi} q(x,\xi)\hat{u}(\xi)\, d\xi \quad \forall u \in C_c^\infty(\mathbb{R}^d)$$

and $\xi \mapsto q(x,\xi)$ is a continuous negative definite function which has Lévy–Khintchine representation (2.32) with x-dependent Lévy triplet $(l(x), Q(x), N(x, dy))$.

If we want to *construct* a Feller generator A it is, therefore, reasonable to **assume that A is defined on the test functions $C_c^\infty(\mathbb{R}^d)$**; consequently, $A|_{C_c^\infty(\mathbb{R}^d)}$ is a pseudo-differential operator with negative definite symbol. We will make this assumption throughout this chapter.

If A is an operator with constant coefficients, i.e. if $q(x,\xi) = \psi(\xi)$ and the Lévy triplet (l, Q, ν) does not depend on x, the corresponding Feller semigroup is translation invariant and the associated stochastic process is a Lévy process (with killing), cf. Sect. 2.1. In this case there is a one-to-one correspondence between continuous negative definite functions ψ, Lévy triplets (l, Q, ν) and Feller semigroups and processes. This means, in particular, that any constant-coefficient generator A with a continuous negative definite symbol $\psi(\xi)$ will generate a translation invariant Feller semigroup and a Lévy process (with killing).

For variable coefficients this is not any longer the case, and it is an important question under which conditions (either for $q(x,\xi)$ or $(l(x), Q(x), N(x, \cdot))$) the operator $A = -q(x, D)$, defined on $C_c^\infty(\mathbb{R}^d)$, has an extension which is a Feller generator. Unfortunately, this is not always the case, see Example 2.26.

In the following sections we review various methods to construct Feller semigroups or processes if the generator $A|_{C_c^\infty(\mathbb{R}^d)}$ is given. Most constructions are quite

B. Böttcher et al., *Lévy Matters III*, Lecture Notes
in Mathematics 2099, DOI 10.1007/978-3-319-02684-8_3,
© Springer International Publishing Switzerland 2013

technical, and we do not attempt to present proofs; we will, however, give precise references to the literature and try to explain the key ideas behind each approach. All constructions differ in methods and assumptions, and there seems to be no all-embracing, general approach. Worse, it is often unclear how the assumptions needed for the various approaches relate among each other.

3.1 The Hille–Yosida Construction

The starting point for most "analytic" approaches to constructing Feller semigroups is the Hille–Yosida–Ray theorem, Theorem 1.30, which we recall here for further discussions.

Theorem 3.1 (Hille–Yosida–Ray). *Let $(A, \mathcal{D})$ be a linear operator on $C_\infty(\mathbb{R}^d)$. $(A, \mathcal{D})$ is closable and the closure $(A, \mathcal{D}(A))$ is the generator of a Feller semigroup if, and only if,*

a) $\mathcal{D} \subset C_\infty(\mathbb{R}^d)$ *is dense;*
b) $(A, \mathcal{D})$ *satisfies the positive maximum principle;*
c) $(\lambda - A)(\mathcal{D}) \subset C_\infty(\mathbb{R}^d)$ *is dense for some (or all) $\lambda > 0$.*

In order to apply Theorem 3.1 we have to verify the conditions (a)–(c). The first two conditions are met in most applications; for instance, Condition (a) is automatically satisfied if we require that $C_c^\infty(\mathbb{R}^d) \subset \mathcal{D}$. Condition (b) is then a structural assumption: If A is a pseudo-differential operator, it is equivalent to saying that the symbol $q(x, \xi)$ is, for every $x \in \mathbb{R}^d$, a negative definite function. This follows directly from Theorem 2.21. It is the third condition which is troublesome. If $A = -q(x, D)$ and $\mathcal{D} = C_c^\infty(\mathbb{R}^d)$, say, it means that we have to solve the following pseudo-differential equation

$$\lambda u(x) + q(x, D)u(x) = f(x) \tag{3.1}$$

for all right-hand sides f from a dense subset of $C_\infty(\mathbb{R}^d)$ and with $u \in C_c^\infty(\mathbb{R}^d)$. A possibility to approach this problem is to analyse the mapping properties of the operator $q(x, D)$ within the scale of L^p-Sobolev spaces:

$$W^{3,p}(\mathbb{R}^d) = \{u \in \mathcal{S}'(\mathbb{R}^d) : \|u\|_{W^{3,p}} < \infty\}, \quad \|u\|_{W^{3,p}} = \sum_{|\alpha| \leq 3} \|\nabla^\alpha u\|_{L^p}. \tag{3.2}$$

We begin with a classical result for local operators and diffusion processes which we take from the monograph by Jacob [158, Theorem 2.1.43, p. 47].

Theorem 3.2. *Let A be a second order differential operator*

$$Au(x) = \frac{1}{2} \sum_{j,k=1}^{d} q_{jk}(x) \frac{\partial^2 u}{\partial x_j \partial x_k}(x) + \sum_{k=1}^{d} l_k(x) \frac{\partial u}{\partial x_k}(x) + c(x)u(x)$$

with variable coefficients $q_{jk} = q_{kj} \in C_b^3(\mathbb{R}^d), l_j \in C_b^2(\mathbb{R}^d)$ *and* $c \in C_b^1(\mathbb{R}^d)$ *such that* $c(x) \leq 0$ *for all* $x \in \mathbb{R}^d$. *Assume that A is uniformly elliptic, i.e. there exists some* $\lambda_0 > 0$ *such that*

$$\lambda_0 |\xi|^2 \leq \sum_{j,k=1}^d q_{jk}(x)\xi_j \xi_k \quad \forall x, \xi \in \mathbb{R}^d. \tag{3.3}$$

Then the operator $(A, W^{3,p}(\mathbb{R}^d))$ *extends for* $p > d$ *to a generator of a Feller semigroup.*

The proof uses two standard ideas. The first is, that for elliptic operators inverses (at least in some sense) exist, and thus one can solve the associated Poisson equation (3.1) in the scale $W^{m,p}$. Then, in a second step, one uses Sobolev's embedding theorem $W^{m,p}(\mathbb{R}^d) \hookrightarrow C_\infty^k(\mathbb{R}^d)$ for $(m-k)p > d$, cf. [158, Theorem 2.1.12, p. 24], to ensure that (3.1) is solved for all right-hand sides f from a dense subset of $C_\infty(\mathbb{R}^d)$.

Theorem 3.2 only includes local operators, i.e. it corresponds to diffusion processes with continuous paths, cf. Theorem 1.40 and Corollary 1.42. For processes with jumps one has to consider integro-differential operators. The following general result which allows discontinuous coefficients, but only treats processes on bounded domains can be found in Taira [314, Theorem 1.2].

Theorem 3.3. *Let* $E \subset \mathbb{R}^d, d \geq 3$, *be a bounded convex set with boundary of class* $C^{1,1}$. *Define*

$$Au(x) = \frac{1}{2} \sum_{j,k=1}^d q_{jk}(x)\frac{\partial^2 u}{\partial x_j \partial x_k}(x) + \sum_{k=1}^d l_k(x)\frac{\partial u}{\partial x_k}(x) + c(x)u(x)$$

$$+ \int_{\bar{E}} \left(u(y) - u(x) - \sum_{k=1}^d (y_k - x_k)\frac{\partial u}{\partial x_k}(x) \right) s(x, dy),$$

where $l_k, c \in L^\infty(E)$ *with* $c(x) \leq 0$ *a.e.,* $q_{jk} \in VMO \cap L^\infty(\mathbb{R}^d),$[1] $q_{jk} = q_{kj}$ *a.e. and a family of Borel measures* $(s(x, dy))_{x \in E} \subset \mathcal{M}^+(\bar{E})$. *Assume that there exists some* $\lambda > 0$ *such that*

---

[1] VMO is the space of functions with *vanishing mean oscillation*:

$$VMO = \left\{ f \in L_{loc}^1(\mathbb{R}^d) : \sup_{r>0} \|f\|_{*r} < \infty, \ \lim_{r \to 0} \|f\|_{*r} = 0 \right\}$$

$$\|f\|_{*r} = \sup_{B \in \{B(x,r) : x \in \mathbb{R}^d\}} \frac{1}{\text{Leb}(B)} \int_B \left| f(x) - \frac{1}{\text{Leb}(B)} \int_B f(x)\, dx \right| dx.$$

$$\lambda^{-1}|\xi|^2 \le \sum_{j,k=1}^{d} q_{jk}(x)\xi_j\xi_k \le \lambda|\xi|^2 \quad \text{for almost all } x \in E \text{ and all } \xi \in \mathbb{R}^d, \quad (3.4)$$

and that $x \mapsto s(x, B)$ is Borel measurable for each Borel subset $B \subset \bar{E}$, $s(x, \{x\}) = 0$ for all $x \in E$ and for every small $\epsilon > 0$ there exists a bounded function Φ such that

$$\sup_{x \in E} \int_{\{y \in \bar{E} : |y-x| \le \epsilon\}} |y - x|^2 s(x, dy) \le \Phi(\epsilon) \quad \text{and} \quad \lim_{\epsilon \to 0} \Phi(\epsilon) = 0$$

and

$$\sup_{x \in E} \int_{\{y \in \bar{E} : |y-x| > \epsilon\}} |y - x| s(x, dy) < \infty.$$

Then $(A, \{u \in W^{2,p}(E) \cap C_\infty(\bar{E}) : Au \in C_\infty(\bar{E})\})$ for $d < p < \infty$ is the generator of a Feller semigroup on $C_\infty(\bar{E}) = C(\bar{E}) = C_b(\bar{E})$.

The proof of Theorem 3.3 requires (3.4) i.e., A has a non-degenerate, uniformly elliptic second-order diffusion part $\frac{1}{2}\sum_{j,k=1}^{d} q_{jk}(x)\frac{\partial^2 u}{\partial x_j \partial x_k}$; this means that A behaves like the Laplacian and that the non-local jump part can be treated as a (lower-order) perturbation. This strong non-degeneracy assumption appears frequently in the literature, for example in Taira [313], Stroock [309] (see also Theorem 3.23) and in the more analytical literature e.g. Bensoussan–Lions [20, 21] or Garroni–Menaldi [120].

In order to construct pure jump processes, i.e. processes with degenerate diffusion part or no diffusion part at all, we need different techniques. The main problem is that there is no dominating second-order principal part. One way out is to develop for the corresponding operators a symbolic calculus which captures their mapping properties within a scale of certain fractional or even anisotropic Sobolev spaces. The main result based on this approach is stated in Jacob [158, Theorem 2.6.9, p. 133] and it is due to Hoh [138, 137]. Here we present a slight extension of it for complex valued symbols which can be found in [39, Theorem 2.2.25]. Observe that $-q(x, D)$ is *not a classical pseudo-differential operator* in the sense of Hörmander [143], see also Kumano-go [194] and Lemma 2.18. Therefore, we cannot employ the usual symbolic calculus for pseudo-differential operators.

Theorem 3.4. Let $\psi : \mathbb{R}^d \to \mathbb{R}$ be a continuous negative definite function such that for every $\alpha \in \mathbb{N}_0^d$ there exists a constant $c_{|\alpha|} \ge 0$ such that

$$|\nabla_\xi^\alpha(1 + \psi(\xi))| \le c_{|\alpha|}(1 + \psi(\xi))^{1 - \frac{|\alpha| \wedge 2}{2}} \quad \forall \xi \in \mathbb{R}^d \qquad (3.5)$$

and, for some constants $c_0, r_0 > 0$

$$\psi(\xi) \ge c_0 \gamma |\xi|^{r_0} \quad \forall \xi \in \mathbb{R}^d. \qquad (3.6)$$

Assume that $q : \mathbb{R}^d \times \mathbb{R}^d \to \mathbb{C}$ is a C^∞-function such that for every $x \in \mathbb{R}^d$ the function $\xi \mapsto q(x, \xi)$ is continuous and negative definite, and that for all $\alpha, \beta \in \mathbb{N}_0^d$ there exist constants $c_{\alpha,\beta} \geq 0$ such that

$$\left| \nabla_\xi^\alpha \nabla_x^\beta q(x, \xi) \right| \leq c_{\alpha,\beta} (1 + \psi(\xi))^{1 - \frac{|\alpha| \wedge 2}{2}} \quad \forall x, \xi \in \mathbb{R}^d$$

and for some $\delta > 0$

$$\operatorname{Re} q(x, \xi) \geq \delta (1 + \psi(\xi)) \quad \forall |\xi| \gg 1, \; x \in \mathbb{R}^d. \tag{3.7}$$

Then $(-q(x, D), C_c^\infty(\mathbb{R}^d))$ is closable and the closure is a Feller generator.

The idea behind the proof of Theorem 3.4 is to replace the classical Sobolev and Bessel potential scale $W^{s,p}(\mathbb{R}^d)$ by a scale of anisotropic spaces which are determined by the fixed reference function ψ.

$$H_p^{\psi,s}(\mathbb{R}^d) = \{ u \in \mathcal{S}'(\mathbb{R}^d) : \|(1 + \psi(D))^{s/2} u\|_{L^p} < \infty \} \tag{3.8}$$

(if we take $\psi(\xi) = |\xi|^2$, we are back in the classical scales). For a detailed account on these anisotropic Bessel potential spaces we refer to Jacob [157, Chap. 3.10, pp. 207–222] for $p = 2$ and, in an L^p setting, to Jacob [158, Chap. 3.3, pp. 270–300] and Farkas et al. [104, 105].

In this framework, (3.7) can be seen as an ellipticity condition, and the minimal growth condition (3.6) for the reference function ψ ensures that we have an embedding $H_p^{\psi,s}(\mathbb{R}^d) \hookrightarrow C_\infty^k(\mathbb{R}^d)$ if $s \cdot p$ is sufficiently large. Still, the operators considered in Theorem 3.4 have a fixed order (determined by the reference function ψ). For operators with variable order, one has the following result of Hoh [138, Theorem 7.1] (for symbols in classical symbol classes we refer to Kikuchi–Negoro [181]).

Theorem 3.5 (Hoh). *Let $\psi(\xi)$ and $q(x, \xi)$ be as in Theorem 3.4. Furthermore, assume that $q(x, \xi)$ is real valued, and $m : \mathbb{R}^d \to (0, 1]$ is a C^∞-function with bounded derivatives such that*

$$\sup_{x \in \mathbb{R}^d} m(x) - \inf_{x \in \mathbb{R}^d} m(x) < \frac{1}{2} \quad \text{and} \quad \inf_{x \in \mathbb{R}^d} m(x) > 0.$$

Then

$$p(x, \xi) := q(x, \xi)^{m(x)}$$

is a negative definite symbol, the operator $(-p(x, D), C_c^\infty(\mathbb{R}^d))$ is closable and the closure is a Feller generator.

The proof of this result uses techniques for degenerated elliptic operators and sectorial forms,[2] cf. Kato [173, VI.2]. For the latter the restrictions on m are essential. In Sect. 3.5 below we will see how we can relax these assumptions by the use of the martingale problem.

The Role of Smoothness. Let us briefly discuss the smoothness requirements on the symbol. As we have mentioned, classical symbol classes require smoothness and decay of the derivatives of the symbol $(x, \xi) \mapsto q(x, \xi)$, both in x and ξ. If we assume this for a negative definite symbol $q(x, \xi)$, we can use the standard Hörmander symbolic calculus to construct Feller processes see, e.g. Boyarchenko–Levendorskiĭ [50,51] and Barndorff-Nielsen–Levendorskiĭ [13]. This approach is somewhat unnatural since, at least in the co-variable $\xi \mapsto q(x, \xi)$, negative definite symbols are, in general, only continuous. Even if we enforce C^∞-smoothness in ξ, e.g. by cutting the support of the Lévy measure, we will not be able to observe the usually required decay improvement for classical symbols (think of the derivatives of homogeneous functions, say): Lemma 2.18 clearly shows that only the first two derivatives will contribute towards a decay.

Jacob [153, Theorem 5.2] considers negative definite symbols which are continuous in ξ and q times differentiable in x where $q = q(d)$ is a multiple of the space dimension d. If $x \mapsto q(x, \xi) - q(x_0, \xi)$ is small (in a suitable sense)—here x_0 is fixed and $q(x_0, \xi)$ is then the characteristic exponent of a Lévy process—then $(-q(x, D), C_c^\infty(\mathbb{R}^d))$ has a closure which is a Feller generator. Using the martingale problem and stopping techniques, Hoh [136] could localize the oscillation condition; in fact, he proved that the existence of *some* solution to the martingale problem only requires continuity in (x, ξ), while it is the well-posedness (hence, the uniqueness of the process and the Markov property of the process) that needs more assumptions, see also Theorem 3.24 below. Hoh's symbolic calculus approach, cf. Theorem 3.4, requires C^∞-smoothness in x and ξ; with a suitable truncation technique for the Lévy kernel, see Lemma 2.18, we can reduce this to C^∞-smoothness in x, if we use the perturbation technique from Hoh [138, Sect. 6.6] and [139]. Potrykus [239, 240] gives a very careful analysis of the degree of smoothness in x which is really needed in Hoh's symbolic calculus.

3.2 Stochastic Differential Equations (SDEs)

For probabilists there is a natural, path-by-path construction for stochastic processes: Stochastic differential equations (SDEs). In our setting, one typically shows that a given SDE has a solution and that this solution is in fact a Feller process. Sometimes it is helpful to understand $x \mapsto X_t^x$ (the superscript indicates the

---

[2]In the theory of Dirichlet forms these are also called coercive closed forms, cf. Ma–Röckner [210, I.2.3].

starting point $X_0 = x$) as a stochastic flow and to study its smoothness properties; for diffusions the classic reference is Kunita's monograph [195] and its recent counterpart [196] for Lévy-driven SDEs. Our standard references for jump-type SDEs are the monographs by Ikeda–Watanabe [146, Chap. IV.9, pp. 244–246], Situ [302] and Protter [243]. We begin with some classical results from the theory of diffusion processes. Consider the SDE

$$dX_t = l(X_t)\, dt + \sigma(X_t)\, dW_t \qquad (3.9)$$

with initial condition $X_0 = x$, $x \in \mathbb{R}^d$ and a d-dimensional Brownian motion $(W_t)_{t \geq 0}$ as driving noise. The coefficients $l : \mathbb{R}^d \to \mathbb{R}^d$ and $\sigma : \mathbb{R}^d \to \mathbb{R}^{d \times d}$ are such that

$$|l(x) - l(y)|^2 + |\sigma(x) - \sigma(y)|^2 \leq L^2 |x - y|^2 \qquad (3.10)$$

$$|l(x)|^2 + |\sigma(x)|^2 \leq M^2 (1 + |x|)^2 \qquad (3.11)$$

for some constants L and M; $|\cdot|$ is any vector or matrix norm. Usually (3.10) is called **Lipschitz condition** and (3.11) is called **linear growth condition**. Note that (3.10) (if the constant L does not depend on $x, y \in \mathbb{R}^d$) implies (3.11), see also (3.17) and (3.18) below. The following is a basic result in this context, see for instance [284, Theorem 19.9, p. 306].

Theorem 3.6. *Assume that the Lipschitz condition* (3.10) *holds for all* $x, y \in \mathbb{R}^d$ *with a global Lipschitz constant* L. *Then the SDE* (3.9) *has for each initial value* $x \in \mathbb{R}^d$ *a unique solution* $(X_t)_{t \geq 0}$. *Moreover this solution is a Feller process with symbol*

$$q(x, \xi) = -i\, l(x) \cdot \xi + \frac{1}{2} \xi \cdot Q(x) \xi \qquad (3.12)$$

where $Q(x) = \sigma(x)\sigma^\top(x)$.

The converse problem, whether for a given symbol (3.12) with variable coefficients $l(\cdot) \in \mathbb{R}^d$ and positive semidefinite $Q(\cdot) \in \mathbb{R}^{d \times d}$ a corresponding process exists has no straightforward answer. Although we can write down the SDE (3.9) since we may define $\sigma(x) \in \mathbb{R}^{d \times d}$ as the unique positive semidefinite square root of $Q(x) \in \mathbb{R}^{d \times d}$, it is by far not clear whether this SDE can be solved. One possibility is to impose conditions on $Q(x)$ such that $\sigma(x)$ is Lipschitz, e.g. we could require that $Q(x)$ is globally Lipschitz continuous and uniformly positive definite, see [284, Corollary 19.11] for a very simple proof of this; the following more general result is from Ikeda–Watanabe [146, Theorem 6.1, Proposition 6.2, p. 215] or Stroock–Varadhan [312, Chap. 5, pp. 122–135].

Theorem 3.7. *Let* $q(x, \xi)$ *be a symbol of the form* (3.12) *where* $x \mapsto l(x) \in \mathbb{R}^d$ *is globally Lipschitz continuous and* $x \mapsto Q(x) \in \mathbb{R}^{d \times d}$ *is of class* $C_b^2(\mathbb{R}^d)$ *such that* $Q(x)$ *is for every* $x \in \mathbb{R}^d$ *a positive definite matrix. Then there exists* *a Feller process with continuous sample paths such that the generator* $(A, \mathcal{D}(A))$ *is a second-order differential operator with the symbol* $q(x, \xi)$.

In the SDE setting (strong, i.e. pathwise) uniqueness plays a major role. This is a quite restrictive assumption, and Stroock and Varadhan developed the martingale problem approach to cope with situations where we only have uniqueness in the sense of finite-dimensional distributions. A short discussion is given in [284, pp. 309–310], the ultimate reference is the monograph by Stroock–Varadhan [312]. The martingale problem is briefly discussed in Sect. 3.5 below.

Since constant drift and Brownian motion are both Lévy processes, Theorem 3.6 is a special case of the next result, taken from [285, Sect. 3], which allows the construction of jump processes.

Theorem 3.8. *Let* $\Phi : \mathbb{R}^d \to \mathbb{R}^{d \times n}$ *be locally Lipschitz continuous and bounded, and let* $(L_t)_{t \geq 0}$ *be an n-dimensional Lévy process with symbol* $\psi : \mathbb{R}^n \to \mathbb{C}$. *Then the solution of the SDE*

$$dX_t = \Phi(X_{t-}) \, dL_t \tag{3.13}$$

exists for every initial condition $X_0 = x \in \mathbb{R}^d$ *and yields a Feller process with symbol* $q(x, \xi) = \psi(\Phi^\top(x)\xi)$.

The boundedness of the coefficient Φ is mainly needed to ensure that the strong solution of the SDE (3.13) is a Feller process. If Φ is not bounded and if the SDE admits a strong solution (e.g. if Φ satisfies a local Lipschitz and linear growth condition), then $(X_t)_{t \geq 0}$ is still a semimartingale admitting a symbol in the sense of Sect. 2.4, cf. [285, Theorem 3.1].

Example 3.9. Let $q(x, \xi) = a(x)|\xi|^\alpha + ib(x) \cdot \xi$, $x, \xi \in \mathbb{R}^d$, where $0 < \alpha \leq 2$ and $a : \mathbb{R}^d \to \mathbb{R}$ and $b : \mathbb{R}^d \to \mathbb{R}^d$ are bounded and Lipschitz continuous functions such that $a(x) > 0$. Then Theorem 3.8 guarantees the existence of a Feller process with generator $A|_{C_c^\infty(\mathbb{R}^d)} = -q(x, D)$. Indeed, let $(L_t^\alpha)_{t \geq 0}$ be a d-dimensional rotationally symmetric α-stable Lévy process and consider the $d + 1$-dimensional Lévy process $(L_t, t)^\top$. Let $\Phi(x)$ be a $d \times d$ diagonal matrix with diagonal entries $a(x)^{1/\alpha}$. Then the SDE

$$dX_t = (\Phi, b)(X_{t-}) \, d\begin{pmatrix} L_t \\ t \end{pmatrix}, \quad X_0 = x$$

yields a Feller process with the symbol $q(x, \xi) = a(x)|\xi|^\alpha + ib(x) \cdot \xi$.   □

The symbols appearing in Theorem 3.8 have a special multiplicative form. To generalize the construction principle, we note that every Feller process $(X_t)_{t \geq 0}$ living on the whole space, i.e. without boundary terms, is the solution to an SDE. We will explain this based on the exposition in [46, Sect. 3].

Assume that we have a Feller process $(X_t)_{t\geq 0}$ with the symbol

$$q(x,\xi) = -il(x)\cdot\xi + \frac{1}{2}\xi\cdot Q(x)\xi$$

$$+ \int_{\mathbb{R}^d\setminus\{0\}} \left(1 - e^{iy\cdot\xi} + i\xi\cdot y\mathbb{1}_{[0,1]}(|y|)\right) N(x,dy). \tag{3.14}$$

Then by [294, Theorem 3.14], see also Sect. 2.5, $(X_t)_{t\geq 0}$ is an Itô process in the sense of Çinlar et al. [71], i.e. it is a strong Markov process which is a semimartingale with respect to every $\mathbb{P}^x$ and its semimartingale characteristics relative to the truncation function $\chi(y) := \mathbb{1}_{[0,1]}(|y|)$ are

$$B_t(\omega) = \int_0^t l(X_s(\omega))\,ds,$$

$$C_t(\omega) = \int_0^t Q(X_s(\omega))\,ds,$$

$$v(\omega, ds, dy) = N(X_s(\omega), dy)\,ds.$$

By Çinlar–Jacod [70, Theorem 3.33] there exists a suitable enlargement of the stochastic basis, the so-called Markov extension, denoted by

$$\left(\tilde{\Omega}, \tilde{\mathscr{F}}, (\tilde{\mathscr{F}}_t)_{t\geq 0}, (X_t)_{t\geq 0}, \tilde{\mathbb{P}}^x\right)_{x\in\mathbb{R}^d}$$

such that the process $(X_t)_{t\geq 0}$ is the solution of the following SDE:

$$X_t = x + \int_0^t l(X_{s-})\,ds + \int_0^t \sigma(X_{s-})\,d\tilde{W}_s$$

$$+ \int_0^t \int_{\mathbb{R}\setminus\{0\}} k(X_{s-},z)\mathbb{1}_{\{|k(X_{s-},z)|\leq 1\}} \left(\tilde{\mu}(\cdot\,; ds, dz) - ds\,\tilde{v}(dz)\right)$$

$$+ \int_0^t \int_{\mathbb{R}\setminus\{0\}} k(X_{s-},z)\mathbb{1}_{\{|k(X_{s-},z)|>1\}} \tilde{\mu}(\cdot\,; ds, dz)$$

where $(\tilde{W}_t)_{t\geq 0}$ is a d-dimensional Brownian motion, $\tilde{\mu}$ is a Poisson random measure on $[0,\infty)\times(\mathbb{R}\setminus\{0\})$ with compensator (or dual predictable projection) $dt\,\tilde{v}(dz)$. The functions $\sigma : \mathbb{R}^d \to \mathbb{R}^{d\times d}$ and $k : \mathbb{R}^d\times(\mathbb{R}\setminus\{0\}) \to \mathbb{R}^d$ are measurable and satisfy

$$\tilde{v}(k(X_s(\omega),\bullet)\in dy)\,ds = v(\omega, ds, dy) \tag{3.15}$$

$\tilde{\mathbb{P}}^x$-a.s. for every $x\in\mathbb{R}^d$ on the Markov extension, cf. Çinlar–Jacod [70, Equation (3.9), remark after Theorem 3.7]. Furthermore, by changing the truncation function and setting $\tilde{l}(x) = l(x) - \int_{\mathbb{R}\setminus\{0\}} k(x,z)\left(\mathbb{1}_{|k(x,z)|\leq 1} - \mathbb{1}_{\{|z|\leq 1\}}\right)\tilde{v}(dz)$, the process solves the SDE

$$X_t = x + \int_0^t \tilde{l}(X_{s-})\,ds + \int_0^t \sigma(X_{s-})\,d\tilde{W}_s$$

$$+ \int_0^t \int_{0<|z|\leq 1} k(X_{s-},z)\left(\tilde{\mu}(\cdot;ds,dz) - ds\,\tilde{v}(dz)\right) \qquad (3.16)$$

$$+ \int_0^t \int_{|z|>1} k(X_{s-},z)\,\tilde{\mu}(\cdot;ds,dz).$$

Thus we have shown the following result.

Proposition 3.10. *Let $(X_t)_{t\geq 0}$ be a Feller process with generator $(A, \mathcal{D}(A))$ such that $C_c^\infty(\mathbb{R}^d) \subset \mathcal{D}(A)$ and $A|_{C_c^\infty(\mathbb{R}^d)}$ is a pseudo-differential operator with negative definite symbol $q(x,\xi)$ given by (3.14). Then $(X_t)_{t\geq 0}$ is a solution of the SDE (3.16).*

Assume now that an SDE of the form (3.16) is given. The standard conditions in order to *solve* such a jump-type SDE are again (local) Lipschitz and linear growth assumptions

$$|l(x) - l(y)|^2 + |\sigma(x) - \sigma(y)|^2 + \int_{|z|\leq 1} |k(x,z) - k(y,z)|^2\,\tilde{v}(dz) \leq L^2|x-y|^2,$$

$$\qquad (3.17)$$

$$|l(x)|^2 + |\sigma(x)|^2 + \int_{|z|\leq 1} |k(x,z)|^2\,\tilde{v}(dz) \leq M^2(1 + |x|)^2.$$

$$\qquad (3.18)$$

The problem is that the relation (3.15) of the pullback k and the measure v is a mere *existence* statement. In most cases, it is not possible (or at least not trivial) to deduce any smoothness property of k, given the symbol of the generator; the converse problem, to get information on the smoothness of the symbol from a given k, is also not simple. In both cases we refer to the discussion in Tsuchiya [326]. Based on this we close this section with a result that anticipates Sect. 3.5: A solution to the martingale problem, formulated in terms of the coefficients in the SDE (3.16), cf. Stroock [311, Theorem 3.1.26, p. 92].

Theorem 3.11. *Let $\tilde{v}$ be a Lévy measure on $\mathbb{R}$,[3] and assume that $k : \mathbb{R}^d \times \mathbb{R} \to \mathbb{R}^d$ is Borel measurable such that*

a) $k(x,0) = 0 \quad \forall x \in \mathbb{R}^d$,

---

[3] Stroock [311, Theorem 3.1.26] considers multidimensional Lévy measures, but for Feller processes it is sufficient to consider $\tilde{v}$ on $\mathbb{R} \setminus \{0\}$, cf. (3.16).

b) $\displaystyle\lim_{r\downarrow 0}\,\sup_{x\in\mathbb{R}^d}(1+|x|)^{-2}\int_{B(0,r)}|k(x,z)|^2\,\tilde{v}(dz)=0,$

c) $\displaystyle\sup_{x\in\mathbb{R}^d}(1+|x|)^{-2}\int_{B(0,R)}|k(x,z)|^2\,\tilde{v}(dz)<\infty\quad\forall R>0,$

d) $\displaystyle\sup_{x\neq y}|x-y|^{-2}\int_{B(0,R)}|k(x,z)-k(y,z)|^2\,\tilde{v}(dz)<\infty\quad\forall R>0,$

and $\sigma:\mathbb{R}^d\to\mathbb{R}^{d\times d}$ and $l:\mathbb{R}^d\to\mathbb{R}^d$ are continuous such that

$$\sup_{x\neq y}\frac{\max(|\sigma(x)-\sigma(y)|,\,|l(x)-l(y)|)}{|x-y|}<\infty.$$

Then for any starting point $x\in\mathbb{R}^d$ there exists a solution to the martingale problem (see Sect. 3.5) for the operator $(-q(x,D),C_c^2(\mathbb{R}^d))$ with the symbol

$$-i\,l(x)\cdot\xi+\frac{1}{2}\xi\cdot\sigma(x)\sigma^\top(x)\xi+\int_{\mathbb{R}^d\setminus\{0\}}\left(1-e^{ik(x,z)\cdot\xi}+i\,\xi\cdot k(x,z)\mathbb{1}_{[0,1]}(|z|)\right)\tilde{v}(dz).$$

3.3 Dirichlet Forms

Dirichlet forms are a powerful tool in order to characterize and construct Markov processes. In contrast to the Feller setting, Dirichlet forms are defined on the space $L^2(E,m)$, and the transition semigroups are strongly contractive sub-Markovian semigroups on $L^2(E,m)$ where $(E,\mathcal{B}(E),m)$ can be a rather general (even infinite-dimensional) measure space. We assume that m has full support; as a consequence, $\|\cdot\|_{L^\infty(m)}$ coincides with the usual (ess)sup-norm $\|\cdot\|_\infty$. We will consider only locally compact, separable metric spaces E and quite often $E\subset\mathbb{R}^d$. Our standard references for (symmetric) Dirichlet forms are the monographs [118] by Fukushima–Oshima–Takeda and [210] by Ma–Röckner. For semi-Dirichlet forms we refer to the recent monograph by Oshima [235] which extends his *Erlangen lecture notes* [234].

A **Dirichlet form** is a densely defined, positive and symmetric bilinear form[4] $\mathcal{E}:\mathcal{D}(\mathcal{E})\times\mathcal{D}(\mathcal{E})\to\mathbb{R}$ on $L^2(E,m)$ which satisfies for $u\in\mathcal{D}(\mathcal{E})$

$(\mathcal{D}(\mathcal{E}),\|\cdot\|_\mathcal{E})$ is complete, where $\|u\|_\mathcal{E}^2:=\mathcal{E}(u,u)+\|u\|_{L^2(m)}$,   **(closed)**

$w:=(0\vee u)\wedge 1\in\mathcal{D}(\mathcal{E})$ and $\mathcal{E}(w,w)\leqslant\mathcal{E}(u,u)$.   **(Markovian)**

As usual, we write $\mathcal{E}_\lambda(u,w):=\mathcal{E}(u,w)+\lambda\langle u,w\rangle_{L^2(m)}$ for all $\lambda>0$.

A **regular Dirichlet form** satisfies additionally

---

[4]That is $\mathcal{D}(\mathcal{E})\subset L^2(E,m)$ is dense, $\mathcal{E}(u,w)=\mathcal{E}(w,u)$, $\mathcal{E}(u,u)\geqslant 0$ and $u\mapsto\mathcal{E}(u,w)$ is linear.

$\mathcal{D}(\mathcal{E}) \cap C_c(E)$ is dense both in $(\mathcal{D}(\mathcal{E}), \| \cdot \|_{\mathcal{E}})$ and $(C_c(E), \| \cdot \|_{\infty})$.   **(regular)**

From the above definition it is not hard to see, cf. [210], that a Dirichlet form automatically satisfies for all $u, w \in \mathcal{D}(\mathcal{E})$

$$\exists \kappa > 0 : |\mathcal{E}_1(u, w)|^2 \leqslant \kappa\, \mathcal{E}_1(u, u)\mathcal{E}_1(w, w), \qquad \textbf{(weak sector condition)}$$

$$w := (0 \vee u) \wedge 1 \in \mathcal{D}(\mathcal{E}) \quad \text{and} \quad \begin{cases} \mathcal{E}(u - w, u + w) \geqslant 0, \\[2mm] \mathcal{E}(u + w, u - w) \geqslant 0. \end{cases} \qquad \textbf{(semi-Dirichlet)}$$

A densely defined, closed (but not necessarily symmetric) bilinear form satisfying the weak sector condition and the semi-Dirichlet property is called a **non-symmetric Dirichlet form**; if only *one* of the semi-Dirichlet conditions is satisfied, $\mathcal{E}$ is a **semi-Dirichlet form**. Obviously, any Dirichlet form is also a semi-Dirichlet form.

There is a one-to-one correspondence between Dirichlet forms and strongly continuous symmetric operator semigroups $(T_t)_{t \geqslant 0}$ on $L^2(E, m)$, cf. Fukushima–Oshima–Takeda [118, Theorem 1.4.1]. If $(A, \mathcal{D}(A))$ is the infinitesimal generator[5] of the semigroup $(T_t)_{t \geqslant 0}$, then A is a (spectrally negative) self-adjoint operator, $\mathcal{D}(A) \subset \mathcal{D}(\mathcal{E}) = \mathcal{D}((-A)^{1/2})$ and

$$\mathcal{E}(u, w) = \langle Au, w \rangle_{L^2(m)} \quad \forall u \in \mathcal{D}(A), \ w \in \mathcal{D}(\mathcal{E}). \tag{3.19}$$

In particular $(A, \mathcal{D}(A))$ and $(\mathcal{E}, \mathcal{D}(\mathcal{E}))$ are in one-to-one correspondence. Recall that the semigroup $(T_t)_{t \geqslant 0}$ is called Markovian, if for all $u \in L^2(E, m)$ with $0 \leqslant u \leqslant 1$ we have $0 \leqslant T_t u \leqslant 1$. The Markov property of the semigroup and the semi-Dirichlet property of the form are equivalent to the fact that A is a **Dirichlet operator**, i.e.

$$\langle Au, (u - 1)^+ \rangle_{L^2(m)} \leqslant 0 \quad \forall u \in \mathcal{D}(A) \tag{3.20}$$

cf. Bouleau–Hirsch [49, Sect. I.3, pp. 12–16] for the symmetric case, and Ma–Röckner [210, Proposition I.4.3, p. 31] and Jacob [157, Sect. 4.6, pp. 364–382] or [281] for the general case. The connection of Dirichlet operators and the positive maximum principle is discussed in Remark 1.29.

Every regular Dirichlet form is given by a Beurling–Deny representation, cf. [118, Theorem 3.2.1, p. 108 and Example 1.2.1, p. 6].

Theorem 3.12 (Beurling–Deny). *Let $(\mathcal{E}, \mathcal{D}(\mathcal{E}))$ be a regular Dirichlet form. Then it has for $u, w \in \mathcal{D}(\mathcal{E}) \cap C_c(E)$ the following **Beurling–Deny representation***

---

[5] All notions are, mutatis mutandis, to be understood in the space $(L^2(E, m), \| \cdot \|_{L^2(m)})$ instead of $(C_{\infty}(E), \| \cdot \|_{\infty})$; they are essentially analogous to those of Sect. 1.4.

$$\mathcal{E}(u, w) = \mathcal{E}^{(c)}(u, w) + \int_E u(x)w(x)\,\kappa(dx)$$

$$+ \iint_{E\backslash\mathrm{diag}} (u(x) - u(y))(w(x) - w(y))\, J(dx, dy) \qquad (3.21)$$

with a strongly local[6] Dirichlet form $\mathcal{E}^{(c)}$ with domain $\mathcal{D}(\mathcal{E}^{(c)}) = \mathcal{D}(\mathcal{E}) \cap C_c(E)$, a symmetric measure $J \in \mathcal{M}^+(E \times E \setminus \mathrm{diag})$ and $\kappa \in \mathcal{M}^+(E)$; moreover, $\mathcal{E}^{(c)}$, J and κ are uniquely determined by $\mathcal{E}$, and vice versa.

If $E \subset \mathbb{R}^d$, then

$$\mathcal{E}^{(c)}(u, w) = \frac{1}{2} \sum_{j,k=1}^d \int_E \frac{\partial u(x)}{\partial x_j} \frac{\partial w(x)}{\partial x_k} v_{jk}(dx) \quad \forall u, w \in C_c^\infty(E) \qquad (3.22)$$

where $v_{jk} \in \mathcal{M}^+(E)$ such that for every compact set $K \subset E$ the matrix $(v_{jk}(K))_{j,k=1}^d$ is positive semidefinite. Moreover, the jump kernel satisfies

$$\iint_{K \times E\backslash\mathrm{diag}} \min(|x - y|^2, 1)\, J(dx, dy) < \infty$$

for every compact set $K \subset E$.

For non-symmetric Dirichlet forms the analogue of the Beurling–Deny formula has been studied by Hu–Ma–Sun [144], see also Fukushima–Uemura [117] for a different approach, as well as [288]. Note that every regular (symmetric) Dirichlet form enjoys a Beurling–Deny representation, but not every quadratic form given by (3.21) gives a Dirichlet form, see e.g. [286, Theorem 1.1]. This is similar to the situation which we have discussed at the beginning of Chap. 3: Every Feller process (whose generator has a sufficiently rich domain) has a negative definite symbol, but not every negative definite symbol yields a Feller process. As for the generators, the constant-coefficient setting is different.

Example 3.13. Let $(T_t)_{t\geq 0}$ be a translation-invariant Feller (i.e. convolution) semigroup, cf. (2.8), with *real-valued* symbol $\psi : \mathbb{R}^d \to \mathbb{R}$. Then $(T_t)_{t\geq 0}$ can be extended to a strongly continuous, *symmetric* semigroup in $L^2(dx)$, see e.g. [282]. Moreover, on $C_c^\infty(\mathbb{R}^d)$ the L^2-generator and the Feller generator coincide, and the corresponding Dirichlet form is given by

$$\mathcal{E}(u, w) = \int_{\mathbb{R}^d} \psi(\xi)\hat{u}(\xi)\overline{\hat{w}(\xi)}\, d\xi \quad \forall u, w \in C_c^\infty(\mathbb{R}^d). \qquad (3.23)$$

---

[6]That is $\mathcal{E}^{(c)}(u, w) = 0$ whenever $u, w \in \mathcal{D}(\mathcal{E}^{(c)})$ and w is constant in an open neighbourhood of $\mathrm{supp}(u)$.

The space $C_c^\infty(\mathbb{R}^d)$ is a core for the form and the generator. Thus we have a Beurling–Deny formula given by

$$\mathcal{E}(u, w) = \frac{1}{2} \sum_{j,k=1}^{d} q_{jk} \int_{\mathbb{R}^d} \frac{\partial u(x)}{\partial x_j} \frac{\partial w(x)}{\partial x_k} \, dx$$

$$+ \iint_{\mathbb{R}^d \times \mathbb{R}^d \setminus \text{diag}} (u(x+h) - u(x))(w(x+h) - w(x)) \, \nu(dh) \, dx$$

where $(0, Q, \nu)$ is the Lévy triplet of the continuous negative definite function ψ.

In this case we do have a one-to-one relation between *symmetric*[7] Lévy processes $(X_t)_{t \geq 0}$, real-valued continuous negative definite functions ψ, Lévy triplets $(0, Q, \nu)$ and constant-coefficient Dirichlet forms where the Beurling–Deny representation is as above. For details we refer to Fukushima–Oshima–Takeda [118, Example 1.4.1, p. 29] and Jacob [157, Example 4.7.28, pp. 407–409].

Berg–Forst [23] show that (3.23) defines a *non-symmetric* Dirichlet form if, and only if, $\psi : \mathbb{R}^d \to \mathbb{C}$ is the symbol of a (non-symmetric) Lévy process satisfying the **sector condition**, i.e. $|\operatorname{Im} \psi(\xi)| \leq \kappa \operatorname{Re} \psi(\xi)$ for a constant κ and all $\xi \in \mathbb{R}^d$.

□

The basic relation of (symmetric) Markov processes and regular (symmetric) Dirichlet forms is the following result which we quote from Fukushima–Oshima–Takeda [118, Theorem 7.2.1, p. 302].

Theorem 3.14 (Fukushima). *Let $(\mathcal{E}, \mathcal{D}(\mathcal{E}))$ be a regular (symmetric) Dirichlet form on $L^2(E, m)$ with semigroup $(T_t)_{t \geq 0}$. Then there exists a Hunt process[8] with state space E such that the transition semigroup $(P_t)_{t \geq 0}$ of the process is symmetric in the space $L^2(E, m)$ and $P_t u$ is, for all $u \in C_c(E)$ a quasi-continuous version of $T_t u$.*

Among the pioneering works in the study of symmetric Markov processes and their study via Dirichlet forms are the monographs by Fukushima [113] and Silverstein [301]. Theorem 3.14 has various extensions to non-symmetric quasi-regular Dirichlet forms (Ma–Röckner [210, Theorem 3.5]), semi-Dirichlet forms (Ma–Overbeck–Röckner [211]), and lower bounded forms (Fukushima–Uemura [117]). Let us point out that the notion of quasi-regularity of a Dirichlet form is

---

[7] That is X_t has the same law as $-X_t$; this is equivalent to ψ being real-valued.

[8] A **Hunt process** is a strong Markov process $(X_t, \mathcal{F}_t)_{t \geq 0}$ with respect to a right-continuous, complete filtration and life-time $\zeta(\omega)$; moreover, $t \mapsto X_t(\omega)$ is a.s. right-continuous on $[0, \infty)$, has left-hand limits on $(0, \zeta(\omega))$ and is quasi-left continuous on $[0, \infty)$.

necessary and sufficient for the existence of a (m-special standard[9]) strong Markov process, cf. Ma–Overbeck–Röckner [211, Theorems 3.8, 3.9].

The connection between Feller processes and Dirichlet forms is not straightforward, unless we are in the constant-coefficient case discussed in Example 3.13. The general case requires deep results from probabilistic potential theory. This is beautifully explained in Fukushima–Oshima–Takeda [118, Introduction to Chap. 7, p. 292]:

> In general, it is hopeless to construct a Feller transition function from the L^2-semigroup T_t associated with the given Dirichlet space *[i.e. the Hilbert space $(\mathcal{F}, \|\cdot\|_\mathcal{E})$ where $\mathcal{F} = \mathcal{D}(\mathcal{E})$ and $\|u\|_\mathcal{E}^2 = \mathcal{E}(u,u)$]*. However, if the Dirichlet space is regular, then the potential theory *[...]* provides us with quasi continuous versions $\widetilde{T_t f}$ and a sequence $t_n \downarrow 0$ such that $\widetilde{T_{t_n} f}(x) \to f(x)$, $n \to \infty$, q.e., *[quasi everywhere]* for sufficiently many f. Going along a similar line as in the case of the Feller transition function, but ignoring successively the sets of capacity zero on which things might go wrong, we finally get a Hunt process outside some set of capacity zero.

Let us begin with the (slightly easier) converse problem, as to whether a Feller process gives a (semi-)Dirichlet form. The following result is a special case of the situation discussed in Sect. 1.5.

Proposition 3.15. *Let $(T_t)_{t \geqslant 0}$ be a Feller semigroup with generator $(A, \mathcal{D}(A))$, let $\mathcal{D} \subset \mathcal{D}(A) \cap L^2(m)$ be a dense subset in $(C_\infty(\mathbb{R}^d), \|\cdot\|_\infty)$ and $(L^2(m), \|\cdot\|_{L^2(m)})$ such that $A(\mathcal{D}) \subset L^2(m)$, and assume that $T_t|_\mathcal{D}$ is symmetric in $L^2(\mathbb{R}^d)$, i.e.*

$$\langle T_t f, u \rangle_{L^2(m)} = \langle f, T_t u \rangle_{L^2(m)} \quad \forall f, u \in \mathcal{D}.$$

Then the operators $T_t|_\mathcal{D}$ have extensions to contractions $T_t : L^2(m) \to L^2(m)$ and $(T_t)_{t \geqslant 0}$ is a strongly continuous sub-Markovian contraction semigroup in $L^2(m)$. Its generator $(A^{(2)}, \mathcal{D}(A^{(2)}))$ satisfies $\mathcal{D} \subset \mathcal{D}(A^{(2)})$ and $A|_\mathcal{D} = A^{(2)}|_\mathcal{D}$.

Moreover, there is a Dirichlet form $(\mathcal{E}, \mathcal{D}(\mathcal{E}))$ with generator $(A^{(2)}, \mathcal{D}(A^{(2)}))$ and semigroup $(T_t^{(2)})_{t \geqslant 0}$.

Proof. We have $|\langle T_t f, u \rangle_{L^2(m)}| \leqslant \|f\|_{L^1(m)} \|T_t u\|_\infty \leqslant \|f\|_{L^1(m)} \|u\|_\infty$ for all $u, f \in \mathcal{D}$. This shows that $T_t|_\mathcal{D}$ has an extension to a contraction operator $\tilde{T}_t : L^1(m) \to L^1(m)$ and $(\tilde{T}_t)_{t \geqslant 0}$ is a strongly continuous contraction semigroup of sub-Markovian operators on $L^1(m)$. By interpolation, there is a further extension $(T_t^{(2)})_{t \geqslant 0}$ of $T_t|_\mathcal{D}$ which is strongly continuous on $L^2(m)$ (and on all spaces $L^p(m)$, $1 < p < \infty$); for details we refer to [104].

If $f \in \mathcal{D}$ and $u \in L^2(m)$, then we get

$$\lim_{t \to 0} \langle t^{-1}(T_t^{(2)} f - f), u \rangle_{L^2(m)} = \lim_{t \to 0} \langle t^{-1}(T_t f - f), u \rangle_{L^2(m)} = \langle Af, u \rangle_{L^2(m)}.$$

---

[9]This means essentially: Càdlàg and quasi-left continuous up to its life-time for all initial distributions $X_0 \sim \mu$.

Since $\mathcal{D} \subset \mathcal{D}(A)$ and $T_t|_{\mathcal{D}} = T_t^{(2)}|_{\mathcal{D}}$, we can use Lemma 1.26 to get

$$\left\| \frac{T_t f - f}{t} \right\|_{L^2(m)} = \left\| \frac{1}{t} \int_0^t T_s A f \, ds \right\|_{L^2(m)} \leq \frac{1}{t} \int_0^t \left\| T_s^{(2)} A f \right\|_{L^2(m)} ds \leq \| A f \|_{L^2(m)}.$$

Thus, Au is for $u \in \mathcal{D}$ the weak limit of $t^{-1}(T_t^{(2)} u - u)$, cf. Yosida [354, Theorem V.1.3, p. 121]. From the standard *weak generator equals strong generator* theorem (e.g. Pazy [236, Theorem II.1.3, p. 43]) we conclude that $\mathcal{D} \subset \mathcal{D}(A^{(2)})$ as well as $A|_{\mathcal{D}} = A^{(2)}|_{\mathcal{D}}$.

Since there is a one-to-one correspondence between strongly continuous symmetric sub-Markovian contraction semigroups and Dirichlet forms, cf. the paragraph before (3.19), the last assertion is clear.   □

The following result, taken from Jacob [154, Sect. 2], extends this to the non-symmetric case. For the definition of a Dirichlet operator we refer to Remark 1.29 and (3.20).

Theorem 3.16. *Let $(A, \mathcal{D}(A))$ be the generator of a Feller process and $\mathcal{D} \subset \mathcal{D}(A)$ be a dense subspace of $L^2(\mathbb{R}^d)$. If $A|_{\mathcal{D}}$ extends to a generator $A^{(2)}$ of a strongly continuous contraction semigroup on $L^2(\mathbb{R}^d)$ such that the set $(\lambda - A^{(2)})^{-1} \mathcal{D}$ is an operator core, then $A^{(2)}$ is a Dirichlet operator, the corresponding semigroup is sub-Markovian, and $\mathcal{E}(u, v) := \langle Au, v \rangle_{L^2}$ can be extended to a semi-Dirichlet form.*

Now we turn to the problem to construct a Feller process using Dirichlet forms. We have seen in Theorem 3.14 that every regular Dirichlet forms defines a Hunt process. Although every Feller process is a Hunt process, see [118, Theorem A.2.2, p. 315], the converse does not hold in general. Thus, to construct Feller processes via Dirichlet forms, one has to overcome some difficulties:

(a) starting from a negative definite symbol $q(x, \xi)$ with x-dependent Lévy triplet $(l(x), Q(x), N(x, dy))$, we have to show that the corresponding operator $-q(x, D)$ defines a regular Dirichlet form;
(b) starting from a Beurling–Deny representation (3.21), (3.22) we have to impose conditions on $(v_{jk}(dx), \kappa(dx), J(dx, dy))$ such that we can perform an "integration by parts" to calculate $-q(x, D)$ by the formula $\mathcal{E}(u, w) = \langle -q(\cdot, D)u, w \rangle_{L^2(m)}$;
(c) the associated Hunt process is defined only for quasi-every starting point. We have to find conditions that guarantee that the exceptional set is actually empty;
(d) we have to show that the Hunt process is, in fact, a Feller process.

To cope with the first problem, the key observation is that the Beurling–Deny formula (3.21), (3.22) resembles the Lévy–Khintchine representation (2.25) of a Feller generator. In the constant-coefficient case, cf. Example 3.13, this is obvious. Without the a priori assumption that the corresponding process exists, this problem has been discussed for pseudo-differential operators $(-q(x, D), C_c^\infty(\mathbb{R}^d))$ with negative definite symbols in [281, Theorem 3.6] and Barlow et al. [12].

The second problem can be approached via (r, p)-capacities, see Kazumi–Shigekawa [174] and [104, 105] and the references given there. Alternatively one can assume that the L^2-semigroup is absolutely continuous, cf. Fukushima [115] and [165] where this problem is discussed in terms of mapping properties of the semigroup, see also Barlow et al. [12].

For the third problem one usually proves first the C_b-Feller property and then, by the usual tightness argument, cf. Theorem 2.49, the Feller property. The following Theorem from [286, Corollary 6.4] is based on this approach.

Theorem 3.17. *Let $\alpha : \mathbb{R}^d \times \mathbb{R}^d \to \mathbb{R}$ and suppose there are constants η, β, γ such that*

$$-\infty < \eta \leqslant \alpha(x, y) \leqslant \beta < 2 \quad \forall |x - y| \leqslant 1$$

$$\alpha(x, y) \equiv \beta \qquad \text{near the diagonal } x = y$$

$$0 < \gamma \leqslant \alpha(x, y) \leqslant \infty \qquad \forall |x - y| > 1$$

and for any compact set $K \subset \mathbb{R}^d$

$$|\alpha(x, y) - \alpha(x, z)| + |\alpha(y, x) - \alpha(z, x)| \leqslant C_K |y - z| \quad \forall x, y, z \in K.$$

Set

$$j(x, y) := |x - y|^{-\alpha(x,y)-d} + |x - y|^{-\alpha(y,x)-d}.$$

Then the operator $(-q(x, D), C_c^\infty(\mathbb{R}^d))$ with the symbol

$$-q(x, \xi) = -\int_{h \neq 0} (1 - \cos h \cdot \xi) j(x, x + h) \, dh$$

$$+ \frac{i}{2} \int_{h \neq 0} \sin h \cdot \xi (j(x, x + h) - j(x, x - h)) \, dh$$

has an extension which generates a Feller semigroup.

Let us briefly sketch the idea how to show the C_b-Feller property in Theorem 3.17; details can be found in [286, p. 420]. First one shows that the quadratic form $\mathcal{E}(u, w) := \langle -q(\cdot, D)u, w\rangle_{L^2(dx)}$, $u, w \in C_c^\infty(\mathbb{R}^d)$ extends to a regular Dirichlet form; this proves the existence of a Hunt process $(X_t)_{t \geqslant 0}$ and a symmetric sub-Markovian semigroup in $L^2(dx)$. Recall that, because of the construction via Dirichlet forms, $(X_t)_{t \geqslant 0}$ is only defined in $\mathbb{R}^d \setminus N$ where N is an exceptional (i.e. capacity-zero) set. Based on the work by Bass–Levin [17], see also Song–Vondraček [306], one shows that any bounded harmonic function[10] is Hölder continuous:

---

[10] That is, the function h satisfies the spherical mean-value property relative to the process $(X_t)_{t \geqslant 0}$: $h(x) = \mathbb{E}^x h(X_{\tau_U})$ where τ_U is the first exit time from $U \subset \mathbb{R}^d$.

$$\exists c, \kappa > 0 \; \forall x_0 \; \forall h \in B_b(\mathbb{R}^d \setminus N) \text{ harmonic in } B_r(x_0) \text{ such that:}$$

$$\forall x, y \in \mathbb{B}(x_0, r/2): \quad |h(x) - h(y)| \leq c\|h\|_\infty |x - y|^\kappa. \tag{3.24}$$

Denote by $R_\lambda f(x) = \mathbb{E}^x \left(\int_0^\infty e^{-\lambda t} f(X_t) \, dt \right)$ the resolvent corresponding to $(X_t)_{t \geq 0}$. By the strong Markov property and Dynkin's formula (1.55) one sees that $R_\lambda f(x)$ is harmonic for every $f \in B_b(\mathbb{R}^d)$. Therefore (3.24) proves that $x \mapsto R_\lambda f(x)$ is Hölder continuous:

$$|R_\lambda f(x) - R_\lambda f(y)| \leq c(1 + \lambda^{-1})\|f\|_\infty |x - y|^{\kappa \wedge \alpha} \quad \forall f \in L^\infty(\mathbb{R}^d).$$

Although this holds only for $x, y \notin N$, this estimate allows us to extend $R_\lambda f$ continuously onto $B_b(\mathbb{R}^d)$. This means that the resolvent is strongly Feller, hence it has the C_b-Feller property, and so does the semigroup.

Based on ideas of Nash [231], similar results were obtained by Komatsu [193] for non-degenerate Lévy kernels of the form $J(dx, dy) = k(x, y)|x - y|^{-\alpha - d} \, dx \, dy$ where $0 < \kappa_1 \leq k(x, y) \leq \kappa_2 < \infty$ using pseudo-differential operator methods and a smoothing technique if $k(x, y)$ is rough.

Chen–Kumagai [62] were able to construct (stable-like) Feller processes associated with non-local jump-type Dirichlet forms on d-sets. Their work uses the parabolic Harnack principle and sharp two-sided heat-kernel estimates.

3.4 Evolution Equations

In his seminal paper [186] (an English translation can be found in [187, pp. 62–198]) Kolmogorov showed that a large class of diffusion processes (which are examples of Feller processes in the sense of this survey) can be described in terms of evolution equations. Later on, Feller [106, 107, 108] extended this to more general Feller processes.

Recall the notion of a fundamental solution of a (time-dependent, differential or pseudo-differential) operator $A = -q(s, x, D_x)$. A **fundamental solution** to the corresponding parabolic problem on the set $[0, T] \times \mathbb{R}^d$ is a measure-valued function $\Gamma(s, x; t, dy)$ defined for all $0 < s < t \leq T$, $x \in \mathbb{R}^d$ such that

$$u(s, t, x) := \int_{\mathbb{R}^d} f(y)\Gamma(s, x; t, dy) \quad \forall f \in C_c(\mathbb{R}^d)$$

satisfies

$$\frac{\partial u(s, t, x)}{\partial s} = q(s, x, D_x)u(s, t, x) \quad \forall x \in \mathbb{R}^d, \; s < t \leq T \tag{3.25}$$

$$\lim_{s \uparrow t} u(s, t, x) = f(x) \quad \forall x \in \mathbb{R}^d. \tag{3.26}$$

The transition function of a (time-homogeneous) Feller process is of the form $p_{t-s}(x, dy) = \Gamma(s, x; t, dy)$ and $T_t f(x) = u(0, t, x)$ is the transition semigroup. The equation (3.25) is usually called **Kolmogorov's backward equation**. In semigroup notation—mind that we differentiate in (3.25) at the left end-point of the interval $[s, t]$—this becomes

$$\frac{d}{dt} T_t u(x) = -q(x, D_x) T_t u(x) \tag{3.27}$$

whenever u is in the domain of the generator $A = -q(x, D)$, cf. also Lemma 1.26.

Using hypersingular integrals, Kochubei [185], see also Jacob [158, Theorem 2.7.16, p. 146], has constructed an example of such a fundamental solution.

Theorem 3.18. *Let $N \geqslant 2d + 3$ and assume that $q_j : \mathbb{R}^d \times \mathbb{R}^d \to \mathbb{C}, 0 \leqslant j \leqslant m$, $m \geqslant 1$, are functions with*

a) *$\operatorname{Re} q_0(x, \xi) \geqslant C_0$ for some $C_0 > 0$ and all $x \in \mathbb{R}^d$, $\xi \in \mathbb{S}^{d-1} \subset \mathbb{R}^d$;*
b) *$\xi \mapsto q_j(x, \xi) \in C^N(\mathbb{R}^d \setminus \{0\})$ is homogeneous of degree γ_j with $\gamma_0 \in (1, 2)$ and $0 < \gamma_j < \gamma_0, \gamma_j \neq 1$ for $0 < j \leqslant m$;*
c) *for all $x \in \mathbb{R}^d, |\alpha| \leqslant N, \xi \in \mathbb{R}^d \setminus \{0\}$*

$$|D_\xi^\alpha q_j(x, \xi)| \leqslant C_N |\xi|^{\gamma_j - |\alpha|};$$

d) *for all $x, y \in \mathbb{R}^d, |\alpha| \leqslant N, \xi \in \mathbb{R}^d \setminus \{0\}$ and some $\lambda \in (0, 1)$*

$$|D_\xi^\alpha(q_j(x, \xi) - q_j(y, \xi))| \leqslant C_N |x - y|^\lambda |\xi|^{\gamma_j - |\alpha|}.$$

Then $q(x, D) = \sum_{j=0}^m q_j(x, D)$ is the sum of hypersingular integral operators, i.e. for $l \geqslant 1$

$$q_j(x, \xi) = c_l \int_{\mathbb{R}^d} \frac{(1 - e^{-ih \cdot \xi})^l}{|h|^{n+\gamma_j}} \Omega_j\left(x, \frac{h}{|h|}\right) dh,$$

where Ω_j is called the characteristic.

If the characteristics are positive and, with respect to the second variable, even functions, then the solution to (3.27) exists and defines a transition function of a Feller semigroup.

A class of Feller processes related to stable processes were constructed in Kolokoltsov [189]; we follow Jacob [158, Theorem 2.7.19].

Theorem 3.19. *Let $\alpha \in (0, 2)$*

$$q(x, \xi) = |\xi|^\alpha \int_{\mathbb{S}^{d-1}} \left|\left\langle \frac{\xi}{|\xi|}, \eta \right\rangle\right|^\alpha \mu(x, d\eta)$$

where the kernel μ satisfies

$$0 < C_1 \le \int_{\mathbb{S}^{d-1}} \left| \left\langle \frac{\xi}{|\xi|}, \eta \right\rangle \right|^{\alpha} \mu(x, d\eta) \le C_2 < \infty,$$

and for $\beta \in \mathbb{N}_0^d$ and $x \in \mathbb{R}^d$ the derivative $\nabla_x^\beta \mu(x, d\eta)$ is a real-valued measure with uniformly (in x) bounded total variation. Then there exists a solution to (3.27) which determines a Feller semigroup.

The proof of Theorem 3.19 uses an asymptotic series representation of the Green's function of the process with symbol $q(x, \xi)$. The guiding idea is to understand the stable-like case (with variable order of differentiability) as a perturbation of the classical stable case and to use Duhamel's formula iteratively to get an asymptotic expansion. This method automatically yields two-sided heat-kernel estimates for the transition densities. More details can be found in the lecture note [188, Chap. 5] by Kolokoltsov; this method also applies to the stable-like processes of Bass, see [188, Chap. 5] and the recent monograph [191, Chap. 7] by Kolokoltsov.

There are further existence results which rely on the classical Hörmander symbolic calculus for pseudo-differential operators with Weyl and Kohn–Nirenberg quantization, cf. Baldus [10], or which use Waldenfels operators, cf. Cancelier [58].

Based on the symbolic calculus for pseudo-differential operators with negative definite symbols one can also obtain the statement of Theorem 3.4 by constructing a fundamental solution, see [40]. We present here a further extension of this result to time-inhomogeneous processes which we take from [41, Theorem 4.2].

Theorem 3.20. *Let ψ be a real-valued continuous negative definite function as in Theorem 3.4, and assume that the function $q : \mathbb{R} \times \mathbb{R}^d \times \mathbb{R}^d \to \mathbb{C}$ is such that for all $\alpha, \beta \in \mathbb{N}_0^d$ and for any compact $K \subset \mathbb{R}$ there are constants $c_{\alpha,\beta,K} \ge 0$ such that*

$$|\nabla_\xi^\alpha \nabla_x^\beta q(t, x, \xi)| \le c_{\alpha,\beta,K}(1 + \psi(\xi))^{\frac{m - (|\alpha| \wedge 2)}{2}}$$

for all $t \in K$, $x \in \mathbb{R}^d$ and $\xi \in \mathbb{R}^d$. Moreover, assume that

a) *$q(\cdot, x, \xi)$ is a continuous function for all $x \in \mathbb{R}^d$, $\xi \in \mathbb{R}^d$,*
b) *$q(t, x, \cdot)$ is continuous negative definite for all $t \in \mathbb{R}$, $x \in \mathbb{R}^d$,*
c) *$q(t, x, 0) = 0$ for all t and x,*
d) *q is elliptic, i.e.*

$$\exists R, c > 0 \quad \forall x, |\xi| \ge R : \operatorname{Re} q(t, x, \xi) \ge c(1 + \psi(\xi))^{\frac{m}{2}}$$

holds uniformly for all t from compact sets.

Then there exists a Markov process $(X_t)_{t \ge 0}$ such that for each t and $f \in C_c^\infty(\mathbb{R}^d)$

$$\lim_{s \downarrow 0} \sup_{x \in \mathbb{R}^d} \left| q(t, x, D) f(x) - \frac{\mathbb{E}(f(X_t) \mid X_s = x) - f(x)}{s} \right| = 0$$

where $q(t, x, D_x)$ is the pseudo-differential operator with symbol $q(t, x, \xi)$.

The proof is based on an extension of the symbolic calculus (for pseudo-differential operators with negative definite symbols) used in Theorem 3.4 to time dependent symbols. This symbolic calculus allows us to solve the evolution equation by mimicking the classical parametrix construction for pseudo-differential operators, see e.g. Kumano-go [194, Chap. 7.4, p. 241].

3.5 The Martingale Problem

The starting point of this technique, which has been developed by Stroock and Varadhan, see e.g. their monograph [312], is the following version of Theorem 1.36 which we take from Jacob [159, Theorem 3.6.5, pp. 109–110].

Theorem 3.21. *Let* $(T_t)_{t \geq 0}$ *be a (conservative) Feller semigroup on* $C_\infty(\mathbb{R}^d)$ *with generator* $(A, \mathcal{D}(A))$ *and* $(X_t, \mathscr{F}_t)_{t \geq 0}$ *be a càdlàg modification of the associated Feller process. Then for every* $u \in \mathcal{D}(A)$ *and every initial distribution* μ

$$M_t^{[u]} = u(X_t) - u(X_0) - \int_0^t Au(X_s)\,ds, \quad t \geq 0, \tag{3.28}$$

is a martingale under the natural filtration $(\mathscr{F}_t^X)_{t \geq 0}$, $\mathscr{F}_t^X = \sigma(X_s : s \leq t)$, *of the process* $(X_t)_{t \geq 0}$ *and under* $\mathbb{P}^\mu$.

This observation raises the question whether a Feller process $(X_t, \mathscr{F}_t^X)_{t \geq 0}$ can be characterized in terms of its martingales $(M_t^{[u]}, \mathscr{F}_t^X)_{t \geq 0}$. This is, in a nutshell, the **martingale problem**. This way of putting things has a few advantages as it enables us to use three powerful techniques:

- weak convergence of processes,
- regular conditional probabilities,
- localization.

It is also tailor-made to avoid strong solutions in the context of SDEs, see also Sect. 3.2. Since our processes have in general sample paths with jumps we work on $\mathbb{D}[0, \infty) = \{\omega : [0, \infty) \to \mathbb{R}^d : \omega \text{ is càdlàg}\}$ the Skorokhod space equipped with the J_1-Skorokhod topology.[11] Equipped with this topology $\mathbb{D}[0, \infty)$ becomes a metrizable space; this is important when studying the convergence of càdlàg processes. Recall Kolmogorov's canonical construction of a stochastic process: Using the canonical projections $X_t : \mathbb{D}[0, \infty) \to \mathbb{R}^d$, $X_t(\omega) = \omega(t)$, every probability measure on $\mathbb{D}[0, \infty)$ defines a (canonical) stochastic process with càdlàg paths (and vice versa).

---

[11] Skorokhod's J_1 topology was introduced in Skorokhod [303], it is discussed e.g. in Billingsley [28, Chap. 3, pp. 109–153], Ethier–Kurtz [100, Chap. 3.5–3.9, pp. 116–147], Jacod–Shiryaev [170, Chap. VI] or Kallenberg [172, Chap. 16, pp. 307–326].

Definition 3.22. Let $A : \mathcal{D} \to B(\mathbb{R}^d)$ denote a linear operator with domain $\mathcal{D} \subset B(\mathbb{R}^d)$. A probability measure $\mathbb{P}$ on the Skorokhod space is called a solution of the **martingale problem** for the operator $(A, \mathcal{D})$, if for every $u \in \mathcal{D}$ the process $(M_t^{[u]})_{t \geqslant 0}$ from (3.28) is a martingale under $\mathbb{P}$ with respect to the canonical filtration $(\mathscr{F}_t^X)_{t \geqslant 0}$.

The martingale problem is said to be **well posed**, if for every probability measure $\mu \in \mathcal{M}^+(\mathbb{R}^d)$ there is a unique solution $\mathbb{P}^\mu$ of the martingale problem such that $X_0 \sim \mu$ under $\mathbb{P}^\mu$.

The principal equivalence of the martingale problem formulation and SDEs is discussed in Kurtz [197]. Stroock [309] seems to be the earliest reference where Feller processes with Lévy-type generators were constructed in this way; note, however, that [309] requires a strictly non-degenerate diffusion part; the following result is adapted from Jacod–Shiryaev [170, Theorem III.2.34, p. 159].

Theorem 3.23 (Stroock). *Let $q : \mathbb{R}^d \times \mathbb{R}^d \to \mathbb{C}$ be a negative definite function such that $\xi \mapsto q(x, \xi)$ is continuous and denote by $(l(x), Q(x), N(x, dy))$ the Lévy triplet relative to the truncation function $\chi(s) = \mathbb{1}_{[0,1]}(s)$. Assume, moreover, that $q(x, D)$ has bounded coefficients (2.33), (2.34) and that $x \mapsto Q(x)$ and $x \mapsto \int_{B \setminus \{0\}} \min(|y|^2, 1)\, N(x, dy)$ are continuous for all Borel sets $B \subset \mathbb{R}^d$. If $Q(x)$ is everywhere invertible, then there exists a unique (in the sense of finite-dimensional distributions) diffusion process with jumps with generator $-q(x, D)$.*

Although $(X_t)_{t \geqslant 0}$ need not be a Feller process, it is a good example of a Markov process with a symbol $q(x, \xi)$. In order to ensure that $(X_t)_{t \geqslant 0}$ is a Feller process, one has to impose further assumptions on $q(x, \xi)$.

The next theorem by Hoh [138, Theorem 3.15] indicates that the martingale problem does have solutions under very general conditions. Well posedness, however, is a completely different matter.

Theorem 3.24. *Let $q : \mathbb{R}^d \times \mathbb{R}^d \to \mathbb{C}$ be a continuous negative definite symbol such that $q(x, 0) = 0$ for all $x \in \mathbb{R}^d$. Moreover, assume that $(x, \xi) \mapsto q(x, \xi)$ is continuous and that $-q(x, D)$ has bounded coefficients, i.e.*

$$|q(x, \xi)| \leqslant c(1 + |\xi|^2) \quad \forall x, \xi \in \mathbb{R}^d.$$

Then for all initial distributions μ there exists a solution to the martingale problem for $(-q(x, D), C_c^\infty(\mathbb{R}^d))$.

Recently some progress has been made to get solutions of the martingale problem for generators with symbols which are discontinuous in x, cf. Imkeller–Willrich [148].

Let us now turn to the question of well posedness. This is an important issue since only a well posed martingale problem gives rise to a Feller process, see Hoh [138, Proposition 5.18].

Theorem 3.25. *Assume that the martingale problem for* $(-q(x, D), C_c^\infty(\mathbb{R}^d))$ *is well posed and that*

$$q(x, D) : C_c^\infty(\mathbb{R}^d) \to C_\infty(\mathbb{R}^d).$$

Then $(-q(x, D), C_c^\infty(\mathbb{R}^d))$ *has an extension which is the generator of a Feller semigroup.*

Theorem 3.25 shows that, in order to get a Feller process, one needs conditions which ensure both the mapping property of the operator and the well-posedness of the corresponding martingale problem. The mapping property is an easy consequence of the following lemma from [44, Lemma 3.1, Remark 3.2].

Lemma 3.26. *Let* $q : \mathbb{R}^d \times \mathbb{R}^d \to \mathbb{C}$ *be a negative definite symbol with an* x-*dependent Lévy triplet* $(l(x), Q(x), N(x, \cdot))$ *for the truncation function* $\chi = \mathbb{1}_{[0,1]}$.

a) $q(x, D)u(x)$ *vanishes at infinity for any* $u \in C_c^\infty(\mathbb{R}^d)$ *if, and only if,*

$$\lim_{|x| \to \infty} N(x, \mathbb{B}(-x, r)) = 0 \quad \forall r > 0. \tag{3.29}$$

b) *Condition* (3.29) *is implied by each of the following conditions*

   1. $\lim\limits_{x \to \infty} \sup\limits_{|\xi| \leq \frac{1}{|x|}} (\operatorname{Re} q(x, \xi) - q(x, 0) - \xi \cdot Q(x)\xi) = 0,$

   2. $\lim\limits_{x \to \infty} \sup\limits_{|\xi| \leq \frac{1}{|x|}} |q(x, \xi) - q(x, 0)| = 0,$

   3. $\lim\limits_{x \to \infty} \sup\limits_{|\xi| \leq \frac{1}{|x|}} |q(x, \xi)| = 0,$

   4. $\xi \mapsto q(x, \xi)$ *is uniformly (in* $x \in \mathbb{R}^d$) *continuous at* $\xi = 0$ *and*

$$\sup_{x \in \mathbb{R}^d} |q(x, \xi)| \leq \kappa_q (1 + |\xi|^2) \quad \forall \xi \in \mathbb{R}^d.$$

c) *If* $x \mapsto q(x, \xi)$ *is a continuous function for every* $\xi \in \mathbb{R}^d$, *then we have* $q(x, D) : C_c^\infty(\mathbb{R}^d) \to C(\mathbb{R}^d)$.

d) *If* $x \mapsto q(x, \xi)$ *is continuous such that* (3.29) *or any of the conditions* b)1–4 *hold, then* $q(x, D) : C_c^\infty(\mathbb{R}^d) \to C_\infty(\mathbb{R}^d)$.

Proof. Let us prove (a). Since u has compact support, there exists some $r > 0$ such that $\operatorname{supp} u \subset \mathbb{B}(0, r)$. As A is a pseudo-differential operator with negative definite symbol, we can use the x-dependent Lévy–Khintchine representation of $q(x, \xi)$ to represent A as an integro-differential operator of the from (2.25):

$$Au(x) = -c(x)u(x) + l(x) \cdot \nabla u(x) + \frac{1}{2} \sum_{j,k=1}^{n} q_{jk}(x) \frac{\partial^2}{\partial x_j \partial x_k} u(x)$$

$$+ \int_{\mathbb{R}^d \setminus \{0\}} \left(u(x+y) - u(x) - \nabla u(x) \cdot y \mathbb{1}_{[0,1]}(|y|) \right) N(x, dy).$$

For all $|x| > r$ this formula becomes

$$|Au(x)| = \left| \int_{\mathbb{R}^d \setminus \{0\}} u(x+y) N(x, dy) \right| \leqslant \|u\|_{\infty} N(x, \mathbb{B}(-x, r))$$

which implies the sufficiency of statement (a).

For the converse note that for every $r > 0$ there exists a positive $u \in C_c^{\infty}(\mathbb{R}^d)$ such that $u \geqslant \mathbb{1}_{\mathbb{B}(0,r)}$ and thus $Au(x) \geqslant N(x, \mathbb{B}(-x, r))$ for $|x| > r$.

For statement (b) note that $4 \Rightarrow 3 \Rightarrow 2 \Rightarrow 1$. So it remains to show that b)1 implies (3.29). We use the elementary relation, cf. [162, Proof of Lemma 5.2],

$$\frac{|y|^2}{1+|y|^2} = \int_{\mathbb{R}^d} (1 - \cos(y \cdot \xi)) \, g(\xi) \, d\xi \quad \forall y \in \mathbb{R}^d \tag{3.30}$$

where $g(\xi) = \frac{1}{2} \int_0^{\infty} (2\pi\lambda)^{-d/2} e^{-|\xi|^2/2\lambda} e^{-\lambda/2} d\lambda$ is integrable and has absolute moments of arbitrary order. This, $\mathbb{B}(-x, r) \subset \mathbb{B}^c(0, |x| - r)$, and the elementary inequality $\frac{1}{2} \leqslant s^2/(1+s^2)$ for $|s| > 1$ yield

$$N(x, \mathbb{B}(-x, r))$$
$$\leqslant N(x, \mathbb{B}^c(0, |x| - r))$$
$$\leqslant 2 \int_{\mathbb{R}^d \setminus \{0\}} \frac{|y/(|x| - r)|^2}{1 + |y/(|x| - r)|^2} N(x, dy)$$
$$\leqslant 2 \int_{\mathbb{R}^d \setminus \{0\}} \int \left(1 - \cos \frac{y \cdot \eta}{|x| - r} \right) g(\eta) \, d\eta \, N(x, dy)$$
$$= 2 \int \left(\operatorname{Re} q \left(x, \frac{\eta}{|x| - r} \right) - q(x, 0) - \frac{1}{(|x| - r)^2} \eta \cdot Q(x)\eta \right) g(\eta) \, d\eta$$
$$\leqslant 2 \sup_{|\xi| \leqslant \frac{1}{|x| - r}} \left(\operatorname{Re} q(x, \xi) - q(x, 0) - \xi \cdot Q(x)\xi \right) \int (1 + |\eta|^2) \, g(\eta) \, d\eta$$
$$\leqslant c_g 2 \sup_{|\xi| \leqslant \frac{1}{|x| - r}} \left(\operatorname{Re} q(x, \xi) - q(x, 0) - \xi \cdot Q(x)\xi \right),$$

since $g(\eta)$ has absolute moments of any order. Finally note that by Proposition 2.17(c) a negative definite function ψ satisfies for $|x| > 2r$

$$\sup_{|\xi| \le \frac{1}{|x|}} |\psi(\xi)| \le \sup_{|\xi| \le \frac{1}{|x|-r}} |\psi(\xi)| \le \sup_{|\xi| \le \frac{2}{|x|}} |\psi(\xi)| = \sup_{|\xi| \le \frac{1}{|x|}} |\psi(2\xi)|$$

$$\le 4 \sup_{|\xi| \le \frac{1}{|x|}} |\psi(\xi)|. \tag{3.31}$$

Hence each of the conditions in (b) implies (3.29).

The Fourier transform $\hat{u}$ of a test function $u \in C_c^\infty(\mathbb{R}^d)$ is in the Schwartz space $\mathcal{S}(\mathbb{R}^d)$ of rapidly decreasing functions. Define

$$Au(x) := -q(x, D)u(x) = -\int q(x, \xi) \, \hat{u}(\xi) \, e^{ix\cdot\xi} \, d\xi \quad \forall u \in C_c^\infty(\mathbb{R}^d)$$

and recall that $\xi \mapsto q(x, \xi)$ grows for all x at most quadratically in ξ (cf. Proposition 2.17(d)). Thus statement (c) is implied by the dominated convergence theorem. The last statement (d) follows from (a)–(c). □

The following rather general result is due to Hoh [138, Theorem 5.24].

Theorem 3.27 (Hoh). *Let* $\psi : \mathbb{R}^d \to \mathbb{R}$ *be a continuous negative definite function and assume that there exist constants* $c_0, r_0 > 0$ *such that*

$$\psi(\xi) \ge c_0 \gamma |\xi|^{r_0} \quad \forall \xi \in \mathbb{R}^d.$$

Set[12] $k := 2\lfloor (\frac{d}{r_0} \vee 2) + d \rfloor + 3 - d$. *Assume that* $q : \mathbb{R}^d \times \mathbb{R}^d \to \mathbb{R}$ *is a continuous negative definite symbol such that* $q(\cdot, \xi) \in C^k(\mathbb{R}^d)$ *and*

$$|\nabla_x^\beta q(x, \xi)| \le c(1 + \psi(\xi)) \quad \forall \beta \in \mathbb{N}_0^d, \ |\beta| \le k$$

and for some strictly positive function $\gamma : \mathbb{R}^d \to (0, \infty)$

$$q(x, \xi) \ge \gamma(x)(1 + \psi(\xi)) \quad \forall x \in \mathbb{R}^d, \ |\xi| \ge 1$$

and

$$\lim_{\xi \to 0} \sup_{x \in \mathbb{R}^d} q(x, \xi) = 0.$$

Then $(-q(x, D), C_c^\infty(\mathbb{R}^d))$ *has an extension that generates a Feller semigroup.*

For special cases one can drop some of the assumptions of Theorem 3.27, e.g. for symbols of the form $\sum_{j=1}^m b_j(x)\psi_j(\xi)$ see Hoh [135, Theorem 6.4].

There are further approaches to solve the martingale problem which pose conditions on the Lévy triplet rather than the symbol; for example, Stroock [311]

---

[12] $\lfloor \cdot \rfloor$ denotes the largest integer which is less or equal than the argument.

describes the state space dependence of the Lévy kernel $N(x, dy)$ by a pullback, Abels and Kassmann [2] assume that $N(x, dy) = n(x, y) \, dy$ and formulate conditions for $n(x, y)$. It is an open problem whether it is possible to re-write such conditions in terms of conditions for the symbol; this applies, in particular, to the setting relying on the pullback, compare Tsuchiya [326] and Stroock [311, Sect. 3.2.2, p. 99].

A very useful feature of the martingale problem is that one can get uniqueness by a localization technique, cf. Hoh [138, Theorem 5.3].

Theorem 3.28. *Let $q, q_k : \mathbb{R}^d \times \mathbb{R}^d \to \mathbb{R}$ ($k \geqslant 1$) be continuous negative definite symbols such that the corresponding pseudo-differential operators map $C_c^\infty(\mathbb{R}^d)$ into $C_b(\mathbb{R}^d)$ and such that*

$$q(x, \xi) = q_k(x, \xi) \quad \forall x \in U_k, \ \xi \in \mathbb{R}^d$$

where $(U_k)_{k \in \mathbb{N}}$ is an open cover of $\mathbb{R}^d$. If the martingale problem for $-q(x, D)$ has for each initial distribution $\mu \in \mathcal{M}^+(\mathbb{R}^d)$ a solution and, if the martingale problem for $-q_k(x, D)$ is well posed for each $k \geqslant 1$, then the martingale problem for $-q(x, D)$ is well posed.

Generators of Variable Order and Stable-Like Processes. Using localization one can improve Theorem 3.5, the existence result for operators of variable order, cf. Hoh [138, Theorem 7.10].

Theorem 3.29. *Let $\psi(\xi)$ and $q(x, \xi)$ be as in Theorem 3.4. Furthermore assume that $q(x, \xi)$ is real-valued, and $m : \mathbb{R}^d \to (0, 1]$ is a C^∞-function. Then*

$$p(x, \xi) := q(x, \xi)^{m(x)}$$

defines an operator $(-p(x, D), C_c^\infty(\mathbb{R}^d))$ which has an extension that is the generator of a Feller semigroup.

Relying only on the martingale problem and localization, Bass [15, Theorem 2.2] obtained another set of general conditions which ensure that a corresponding pure jump processes exist. The following two theorems are stated for $d = 1$ only, but Bass mentions in [15] that the proofs carry over to $d > 1$. We re-state Bass' results in terms of symbols.

Theorem 3.30. *Let $q(x, \xi)$, $x, \xi \in \mathbb{R}$, be a negative definite symbol of the form*

$$q(x, \xi) = \int_{\mathbb{R} \setminus \{0\}} \left(1 - e^{ih\xi} + i\xi \cdot h \mathbb{1}_{[0,1]}(|h|)\right) v(x, dh),$$

and define for $\lambda > 0$

$$Q_\lambda(x, y, \xi) := \frac{\lambda + q(x, \xi)}{\lambda + q(y, \xi)}.$$

Assume that

a) *there exists* $\delta \in (0, 1)$ *such that* $\inf\limits_{x \in \mathbb{R}} |q(x, \xi)| \geq c|\xi|^{\delta}$ *for* $|\xi| \geq 1$;

b) *there exists a constant c such that* $\sup\limits_{x \in \mathbb{R}} |q(x, \xi)| \leq c(1 + |\xi|^2)$;

c) *there exists* $\rho > 0$ *such that* $\sup\limits_{|x-y|<\rho} \xi^2 \left| \dfrac{d^2}{d\xi^2} Q_\lambda(x, y, \xi) \right| \leq c_\lambda |\xi|^\delta$;

d) *for all* $r < 1$ *one has* $\overline{\lim\limits_{\epsilon \downarrow 0}} \sup\limits_{|x-y| \leq \epsilon} \sup\limits_{r \leq |\xi| \leq 1/r} \left| \dfrac{d^2}{d\xi^2} Q_\lambda(x, y, \xi) \right| = 0$;

e) $\overline{\lim\limits_{\epsilon \downarrow 0}} \sup\limits_{|\xi| \in \left[\sqrt{\epsilon}, \epsilon^{-1}\right]} \sup\limits_{x \in \mathbb{R}, |y| \leq 1} \left| Q_\lambda\left(x, x + \epsilon y, \epsilon^{-1} u\right) - 1 \right| = 0$;

f) $\overline{\lim\limits_{\eta \downarrow 0}} \int_{-\eta}^{\eta} \sup\limits_{x \in \mathbb{R}} \left[\dfrac{1}{|z|^2} \int_{\mathbb{R}} \min(\xi^2 z^2, 1) \left| \dfrac{d^2}{d\xi^2} Q_\lambda(x, x+z, \xi) \right| d\xi \right] dz = 0.$

Then $(-q(x, D), C_c^\infty(\mathbb{R}))$ *extends uniquely to the generator of a Feller semigroup.*

Proof. By Theorem 2.31, Assumption (b) is equivalent to

$$\sup_{x \in \mathbb{R}} \int \min(|y|^2, 1)\, \nu(x, dy) < \infty.$$

Then, cf. [15, Theorem 2.2, Proposition 6.2], the operator $(q(x, D), C_c^\infty(\mathbb{R}))$ extends uniquely to the generator of a C_b-Feller process. For the extension of $q(x, D)$ we use Theorem 2.37(a). Moreover, by dominated convergence (2.45) holds, and Theorem 2.49 implies the result. □

The above theorem simplifies considerably for stable-like processes, see Bass [15, Theorem 2.2].

Theorem 3.31. *Let* $q(x, \xi)$, $x, \xi \in \mathbb{R}$, *be a negative definite symbol of the form*

$$q(x, \xi) = c_{\alpha(x)} \int_{\mathbb{R} \setminus \{0\}} \left(1 - e^{i\xi h} + i\xi h \mathbb{1}_{[0,1]}(|h|)\right) \frac{dh}{|h|^{1+\alpha(x)}}$$

where $c_{\alpha(x)} = \alpha(x) 2^{\alpha(x)-1} \pi^{-d/2} \Gamma\left(\frac{\alpha(x)+d}{2}\right) / \Gamma\left(1 - \frac{\alpha(x)}{2}\right)$ *is a constant such that*

$$e^{-ix\xi} q(x, D_x) e^{ix\xi} = q(x, \xi) = |\xi|^{\alpha(x)},$$

cf. (1.34). *Assume that*

a) $0 < \inf\limits_{x \in \mathbb{R}} \alpha(x) \leq \sup\limits_{x \in \mathbb{R}} \alpha(x) < 2$;

b) $\sup\limits_{|x-y| \leq z} |\alpha(x) - \alpha(y)| = o(1/|\ln z|)$ *as* $z \to 0$;

c) $\int_0^1 \sup\limits_{|x-y| \leq z} |\alpha(x) - \alpha(y)| \dfrac{dz}{z} < \infty.$

Then $(-q(x, D), C_c^\infty(\mathbb{R}))$ *extends uniquely to the generator of a Feller process.*

Let us close this section with a word of caution. Bass' stable-like processes are generated by fractional Laplacians of variable order, to wit $q(x, D) = (-\Delta)^{\alpha(x)}$, and these operators are, in general, not symmetric for the normal scalar product in the space $L^2(dx)$. On the other hand, there are (symmetric) stable-like processes associated with certain stable-like *symmetric* Dirichlet forms, cf. Uemura [328,329] where the Beurling–Deny representation of the Dirichlet form has the jump kernel $N(x, dy) dx = C_{\alpha(x)} |y|^{-d-\alpha(x)} dy dx$, cf. Sect. 3.3; their generators $L_{\alpha(x)}(x, D)$ are, by construction, symmetric in $L^2(dx)$. The relation between $(-\Delta)^{\alpha(x)}$ and $L_{\alpha(x)}(x, D)$ is discussed in [286] in the symmetric case (see also Theorem 3.17) and in Fukushima–Uemura [117] as well as in [288] for the non-symmetric case.

3.6 Unbounded Coefficients

All methods to construct a Feller process which we have discussed so far assume that the coefficients of the generator are bounded, i.e.

$$|q(x, \xi)| \leq c(1 + |\xi|^2) \quad \forall x, \xi \in \mathbb{R}^d$$

where $c = 2 \sup_{y \in \mathbb{R}^d} \sup_{|\eta| \leq 1} |q(y, \eta)|$, cf. Theorem 2.31. In some cases, this technical restriction can be relaxed, for instance it is shown in Hoh [138, Theorem 9.4] that Theorem 3.24 can be extended to the following statement.

Theorem 3.32 (Hoh). *Let $\psi : \mathbb{R}^d \to \mathbb{R}$ be a continuous negative definite function, such that $\psi \not\equiv 0$ and $\psi(0) = 0$. Let $q : \mathbb{R}^d \times \mathbb{R}^d \to \mathbb{R}$ be a negative definite symbol such that $(x, \xi) \mapsto q(x, \xi)$ is continuous, $q(x, 0) = 0$ for all $x \in \mathbb{R}^d$ and*

$$q(x, \xi) \leq \frac{c \psi(\xi)}{\sup_{|\eta| \leq 1/|x|} \psi(\eta)} \quad \forall |x| \geq 1, \; \xi \in \mathbb{R}^d. \tag{3.32}$$

Then for any initial distribution μ there is a solution of the martingale problem for $(-q(x, D), C_c^\infty(\mathbb{R}^d))$.

On the basis of Theorem 3.32, Hoh [138, Theorem 9.5] was able to prove an analogous extension of Theorem 3.27 to generators with unbounded coefficients. It is interesting to note that the condition (3.32) represents, essentially, a trade-off between the growth in x as $|x| \to \infty$ and the decay in ξ as $|\xi| \to 0$. This is already familiar from the conservativeness condition from Theorem 2.34.

Remark 3.33. For the martingale problem approach, there is also a Lyapunov function technique for the construction of processes with unbounded coefficients, cf. Kolokoltsov [190, Chap. 5]. This method adapts the underlying function space taking into account the growth of the coefficients, i.e. the semigroups are not considered on $C_\infty(\mathbb{R}^d)$ but on weighted function spaces. The functional analytic basics can be found in the monograph by van Casteren [331, Chap. 2].   □

The solution of a Lévy-driven stochastic differential equation of the form

$$dX_t = \Phi(X_{t-})\, dL_t, \qquad X_0 = x \in \mathbb{R}^d,$$

where $\Phi : \mathbb{R}^d \to \mathbb{R}^{d \times n}$ is a suitable coefficient and $(L_t)_{t \geq 0}$ is a Lévy process (with killing) and with characteristic exponent $\psi : \mathbb{R}^n \to \mathbb{C}$, yields a strong Markov processes admitting a symbol $q(x, \xi) = \psi(\Phi(x)^\top \xi)$, see Example 2.42(e). For the existence and uniqueness one usually has to assume a local Lipschitz and linear growth condition on Φ. In principle, this allows to construct Markov processes with unbounded symbols; if Φ is not bounded (boundedness was assumed in Theorem 3.8), the Feller property of the process needs to be checked on a case-by-case basis.

Example 3.34. Here are two typical examples which illustrate this approach.

a) (*Ornstein–Uhlenbeck processes*) Consider the following Lévy-driven SDE

$$dX_t = b(X_t)\, dt + dL_t, \qquad X_0 = x \in \mathbb{R}^d,$$

where $(L_t)_{t \geq 0}$ is a d-dimensional Lévy process and $b : \mathbb{R}^d \to \mathbb{R}^d$. If for some constant $c > 0$

$$\langle b(x), x \rangle \leq c|x|^2 \quad \forall |x| \gg 1 \quad \text{and} \quad \lim_{|x| \to \infty} \frac{|b(x)|}{|x|} = 0,$$

then $(X_t)_{t \geq 0}$ is a Feller process. The proof depends on the fact that one can find an explicit Lyapunov function, cf. [345].

b) (*generalized Ornstein–Uhlenbeck processes*) Consider the following Lévy-driven SDE

$$V_t^x = x + \int_0^t V_{s-}^x\, dX_s^{(1)} + X_t^{(2)} = x + \int_0^t \phi(V_{s-}^x)\, dX_s \quad \forall x \in \mathbb{R}$$

where $\phi(y_1, y_2) = (y_1, 1)$ and $X_t = (X_t^{(1)}, X_t^{(2)})^\top$, $t \geq 0$, is a two-dimensional Lévy process with Lévy triplet (l, Q, ν) such that $\nu(\{-1\} \times dy_2) = 0$. Behme–Lindner [18, Theorem 3.1] show that $(V_t^x)_{t \geq 0}$ is a one-dimensional Feller process.   □

The following general construction principle by a limit procedure, which also allows unbounded coefficients, can be found in [44, Theorem 1.1].

Theorem 3.35. *Let* $q : \mathbb{R}^d \times \mathbb{R}^d \to \mathbb{C}$ *be a function such that*

$$\lim_{r \to \infty} \sup_{|y| \leq r} \sup_{|\xi| \leq \frac{1}{r}} |q(y, \xi)| = 0. \tag{3.33}$$

Assume that for each $k \geq 0$ the process $(X_t^k)_{t\geq0}$ is a Feller process with semigroup $(T_t^k)_{t\geq0}$ such that its generator A_k satisfies $C_c^\infty(\mathbb{R}^d) \subset \mathcal{D}(A_k)$ and the symbol $q_k(x, \xi)$ of $A_k\big|_{C_c^\infty(\mathbb{R}^d)}$ satisfies

$$|q_k(x,\xi)| \leq |q(x,\xi)| \qquad \forall x, \xi \in \mathbb{R}^d, \tag{3.34}$$

$$q_k(x,\xi) = q(x,\xi) \qquad \forall |x| \leq k, \ \xi \in \mathbb{R}^d \tag{3.35}$$

and

$$\left(X_{t\wedge\tau_{\mathbb{B}(0,k)}^k}^k\right)_{t\geq0} \overset{\text{fdd}}{\sim} \left(X_{t\wedge\tau_{\mathbb{B}(0,k)}^l}^l\right)_{t\geq0} \quad \forall l \geq k,$$

where the symbol $\overset{\text{fdd}}{\sim}$ refers to equality of all finite-dimensional distributions, and $\tau_{\mathbb{B}(0,k)}^l$ denotes the first exit time of the process X^l from the ball $\mathbb{B}(0,k)$.

Then the operator $(-q(x, D), C_c^\infty(\mathbb{R}^d))$ has an extension which generates a Feller process and the corresponding semigroup is given by

$$T_t u = \lim_{k\to\infty} T_t^k u$$

for $u \in C_\infty(\mathbb{R}^d)$ where the limit is meant in the strong sense, i.e. with respect to $\|\cdot\|_\infty$.

The proof of this theorem is based on the constructive approximation argument which we will discuss in Sect. 7.1.

Example 3.36. Consider the negative definite symbol $q(x,\xi) = (1+|x|^\beta)|\xi|^\alpha$ with $0 \leq \beta < \alpha \leq 2$. Since

$$\sup_{|y|\leq r} \sup_{|\xi|\leq 1/r} |q(y,\xi)| = (1 + r^\beta)r^{-\alpha},$$

Theorem 3.35 shows that there exists a Feller process with symbol $q(x,\xi)$.   □

Chapter 4
Transformations of Feller Processes

A Markov process can be transformed in many ways: For example by

- pinching and twisting, cf. Evans–Sowers [102];
- pasting, cf. Nagasawa [230];
- piecing out, cf. Ikeda–Nagasawa–Watanabe [147];
- killing and creation, cf. Meyer–Smyth–Walsh [222];
- adding and removing jumps (interlacing), cf. Meyer [221];
- restarting after random times, cf. Sawyer [270];
- censoring and resurrecting, cf. Bogdan–Burdzy–Chen [37].

Most of these constructions are path-by-path transformations of the process. Here we will focus on transformations of the generators, in particular on time changes and perturbations.

Throughout this chapter we assume that $(T_t)_{t \geq 0}$ is a Feller semigroup on $C_\infty(\mathbb{R}^d)$.

4.1 Random Time Changes

Given a Feller process $(X_t)_{t \geq 0}$ one can construct a new Feller process by changing the intrinsic clock of the process. The simplest method is to change the speed. To illustrate this, consider a Lévy process $(X_t)_{t \geq 0}$ with symbol $\psi(\xi)$. It is easy to check that $(X_{ct})_{t \geq 0}$ is for any $c > 0$ again a Lévy process, hence a Feller process, and we find

$$\mathbb{E} \, e^{i\xi \cdot X_{ct}} = e^{-(ct)\psi(\xi)} = e^{-t(c\psi)(\xi)}.$$

Thus, $(X_{ct})_{t \geq 0}$ has the symbol $c\psi(\xi)$ and the generator $-c\psi(D)$, i.e. a time-change results in a multiplicative perturbation of the generator. We can even change the time

B. Böttcher et al., *Lévy Matters III*, Lecture Notes
in Mathematics 2099, DOI 10.1007/978-3-319-02684-8_4,
© Springer International Publishing Switzerland 2013

depending on the current position of the process. The following result is adapted from Lumer [209, Theorem 2] extending an earlier result by Dorroh [87].

Theorem 4.1. *Let $(A, \mathcal{D}(A))$ be a Feller generator and $s(\cdot) \in C_b(\mathbb{R}^d)$ be real valued and strictly positive. Then the closure of $(s(\cdot)A, \mathcal{D}(A))$ is also a Feller generator.*

Using additive functionals we obtain the corresponding transformation at the level of sample paths. The following result, with a different proof and in the context of martingale problems, can be found in Ethier–Kurtz [100, Chap. 6].

Corollary 4.2. *Let $(X_t, \mathcal{F}_t)_{t \geq 0}$ be a Feller process with generator $(A, \mathcal{D}(A))$ and $s \in C_b(\mathbb{R}^d)$ be real valued and strictly positive. Denote by*

$$\alpha(t, \omega) := \int_0^t \frac{dr}{s(X_r(\omega))} \quad and \quad \alpha(\infty, \omega) := \int_0^\infty \frac{dr}{s(X_r(\omega))} \qquad (4.1)$$

and by $\tau(t)$ the (generalized, right-continuous) inverse

$$\tau(t, \omega) := \inf\{u > 0 : \alpha(u, \omega) > t\}, \quad \inf \emptyset = +\infty. \qquad (4.2)$$

Then the time-changed process $(\hat{X}_t, \hat{\mathcal{F}}_t)_{t \geq 0}$, $\hat{X}_t := X_{\tau(t)}$, $\hat{\mathcal{F}}_t := \mathcal{F}_{\tau(t)}$, is again a Feller process and the infinitesimal generator is the closure of $(s(\cdot)A, \mathcal{D}(A))$.

Proof. Since s is strictly positive and bounded, $0 < s(x) < c$ for some $c < \infty$; this ensures that we have $\alpha(\infty, \omega) = \int_0^\infty s(X_r(\omega))^{-1} \, dr \geq \int_0^\infty c^{-1} \, dr = \infty$; therefore, $\tau(\infty, \omega) = \infty$, and the time-changed process $\hat{X}$ has a.s. infinite life-time whenever the original process X has infinite life-time.

In view of Theorem 4.1 we only have to check that $s(\cdot)A$ is the generator of $\hat{X}$. For this we can restrict ourselves to the core $\mathcal{D}(A)$ of $s(\cdot)A$. Let $u \in \mathcal{D}(A)$. Then

$$u(X_t) - u(X_0) - M_t = \int_0^t Au(X_r) \, dr$$

where $(M_t)_{t \geq 0}$ is a martingale (with respect to the natural filtration $(\mathcal{F}_t^X)_{t \geq 0}$ of $(X_t)_{t \geq 0}$ and under all $\mathbb{P}^x$, $x \in \mathbb{R}^d$), cf. Theorem 1.36. From the definition of $(\alpha(r))_{r \geq 0}$ we find

$$\begin{aligned}
u(X_{\tau(t) \wedge n}) - u(X_0) - M_{\tau(t) \wedge n} &= \int_0^{\tau(t) \wedge n} Au(X_r) \, dr \\
&= \int_0^{\tau(t) \wedge n} s(X_r) Au(X_r) \, d\alpha(r) \\
&= \int_0^{t \wedge \alpha(n)} s(X_{\tau(r)}) Au(X_{\tau(r)}) \, dr.
\end{aligned}$$

Since $\tau(t)$, $t \geq 0$, is a family of $(\mathscr{F}_t^X)_{t \geq 0}$ stopping times, we can use optional stopping to see that $(M_{\tau(t) \wedge n})_{t \geq 0}$ is a martingale for every $n \geq 0$. Thus,

$$\mathbb{E}^x u(X_{\tau(t) \wedge n}) - u(x) = \mathbb{E} \left(\int_0^{t \wedge \alpha(n)} s(X_{\tau(r)}) A u(X_{\tau(r)}) \, dr \right)$$

and, as $n \to \infty$, we find from dominated convergence that

$$\mathbb{E}^x u(\hat{X}_t) - u(x) = \mathbb{E} \left(\int_0^t s(\hat{X}_r) A u(\hat{X}_r) \, dr \right).$$

Using the fact that $\mathcal{D}(A)$ is an operator core, we conclude that $s(\cdot) A$ is the generator of $(\hat{X}_t)_{t \geq 0}$. $\square$

Formally related to this time change is **Doob's h-transformation**. Let $(T_t)_{t \geq 0}$ be the transition semigroup associated with a Markov process and set

$$T_t^h u(x) := \frac{1}{h(x)} T_t[hu](x), \quad u \in B_b(\mathbb{R}^d), \ x \in \mathbb{R}^d.$$

This definition requires that h is not zero and measurable. Moreover, $(T_t^h)_{t \geq 0}$ defines only a (sub-)Markovian semigroup, if $h > 0$ and $T_t h \leq h$ for all $t > 0$, i.e. if h is a strictly positive **supermedian** function. If $(T_t)_{t \geq 0}$ is a Feller semigroup, it is a natural question to ask under which conditions $(T_t^h)_{t \geq 0}$ is again a Feller semigroup. A necessary condition is that $hu \in C_\infty(\mathbb{R}^d)$ for any $u \in C_\infty(\mathbb{R}^d)$, and this is equivalent to saying that $h \in C_b(\mathbb{R}^d)$. Moreover, $h^{-1} T_t(hu)$ must vanish at infinity and, unless we have more information on the mapping properties of the semigroup $(T_t)_{t \geq 0}$, this usually means that $h > 0$ and $1/h$ is bounded as $|x| \to \infty$. With a bit more effort one can prove the following result.

Lemma 4.3. Let $(T_t)_{t \geq 0}$ be a Feller semigroup with generator $(A, \mathcal{D}(A))$. Assume that $h > 0$ is a strictly positive, bounded, continuous and supermedian[1] function satisfying $\underline{\lim}_{|x| \to \infty} |h(x)| > 0$. Then the h-transformed semigroup $T_t^h u := h^{-1} T_t[hu]$ is again a Feller semigroup, and the generator is of the form

$$A^h u = h^{-1} A[hu] \quad and \quad \mathcal{D}(A^h) = \{u \in C_\infty(\mathbb{R}^d) : hu \in \mathcal{D}(A)\}.$$

Proof. Only the assertion about $\mathcal{D}(A^h)$ requires further proof. Denote by $R(\alpha, A)$ and $R(\alpha, A^h)$ the resolvent of the original and of the h-transformed semigroup, respectively. Then, cf. Definition 1.21 and Lemma 1.27,

$$R(\alpha, A^h) u = \int_0^\infty e^{-\alpha s} T_s^h u \, ds = \frac{1}{h} \int_0^\infty e^{-\alpha s} T_s[hu] \, ds = \frac{1}{h} R(\alpha, A)[hu]$$

---

[1] Some authors require h to be **excessive**, i.e. supermedian and $\sup_{t > 0} T_t h = h$; the last condition is, however, not needed for the proof of the lemma.

holds for all $u \in C_\infty(\mathbb{R}^d)$. Since $\mathcal{D}(A^h) = R(\alpha, A^h)[C_\infty(\mathbb{R}^d)]$ for any $\alpha > 0$, cf. Lemma 1.27, and since multiplication by h preserves the set $C_\infty(\mathbb{R}^d)$, we get

$$\mathcal{D}(A^h) = R(\alpha, A^h)[C_\infty(\mathbb{R}^d)] = \frac{1}{h} R(\alpha, A)[hC_\infty(\mathbb{R}^d)] \subset \frac{1}{h} R(\alpha, A)[C_\infty(\mathbb{R}^d)]$$

$$= \frac{1}{h} \mathcal{D}(A).$$

The converse inclusion is trivial. $\square$

The h-transform is frequently used in probabilistic potential theory in order to construct Markov processes conditioned to hit certain sets, e.g. a Brownian bridge. More information can be found in Sharpe [298, p. 298] and Rogers–Williams [254, Sect. IV. 39, p. 83]. In the context of Dirichlet forms time changes by additive functionals are discussed in Fukushima–Oshima–Takeda [118, Chap. 6.2, pp. 265–279, and Chap. A.3, pp. 331–344].

4.2 Subordination in the Sense of Bochner

In the previous section we considered a time-change of the process $(X_t)_{t\geq 0}$ by an increasing stochastic process $\tau(t)$ which was a functional of the process, hence *not independent*. **Subordination (in the sense of Bochner)** is a random time-change by an *independent* increasing Lévy process (subordinator) $(S_t)_{t\geq 0}$. While the *dependent* time-change of Sect. 4.1 led to a multiplicative perturbation of the generator $(A, \mathcal{D}(A))$ of $(X_t)_{t\geq 0}$, subordination yields a *function of the generator* in an appropriate sense.

Recall from Examples 1.17(g) and 2.4(g) that a subordinator can be characterized in terms of its Laplace exponent $f(\lambda) = -\log \mathbb{E}\, e^{-\lambda X_t}$; any such f is a **Bernstein function** and every Bernstein functions is uniquely given by its Lévy–Khintchine representation

$$f(\lambda) = a + b\lambda + \int_{(0,\infty)} (1 - e^{-\lambda t})\, \mu(dt) \tag{4.3}$$

where $a, b \geq 0$ and μ is a measure on $(0, \infty)$ such that $\int_{(0,\infty)} \min(t, 1)\, \mu(dt) < \infty$. Moreover, $\mu_t(ds) := \mathbb{P}(X_t \in ds)$ is a vaguely continuous convolution semigroups of sub-probability measures on the half-line. Although the following definition holds for abstract spaces E, we state it only for $E = \mathbb{R}^d$.

Definition 4.4. Let $(X_t)_{t\geq 0}$ be a Feller process with transition semigroup $(T_t)_{t\geq 0}$ and let $(S_t)_{t\geq 0}$ be a subordinator with Bernstein function f and convolution semigroup $(\mu_t)_{t\geq 0}$. Assume that $(X_t)_{t\geq 0}$ and $(S_t)_{t\geq 0}$ are stochastically independent. Then the stochastic process given by

$$X_t^f(\omega) := X_{S_t(\omega)}(\omega) \quad \forall t \geq 0 \tag{4.4}$$

is said to be **subordinate** to $(X_t)_{t \geqslant 0}$ (with respect to f or $(S_t)_{t \geqslant 0}$). The family of operators

$$T_t^f u := \int_{[0,\infty)} T_s u \, \mu_t(ds) \quad \forall t \geqslant 0, \ u \in C_\infty(\mathbb{R}^d) \tag{4.5}$$

is said to be **subordinate** to $(T_t)_{t \geqslant 0}$ (with respect to f or $(\mu_t)_{t \geqslant 0}$).

For semigroups, subordination was introduced by Bochner [34] while the time-change aspect was rigorously worked out by Nelson [232]. Our standard reference is the monograph [293, Chaps. 5 and 13] where the following simple result can be found.

Lemma 4.5. *Let $(X_t)_{t \geqslant 0}$ be a Feller process and f a Bernstein function. Then the subordinate process $(X_t^f)_{t \geqslant 0}$ is again a Feller process and its transition semigroup is the subordinate semigroup $(T_t^f)_{t \geqslant 0}$.*

The structure of the subordinate generator is given by Phillips' theorem [293, Theorem 13.5].

Theorem 4.6 (Phillips). *Let $(A, \mathcal{D}(A))$ be a generator of the Feller semigroup $(T_t)_{t \geqslant 0}$ and f a Bernstein function given by (4.3). Then the subordinate generator A^f is given by*

$$A^f u := -au + bAu + \int_{(0,\infty)} (T_s u - u) \, \mu(ds) \quad \forall u \in \mathcal{D}(A) \tag{4.6}$$

and $\mathcal{D}(A)$ is an operator core for $(A^f, \mathcal{D}(A^f))$.

Formally, $T_t = e^{tA}$ and inserting this into (4.6), we see that $A^f = -f(-A)$. This can be made rigorous if we understand $-f(-A)$ in a proper sense. No problem arises in Hilbert space where we have the usual spectral calculus at our disposal. In a C_∞-context we can use a variant of the unbounded Dunford–Riesz integral to define $-f(-A)$. This has been done in [272, 278], see also Berg–Boyadzhiev–de Laubenfels [26], Pustyl'nik [246] and, quite recently, as an extended Hille–Phillips calculus by Gomilko–Haase–Tomilov [124]. These approaches generalize Balakrishnan's formula for fractional powers, cf. [293, Chap. 13] for a detailed discussion. Let us briefly note that subordination leads to an unbounded functional calculus for the infinitesimal generators, cf. [293, Theorem 13.23, p. 222].

Theorem 4.7. *Let $(A, \mathcal{D}(A))$ be the generator of a Feller semigroup $(T_t)_{t \geqslant 0}$ on the space $C_\infty(\mathbb{R}^d)$, and let f, g be any two Bernstein functions. Then we have*

a) $A^{cf} = cA^f$ *for all $c > 0$;*

b) $A^{f+g} = \overline{A^f + A^g}$;

c) $A^{f \circ g} = (A^g)^f$;

d) $A^{c+\mathrm{id}+f} = -c + A + A^f$ *for all $c \geqslant 0$;*

e) *if fg is again a Bernstein function, then $A^{fg} = -A^f A^g = -A^g A^f$.*

The equalities a)–e) are identities in the sense of closed operators, including their domains which are the usual domains for sums, compositions etc. of closed operators.

This functional calculus can also be extended to the algebra generated by the family of Bernstein functions. Various characterizations of the domain $\mathcal{D}(A^f)$ can be found in [293, Corollary 13.20, p. 219].

Let us discuss what is going on at the level of symbols. For constant coefficients things are straightforward. Denote by $\mu_t(ds) = \mathbb{P}(S_t \in ds)$, $t \geq 0$, the family of transition probabilities of the subordinator. Then

$$\widehat{T_t^f u} = \int_0^\infty \widehat{T_s u}\, \mu_t(ds) = \int_0^\infty e^{-s\psi}\, \mu_t(ds) = e^{-tf\circ\psi}.$$

Thus, the symbol of the subordinate semigroup $(T_t^f)_{t\geq 0}$ or the subordinate Lévy process $(X_t^f)_{t\geq 0}$ is $\xi \mapsto f(\psi(\xi))$. Incidentally, this calculation can be used to show that a subordinate Lévy process is again a Lévy process. This shows that for constant-coefficient symbols and translation invariant semigroups the notions of *symbolic calculus* and *functional calculus* coincide. For general Feller processes generated by pseudo-differential operators this is no longer true:

$$-q^f(x,\xi) := \text{Symbol of}\{(-q(x,D))^f\}(x,\xi)$$

$$= \text{Symbol of}\{-f(q(x,D))\}(x,\xi)$$

$$\neq \text{Symbol of}\{-(f\circ q)(x,D)\}(x,\xi) = -f(q(x,\xi)).$$

In fact, we still have

$$\text{Symbol of}\{(-q(x,D))^f\}(x,\xi) = -f(q(x,\xi)) + \text{lower order perturbation.}$$

The *perturbation* is usually measured in a suitable scale of anisotropic function spaces. In a Feller, i.e. $C_\infty(\mathbb{R}^d)$, context we have the following estimates which generalize earlier results from [161, 160], see also [163, Theorem 4.3]. As usual, we denote by $H^n(\mathbb{R}^d) = W^{n,2}(\mathbb{R}^d)$ the classical L^2-Sobolev spaces.

Theorem 4.8. *Let $(T_t)_{t\geq 0}$ be a Feller semigroup with generator $(A, \mathcal{D}(A))$ such that $C_c^\infty(\mathbb{R}^d) \subset \mathcal{D}(A)$ and denote by $q(x,\xi)$ the corresponding negative definite symbol. We assume that $q(x,\xi)$ has bounded coefficients, cf. Theorem 2.31. Moreover, let f be a Bernstein function given by (4.3). Then the following estimates hold*

$$\left\| T_t^f u - e^{-(f\circ q)(x,D)} u \right\|_\infty \leq c \min(t,1)\, \|u\|_{H^{d+3}}$$

$$\left\| f(q(x,D))u - (f\circ q)(x,D)u \right\|_\infty \leq c' \|u\|_{H^{d+3}}.$$

Proof. As in [160, Lemma 4.1] we get from Taylor's theorem (in the variable t) combined with the formulae (1.38) and $-q(x, \xi) = \frac{d}{dt} \lambda_t(x, \xi)|_{t=0}$ from Theorem 2.36

$$\left| \lambda_t(x, \xi) - e^{-tq(x,\xi)} \right| = t \left| \int_0^1 \left(-e^{-ix\cdot\xi} T_{\theta t}[e_\xi q(\cdot, \xi)](x) + q(x, \xi) e^{-\theta tq(x,\xi)} \right) d\theta \right|$$

$$\leqslant 2t \sup_{x \in \mathbb{R}^d} |q(x, \xi)|$$

$$\leqslant c_q t (1 + |\xi|^2).$$

Since we have, trivially, $\left| \lambda_t(x, \xi) - e^{-tq(x,\xi)} \right| \leqslant 2$, we get

$$\left| \lambda_t(x, \xi) - e^{-tq(x,\xi)} \right| \leqslant c \min(t, 1)(1 + |\xi|^2) \quad \forall t > 0, \; x, \xi \in \mathbb{R}^d.$$

Consequently we find for all $u \in C_c^\infty(\mathbb{R}^d)$

$$\left| \lambda_t(x, D)u(x) - e^{-tq(\cdot,\cdot)}(x, D)u(x) \right| \leqslant c \min(t, 1) \int_{\mathbb{R}^d} (1 + |\xi|^2) |\hat{u}(\xi)| \, d\xi$$

$$\leqslant c_d \min(t, 1) \|u\|_{H^{d+3}} \quad \forall t > 0.$$

For the second estimate we used a standard argument from the theory of Sobolev spaces, cf. [160, Corollary 4.3]. Adding and subtracting u on the left-hand side we see

$$\left| (T_t u(x) - u(x)) - \left(e^{-tq(\cdot,\cdot)}(x, D)u(x) - u(x) \right) \right| \leqslant c_d \min(t, 1) \|u\|_{H^{d+3}} \quad \forall t > 0.$$

By construction, $\frac{d}{dt} T_t u \big|_{t=0} = \frac{d}{dt} \left(e^{-tq(\cdot,\cdot)}(\cdot, D)u \right) \big|_{t=0}$. Therefore, we can use Phillips' formula for the subordinate generator (4.6) to get

$$\left| A^f u(x) - (f \circ q)(x, D)u(x) \right|$$

$$= \left| \int_{(0,\infty)} (T_t u(x) - u(x)) \, \mu(dt) - \int_{\mathbb{R}^d} e^{ix\cdot\xi} \int_{(0,\infty)} \left(e^{-tq(x,\xi)} - 1 \right) \mu(dt) \hat{u}(\xi) \, d\xi \right|$$

$$= \left| \int_{(0,\infty)} (T_t u(x) - u(x)) \, \mu(dt) - \int_{(0,\infty)} \left(e^{-tq(\cdot,\cdot)}(x, D)u(x) - u(x) \right) \mu(dt) \right|$$

$$\leqslant c_d \int_{(0,\infty)} \min(t, 1) \, \mu(dt) \|u\|_{H^{d+3}}.$$

Since x is arbitrary, the proof is finished. $\qquad\square$

The concept of subordination has also been used for state space dependent subordinators [101], leading to technical results in the spirit of Theorem 3.4.

4.3 Perturbations

One can understand a Feller process $(X_t)_{t\geq 0}$ as a perturbation of a Lévy process $(L_t)_{t\geq 0}$, if its symbol $q(x,\xi)$ is comparable to the symbol $\psi(\xi)$ of the Lévy process. The following existence result from Jacob [158, Theorem 2.6.4, p. 131] illustrates this point, its proof relies on the Hille–Yosida construction, cf. Sect. 3.1.

Theorem 4.9. *Let* $\psi : \mathbb{R}^d \to \mathbb{R}$ *be a real-valued continuous negative definite function such that for some* $c_0, r_0 > 0$

$$\psi(\xi) \geq c_0 \gamma |\xi|^{r_0} \quad \forall \xi \in \mathbb{R}^d.$$

Let $q_1 : \mathbb{R}^d \to \mathbb{C}$ *be a continuous negative definite function such that there exist constants* $\gamma_0, \gamma_1, \gamma_2 > 0$ *with*

$$\gamma_0 \psi(\xi) \leq \operatorname{Re} q_1(\xi) \leq \gamma_1 \psi(\xi) \quad \forall |\xi| \geq 1,$$

$$|\operatorname{Im} q_1(\xi)| \leq \gamma_2 \operatorname{Re} q_1(\xi) \quad \forall \xi \in \mathbb{R}^d.$$

Set $m = \lfloor d/r_0 \rfloor + d + 3$ *and let* $q_2 : \mathbb{R}^d \times \mathbb{R}^d \to \mathbb{C}$ *be such that for all* $\alpha \in \mathbb{N}_0^d$, $|\alpha| \leq m$, *there are functions* $\phi_\alpha \in L^1(\mathbb{R}^d, dx)$ *with*

$$|\nabla_x^\alpha q_2(x,\xi)| \leq \phi_\alpha(x)(1 + \psi(\xi)) \quad \forall \xi \in \mathbb{R}^d$$

(the existence of the derivatives is implicitly assumed), and

$$\sum_{|\alpha|\leq m} \|\phi_\alpha\|_{L^1(dx)} \quad \text{is small.}$$

Then the operator $-q(x,D)$ *with the symbol* $q(x,\xi) = q_1(\xi) + q_2(x,\xi)$ *extends to a generator of a Feller semigroup.*

The smallness condition on $\sum_{|\alpha|\leq m} \|\phi_\alpha\|_{L^1}$ appearing in Theorem 4.9 is quite technical, and for details we refer to Jacob [158, Assumption 2.6.3, p. 130]. One should mention that such conditions are typical for a perturbation approach; its true meaning lies in the fact that it guarantees a Gårding inequality

$$|\langle q(x,D)u, u\rangle_{L^2(dx)}| \geq c_3 \|u\|_{H^{\psi,1}}^2 - c_4 \|u\|_{L^2(dx)}^2$$

as well as *lower* estimates of the type

$$\|q(x,D)u\|_{H^{\psi,s}} \geq c_1 \|u\|_{H^{\psi,s+2}} - c_2 \|u\|_{L^2(dx)}$$

relative to the scale of anisotropic L^2-Bessel potential spaces $H^{\psi,s}(\mathbb{R}^d)$, $s \in \mathbb{R}$, which are tailor-made for the control function ψ, cf. (3.8).

In semigroup theory, see e.g. Ethier–Kurtz [100] or Pazy [236], perturbations are connected with the notion of A-boundedness.

Definition 4.10. Let $(A, \mathcal{D})$ and $(B, \mathcal{H})$ be linear operators with $\mathcal{D} \subset \mathcal{H}$. Then B is A-**bounded** if there exist $\alpha \in [0, 1)$ and $\beta \geq 0$ such that

$$\|Bu\|_\infty \leq \alpha \|Au\|_\infty + \beta \|u\|_\infty \quad \forall u \in \mathcal{D}. \tag{4.7}$$

Example 4.11. a) Every bounded operator is also A-bounded for any A.
b) Let A be the generator of a Feller process, f a Bernstein function given by (4.3) such that $b = \lim_{x\to\infty} f(x)/x < 1$, and A^f the generator of the subordinate process in the sense of Theorem 4.6. Then A^f is A-bounded, see [293, (13.17), p. 208]. $\qquad\square$

For perturbations by A-bounded operators we have the following existence result, cf. Pazy [236, Corollary 3.3, p. 82] or Ethier–Kurtz [100, Theorem 7.1, p. 37].

Theorem 4.12. *Let the closure of $(A, C_c^\infty(\mathbb{R}^d))$ be the generator of a Feller semigroup and B be an A-bounded linear operator defined on $C_c^\infty(\mathbb{R}^d)$ such that B is a pseudo-differential operator $p(x, D)$ with the negative definite symbol $p(x, \xi)$. Then the closure of $(A + B, C_c^\infty(\mathbb{R}^d))$ generates a Feller semigroup.*

If in the definition of A-boundedness the constant α is allowed to be equal to 1, then the statement of Theorem 4.12 remains true if B has a densely defined adjoint, see Pazy [236, Theorem 3.4, p. 83].

In view of Example 4.11(b) we get the following result as a special case of Theorem 4.12, see also Jacob [158, Corollary 2.8.2, Remark 2.8.3.B, p. 153].

Theorem 4.13. *Assume that the closure of $(-q(x, D), C_c^\infty(\mathbb{R}^d))$ generates a Feller semigroup, and let f be a Bernstein function satisfying*

$$\lim_{x\to\infty} \frac{f(x)}{x} < 1.$$

Denote by $-q^f(x, D)$ the generator of the subordinate semigroup $(T_t^f)_{t\geq 0}$, cf. Theorem 4.6. Then the closure of $(-q(x, D) - q^f(x, D), C_c^\infty)$ generates a Feller semigroup.

The last result of this section, due to Hoh [138, Theorem 6.33] and [139, Theorem 3.5], shows that the existence of a Feller process for a given generator depends essentially on the behaviour of the Lévy kernel near the origin, i.e. the small jumps of the corresponding stochastic process. This situation is familiar from interlacing, cf. Meyer [221], and the solution of jump-type SDEs, cf. Ikeda–Watanabe [146, Theorem IV.9.1, pp. 245–246], where the large jumps are added in afterwards.

Theorem 4.14. *Let $q : \mathbb{R}^d \times \mathbb{R}^d \to \mathbb{R}$ be a real-valued negative definite symbol*

$$q(x, \xi) = \frac{1}{2}\xi \cdot Q(x)\xi + \int_{\mathbb{R}^d\setminus\{0\}} (1 - \cos y \cdot \xi)\, N(x, dy),$$

which has bounded coefficients (cf. Theorem 2.31) and is uniformly continuous at $\xi = 0$, i.e.

$$\lim_{|\xi| \to 0} \sup_{x \in \mathbb{R}^d} (q(x, \xi) - q(x, 0)) = 0.$$

Moreover, for $\phi \in C_c^\infty(\mathbb{R}^d)$ with $0 \leqslant \phi \leqslant 1$ and $\phi = 1$ in a neighbourhood of 0, set

$$q_1(x, \xi) := \frac{1}{2}\xi \cdot Q(x)\xi + \int_{\mathbb{R}^d \setminus \{0\}} (1 - \cos y \cdot \xi)\, \phi(y) N(x, dy).$$

Then $(-q(x, D), C_c^\infty(\mathbb{R}^d))$ has an extension which generates a Feller semigroup if $(-q_1(x, D), C_c^\infty(\mathbb{R}^d))$ has an extension which generates a Feller semigroup.

4.4   Feynman–Kac Semigroups

We start with some heuristic considerations. Let $(T_t)_{t \geqslant 0}$ be a Feller semigroup with generator $(A, \mathcal{D}(A))$ and write $(X_t)_{t \geqslant 0}$ for the corresponding Feller process. Assume that $V : \mathbb{R}^d \to (-\infty, 0)$ is a continuous and strictly negative function. Then

$$T_t^V u(x) := \mathbb{E}^x \left(e^{\int_0^t V(X_s)\,ds} u(X_t) \right) \quad \forall u \in B_b(\mathbb{R}^d), \ t \geqslant 0 \qquad (4.8)$$

is well-defined and gives, because of the Markov property, a sub-Markovian operator semigroup. Using the elementary estimate $|e^{-t} - 1| \leqslant t \wedge 1$ for $t \geqslant 0$ we see

$$|T_t^V u - T_t u| \leqslant \mathbb{E}^x \left(1 \wedge \int_0^t |V(X_s)|\,ds \right) \|u\|_\infty.$$

Thus, $(T_t^V)_{t \geqslant 0}$ is strongly continuous for $\| \cdot \|_\infty$ if, and only if, $(T_t)_{t \geqslant 0}$ is strongly continuous. Therefore, it makes sense to talk about the generator A^V, and it is not hard to see that $\mathcal{D}(A^V) = \mathcal{D}(A)$ and

$$A^V u(x) = Au(x) - V(x)u(x).$$

This sets the scene for the type of perturbations we are going to encounter.

Speaking about the Feller property is a different matter, and one would expect further conditions on the **potential** V, e.g. $V \in C_b(\mathbb{R}^d)$ as in Applebaum [5, Theorem 6.7.9, p. 406]. We adopt the approach by Chung, cf. Chung–Zhao [68], and drop, from now on, the assumption that $V < 0$.

Definition 4.15. Let $(T_t)_{t \geq 0}$ be a Feller semigroup and $V : \mathbb{R}^d \to \mathbb{R}$ be a measurable function. The function V satisfies an **abstract Kato condition**, if

$$\lim_{t \to 0} \sup_{x \in \mathbb{R}^d} \int_0^t T_s |V|(x) \, ds = 0. \tag{4.9}$$

This condition, called class J in Chung–Zhao [68, Theorem 3.6, p. 69] or $K(E)$ in Demuth–van Casteren [85, Definition 2.1, p. 57], is pretty much optimal for our purposes.

Let $(T_t)_{t \geq 0}$ and $(X_t)_{t \geq 0}$ be a Feller semigroup and a Feller process as before. Define

$$e_V(t) := \exp\left(\int_0^t V(X_s) \, ds \right).$$

If (4.9) holds, one can show the following properties, cf. Chung–Zhao [68, Propositions 3.8 and 3.9, pp. 72–74],

$$\lim_{t \to 0} \sup_{x \in \mathbb{R}^d} \mathbb{E}^x [e_{|V|}(t)] = 1; \tag{4.10}$$

$$\sup_{x \in \mathbb{R}^d} \mathbb{E}^x [e_{|V|}(t)] \leq e^{\gamma_0 + \gamma_1 t} \quad \forall t > 0; \tag{4.11}$$

$$\lim_{t \to 0} \sup_{x \in \mathbb{R}^d} \mathbb{E}^x \left[|e_V(t) - 1|^r \right] = 0 \quad \forall r \geq 1. \tag{4.12}$$

The fact that condition (4.9) implies (4.11) is often called **Has'minskii's lemma**; (4.11) ensures that T_t^V, defined by (4.8), is continuous on $(B_b(\mathbb{R}^d), \| \cdot \|_\infty)$; (4.12) guarantees strong continuity of $t \mapsto T_t^V u$. The Feller property requires further input. The following Lemma is adapted from Chung–Zhao [68, Propositions 3.11 and 3.12, p. 78].

Lemma 4.16. *Let $(T_t)_{t \geq 0}$ be a Feller semigroup and assume that V satisfies (4.9). Then $\lim_{h \to 0} \|T_h T_{t-h}^V u - T_t^V u\|_\infty = 0$ for all $u \in B_b(\mathbb{R}^d)$ and $t > 0$.*

In particular, if T_t enjoys the strong Feller property $T_t : B_b(\mathbb{R}^d) \to C_b(\mathbb{R}^d)$ (or $T_t : B_b(\mathbb{R}^d) \to C_\infty(\mathbb{R}^d)$), then T_t^V has this property, too.

Proof. The second part of the lemma follows immediately from the uniform convergence assertion since $T_{t-h}^V u \in B_b(\mathbb{R}^d)$ for all $u \in B_b(\mathbb{R}^d)$, thus $T_h T_{t-h}^V u$ is contained in $C_b(\mathbb{R}^d)$ or $C_\infty(\mathbb{R}^d)$.

In order to prove the first part of the lemma, we use the Markov property to get

$$T_h T_{t-h}^V u(x) = \mathbb{E}^x \left[\mathbb{E}^{X_h} \left(e_V(t-h) u(X_{t-h}) \right) \right] = \mathbb{E}^x \left[e_V(t) e_{-V}(h) u(X_t) \right].$$

By the Cauchy–Schwarz inequality and with (4.11), (4.12) we get

$$\|T_t^V u - T_h T_{t-h}^V u\|_\infty \leq \sup_{x \in \mathbb{R}^d} \mathbb{E}^x \big[e_V(t)\big(1 - e_{-V}(h)\big)\big] \|u\|_\infty$$

$$\leq e^{c_0 + c_1 t} \sup_{x \in \mathbb{R}^d} \sqrt{\mathbb{E}^x \big[|1 - e_{-V}(h)|^2 \big]} \, \|u\|_\infty \xrightarrow{h \to 0} 0. \qquad \square$$

Lemma 4.16 immediately implies the next result.

Theorem 4.17. *Assume that V is bounded with $v := \sup_{x \in \mathbb{R}^d} |V(x)|$ and satisfies the abstract Kato condition (4.9).*

If $(T_t)_{t \geq 0}$ is a strong Feller semigroup with generator A, then $(e^{-tv} T_t^V)_{t \geq 0}$ is a strong Feller semigroup with generator $A^V = A - V + v$.

If $(T_t)_{t \geq 0}$ is a Feller semigroup with generator A and $T_t(B_b(\mathbb{R}^d)) \subset C_\infty(\mathbb{R}^d)$, then $(e^{-tv} T_t^V)_{t \geq 0}$ is a Feller semigroup with generator $A - V + v$.

If we do not assume that V is bounded, the semigroup $(T_t^V)_{t \geq 0}$ inherits the strong continuity and mapping properties of $(T_t)_{t \geq 0}$, but it need not be sub-Markovian or contractive. This can be seen from (4.8). A close analysis of the above arguments show that it is enough to assume that $v := \sup_{x \in \mathbb{R}^d} V^+(x) < \infty$.

If we know that T_t is of the form $T_t u(x) = \int u(y) p_t(x, y) \, m(dy)$ with kernels satisfying

$$p_t(x, y) = p_t(y, x) \quad \text{and} \quad \sup_{x, y \in \mathbb{R}^d} p_t(x, y) < \infty, \qquad (4.13)$$

then we can avoid the strong Feller property in Theorem 4.17. Note, however, that this absolute continuity property is already very close to the strong Feller property, see the discussion on pp. 10–13. Variants of the following theorem can be found in Demuth–van Casteren [85, Theorem 2.5, pp. 61–62] or Chung–Zhao [68, Theorem 3.10, pp. 74–75].

Theorem 4.18. *Let $(T_t)_{t \geq 0}$ be a Feller semigroup such that the operators T_t are integral operators $T_t u(x) = \int u(y) p_t(x, y) \, m(dy)$ whose kernels satisfy (4.13). Assume that $V = V^+ - V^- \in B(\mathbb{R}^d)$ such that both V^- and $V^+ \mathbb{1}_{B(0,r)}$ satisfy the abstract Kato condition (4.9) for all $r > 0$.*

Then $(T_t^V)_{t \geq 0}$ is a strongly continuous, bounded, positivity preserving semigroup in $(C_\infty(\mathbb{R}^d), \|\cdot\|_\infty)$. The operators T_t^V are integral operators with kernels

$$p_t^V(x, y) = \lim_{s \to t} \mathbb{E}^x \big[e_V(s) p_{t-s}(X_s, y) \big]$$

with respect to the measure m.

A much more thorough discussion of Feynman–Kac semigroups can be found in the classic book Chung–Zhao [68]; in a Feller context we refer to Demuth–van Casteren [85] and the recent monograph by van Casteren [331]. In an L^2/Dirichlet form context Feynman–Kac semigroups are discussed in Fukushima–Oshima–Takeda [119, Chaps. 6.3 and 6.4, pp. 332–368] and [118, Chap. 6.3, pp. 280–291].

Chapter 5
Sample Path Properties

In this chapter we will study certain properties of the sample paths of a Feller process. Throughout we assume that $E = \mathbb{R}^d$ and that the Feller process $(X_t)_{t \geqslant 0}$ admits a symbol $q(x, \xi)$; by Theorem 2.21, a sufficient condition for the existence of a symbol is that $C_c^\infty(\mathbb{R}^d) \subset \mathcal{D}(A)$ where $(A, \mathcal{D}(A))$ is the generator of $(X_t)_{t \geqslant 0}$.

The overall philosophy is like this: $(X_t)_{t \geqslant 0}$ is a semimartingale whose differential characteristics coincide with the x-dependent Lévy triplet of the symbol $q(x, \xi)$, cf. Sect. 2.4. Therefore, a Feller process behaves, close to its starting point $X_0 = x_0$, like a Lévy process with characteristic exponent $\psi(\xi) = q(x_0, \xi)$ and generator $-\psi(D_x) = -q(x_0, D_x)$—note that we treat the starting point x_0 as a parameter and freeze the coefficients of the Feller generator $-q(x, D_x)$. This indicates that the path behaviour (of a Feller process) which only depends on short-time increments should be similar to the behaviour of a Lévy process. As we will see, this is indeed true, but unfortunately all proofs in the Lévy setting break down as they often rely on independence and translation invariance. Much has been discovered about this special class of Markov processes in the last 20 years so that a complete account is not possible. Instead we concentrate on a few areas where significant progress has been made in the recent past. In particular, we want to outline how one can move from Lévy to Lévy-type processes. We also take the opportunity to streamline and improve some results in the literature; in these cases we include complete proofs or point out how to modify the known arguments in order to get stronger results.

Most of the results in this section remain valid if we consider **any strong Markov process admitting a symbol** (in the sense of Definition 2.41, see Example 2.42). In order to keep things simple, we restrict ourselves to Feller processes.

It might be instructive to have a look at sample path properties of Lévy processes. The classic references are the survey papers by Fristedt [112] and Taylor [318] as well as the monograph by Sato [267]. Bertoin [27] is the best source for potential theory and the fluctuation theory of Lévy processes, and Xiao [353] contains an up-to-date survey on all aspects of fractal dimensions.

B. Böttcher et al., *Lévy Matters III*, Lecture Notes
in Mathematics 2099, DOI 10.1007/978-3-319-02684-8_5,
© Springer International Publishing Switzerland 2013

5.1 Probability Estimates

Probability estimates, e.g. for the tails of the distribution or for the maximum of a stochastic process, are important tools if one wants to consider sample path properties. In this section we derive some (maximal) estimates for Feller processes and their exponential moments which we will use later on.

Let $(X_t)_{t \geq 0}$ be a Feller process with generator $(A, \mathcal{D}(A))$, $C_c^\infty(\mathbb{R}^d) \subset \mathcal{D}(A)$ and symbol $q(x, \xi)$. By Theorem 2.27(d) the symbol is locally bounded,[1] i.e. for any compact $K \subset \mathbb{R}^d$, there exists a constant $C_K > 0$ such that

$$\sup_{x \in K} |q(x, \xi)| \leq C_K (1 + |\xi|^2) \quad \forall \xi \in \mathbb{R}^d. \tag{5.1}$$

Theorem 2.31 gives equivalent conditions for (5.1) in terms of the Lévy triplet.

As usual, we denote by $\mathbb{P}^x$ and $\mathbb{E}^x$ the probability measure $\mathbb{P}(\cdot \mid X_0 = x)$ and the corresponding expectation, respectively. Recall from Definition 1.35 and Proposition 1.36 that

$$\left(f(X_t) - \int_0^t Af(X_s)\, ds, \ \mathscr{F}_t^X \right)_{t \geq 0} \tag{5.2}$$

is for any $x \in \mathbb{R}^d$ and for all $f \in C_c^\infty(\mathbb{R}^d)$ a local martingale under $\mathbb{P}^x$ with respect to the natural filtration $\mathscr{F}_t^X = \sigma(X_s : s \leq t)$ of the process.

Maximal Estimates for the Transition Probability. For any $x \in \mathbb{R}^d$ and $r > 0$ we denote by $\tau_r^x := \tau_{\overline{\mathbb{B}}(x,r)}$ the first exit time of $(X_t)_{t \geq 0}$ from the ball $\overline{\mathbb{B}}(x, r)$,

$$\tau_r^x = \tau_{\overline{\mathbb{B}}(x,r)} := \inf \left\{ t > 0 : X_t \in \overline{\mathbb{B}}^c (x, r) \right\}.$$

Note that τ_r^x is intimately connected with the supremum of the process,

$$\{\tau_r^x < t\} \subset \left\{ \sup_{s \leq t} |X_s - x| > r \right\} \subset \{\tau_r^x \leq t\} \subset \left\{ \sup_{s \leq t} |X_s - x| \geq r \right\}, \tag{5.3}$$

and we will use these relations to obtain maximal inequalities for $(X_t)_{t \geq 0}$ in terms of its symbol. For a general class of Lévy processes, such estimates were obtained for the first time by Pruitt [245], and for Feller processes in [277]. Our exposition follows the ideas developed in [277, Sect. 6], resulting in both simpler proofs and stronger results; therefore, we will include complete proofs.

The following upper estimate is implicitly contained in [277, Lemma 6.1, Corollary 6.2 and Lemma 5.1] under the assumption that the generator A has bounded coefficients, i.e. (5.2) holds with $K = \mathbb{R}^d$, cf. Theorem 2.31. The following version is valid for all processes with *locally* bounded coefficients where (5.1) holds for all compact sets $K \subset \mathbb{R}^d$.

---

[1]Caution: If we consider more general processes admitting a symbol, one has to *assume* (5.1).

Theorem 5.1. *Let $(X_t)_{t \geqslant 0}$ be a Feller process with generator $(A, \mathcal{D}(A))$, symbol $q(x, \xi)$ and $C_c^\infty(\mathbb{R}^d) \subset \mathcal{D}(A)$. Then*

$$\mathbb{P}^x(\tau_r^x \leqslant t) \leqslant ct \sup_{|y-x| \leqslant r} \sup_{|\xi| \leqslant 1/r} |q(y, \xi)| \quad \forall x \in \mathbb{R}^d, \ r, t > 0 \quad (5.4)$$

holds with an absolute constant $c > 0$.

Proof. Pick $u \in C_c^\infty(\mathbb{R}^d)$ such that $\operatorname{supp} u \subset \mathbb{B}(0, 1)$ and $0 \leqslant u \leqslant 1 = u(0)$. For any $x \in \mathbb{R}^d$ and $r > 0$ we set $u_r^x(\cdot) := u((\cdot - x)/r)$. Then, $u_r^x \in C_c^\infty(\mathbb{R}^d) \subset \mathcal{D}(A)$. By (5.2),

$$M_t := 1 - u_r^x(X_{t \wedge \tau_r^x}) + \int_0^{t \wedge \tau_r^x} (-q(X_s, D)) u_r^x(X_s) \, ds, \quad t \geqslant 0$$

is a martingale for the canonical filtration $(\mathscr{F}_t^X)_{t \geqslant 0}$ under $\mathbb{P}^x$; moreover $M_0 = 0$, hence $\mathbb{E}^x(M_t) = 0$ for all $t \geqslant 0$. Then,

$$\mathbb{E}^x \left(1 - u_r^x(X_{t \wedge \tau_r^x}) \right) = \mathbb{E}^x \left(\int_0^{t \wedge \tau_r^x} q(X_s, D) u_r^x(X_s) \, ds \right).$$

Therefore,

$$
\begin{aligned}
\mathbb{P}^x(\tau_r^x \leqslant t) &\leqslant \mathbb{E}^x \left(1 - u_r^x(X_{t \wedge \tau_r^x}) \right) \\
&= \mathbb{E}^x \int_0^{t \wedge \tau_r^x} q(X_s, D) u_r^x(X_s) \, ds \\
&= \mathbb{E}^x \int_0^{t \wedge \tau_r^x} q(X_{s-}, D) u_r^x(X_{s-}) \, ds \\
&= \mathbb{E}^x \int_0^{t \wedge \tau_r^x} \mathbb{1}_{\{|X_{s-} - x| < r\}} \, q(X_{s-}, D) u_r^x(X_{s-}) \, ds \\
&= \mathbb{E}^x \int_0^{t \wedge \tau_r^x} \left. \left(\mathbb{1}_{\{|y-x| < r\}} \int e^{iy \cdot \xi} q(y, \xi) \, \hat{u}_r^x(\xi) \, d\xi \right) \right|_{y = X_{s-}} ds \\
&\leqslant \mathbb{E}^x \int_0^{t \wedge \tau_{\mathbb{B}(x,r)}} ds \sup_{|z-x| < r} \int |q(z, \xi)| |\hat{u}_r^x(\xi)| \, d\xi \\
&\leqslant \mathbb{E}^x \left(t \wedge \tau_{\mathbb{B}(x,r)} \right) \sup_{|z-x| < r} \int |q(z, \xi)| |\hat{u}_r^x(\xi)| \, d\xi \quad (5.5) \\
&\leqslant t \int \sup_{|z-x| < r} |q(z, \xi)| \, r^d \, |\hat{u}(r\xi)| \, d\xi \\
&= t \int \sup_{|y-x| < r} |q(y, \xi/r)| |\hat{u}(\xi)| \, d\xi,
\end{aligned}
$$

where the second equality (line three) follows since we are integrating with respect to Lebesgue measure and since $t \mapsto X_t(\omega)$ is ($\mathbb{P}^x$-a.s.) a càdlàg function which has at most countably many jumps on $[0, t]$, i.e. $\{s \in [0, t] : X_s \neq X_{s-}\}$ is almost surely a set of Lebesgue measure zero. Using (2.34) from Theorem 2.31 for the function $\xi \mapsto q(y, \xi/r)$, see also [274, Lemma 2.3], we get

$$\int \sup_{|y-x| \leq r} |q(y, \xi/r)| |\hat{u}(\xi)| \, d\xi \leq 2 \sup_{|y-x| \leq r} \sup_{|\xi| \leq 1/r} |q(y, \xi)| \int (1 + |\eta|^2) |\hat{u}(\eta)| \, d\eta$$

$$= c_u \sup_{|y-x| \leq r} \sup_{|\xi| \leq 1/r} |q(y, \xi)|,$$

where $c_u = 2 \int (1 + |\eta|^2) |\hat{u}(\eta)| \, d\eta < \infty$ is an absolute constant. Thus,

$$\mathbb{P}^x(\tau_r^x \leq t) \leq c_u t \sup_{|y-x| \leq r} \sup_{|\xi| \leq 1/r} |q(y, \xi)|$$

which proves the assertion. $\square$

In view of (5.3) we get, without proof, an upper maximal inequality.

Corollary 5.2. *Let $(X_t)_{t \geq 0}$ be a d-dimensional Feller process with symbol $q(x, \xi)$ as in Theorem 5.1. Then*

$$\mathbb{P}^x \left(\sup_{s \leq t} |X_s - x| > r \right) \leq ct \sup_{|y-x| \leq r} \sup_{|\xi| \leq 1/r} |q(y, \xi)| \quad \forall x \in \mathbb{R}^d, \, r, t > 0 \quad (5.6)$$

holds with an absolute constant $c > 0$.

The proof of Theorem 5.1 immediately gives the following lower bound for the expectation of τ_r^x.

Corollary 5.3. *Let $(X_t)_{t \geq 0}$ be a d-dimensional Feller process with symbol $q(x, \xi)$ as in Theorem 5.1. Then*

$$\mathbb{E}^x(\tau_r^x) \geq \frac{c}{\sup_{|y-x| \leq r} \sup_{|\xi| \leq 1/r} |q(y, \xi)|} \quad \forall x \in \mathbb{R}^d, \, r > 0$$

where c is the absolute constant from (5.4).

Proof. From (5.5) we see

$$\mathbb{P}^x(\tau_r^x \leq t) \leq \mathbb{E}^x(t \wedge \tau_r^x) \sup_{|z-x| < r} \int |q(z, \xi)| |\hat{u}_r^x(\xi)| \, d\xi.$$

Letting $t \to \infty$ gives

$$1 \leq \mathbb{E}^x(\tau_r^x) \sup_{|z-x| < r} \int |q(z, \xi)| |\hat{u}_r^x(\xi)| \, d\xi,$$

and we can estimate the right-hand side of this inequality as in the last part of the proof of Theorem 5.1. □

Example 5.4. The function $q(x, \xi) = (1 + |x|^\beta)|\xi|^\alpha$ with $\alpha \in (0, 2)$ and $\beta < \alpha$ is a negative definite symbol. It may be seen as the symbol of a symmetric α-stable Lévy process which is perturbed by the unbounded coefficient $(1 + |x|^\beta)$. Example 3.36 and Theorem 3.35 guarantee the existence of a Feller process with this symbol. Then, for all $x \in \mathbb{R}^d$ and $r, t > 0$,

$$\mathbb{P}^x(\tau_r^x \leq t) \leq ctr^{-\alpha} \sup_{|y-x| \leq r} (1 + |y|^\beta).$$

If $\alpha = \beta$ one can use the argument following Theorem 3.8 to guarantee the existence of a process with the negative definite symbol $q(x, \xi) = (1 + |x|^\alpha)|\xi|^\alpha$. □

The following counterpart of Theorem 5.1 presents an upper bound for the tail probability of the first exit time from a ball.

Theorem 5.5. *Let $(X_t)_{t \geq 0}$ be a Feller process with generator $(A, \mathcal{D}(A))$, symbol $q(x, \xi)$ and $C_c^\infty(\mathbb{R}^d) \subset \mathcal{D}(A)$. Then*

$$\mathbb{P}^x(\tau_r^x \geq t) \leq c \left(t \sup_{|\xi| \leq 1/(rk(x,r))} \inf_{|y-x| \leq r} \operatorname{Re} q(y, \xi) \right)^{-1} \quad \forall x \in \mathbb{R}^d, \ r, t > 0 \tag{5.7}$$

where $c = 4/\cos \sqrt{2/3}$ and

$$k(x, r) := \inf \left\{ k \geq \left(\arccos \sqrt{2/3} \right)^{-1} : \sup_{|\xi| \leq 1/(kr)} \sup_{|y-x| \leq r} \frac{\operatorname{Re} q(y, \xi)}{|\xi| \, |\operatorname{Im} q(y, \xi)|} \geq 2r \right\}.$$

From the very definition of $k(x, r)$ we find that for all $|\xi| \leq 1/(rk(x, r))$ and $|y - x| < r$,

$$\operatorname{Re} q(y, \xi) \geq 2r|\xi| \, |\operatorname{Im} q(y, \xi)|. \tag{5.8}$$

Conversely, if we fix $r > 0$ and if (5.8) holds only for small values of $|\xi|$, then $k(x, r)$ will be large but still stay finite. This observation allows us to use Theorem 5.5 in many situations, e.g. for Feller generators whose symbol satisfies locally uniformly $\operatorname{Re} q(x, \xi) \succeq |\xi|^\alpha$ with $\alpha \in (0, 2)$ and $|\operatorname{Im} q(x, \xi)| \asymp |\xi|$ as $|\xi| \to 0$.[2]

---

[2]Recall that $f \asymp g$ is a shorthand for $cf(t) \leq g(t) \leq Cf(t)$ for all t satisfying the specified condition and with absolute constants $0 < c < C < \infty$. Similarly, $f \succeq g$ means that we have $Cf(t) \geq g(t)$ for some $C < \infty$.

Theorem 5.5 becomes particularly simple, if the symbol $q(x, \xi)$ satisfies a **sector condition**. This means that there exists a constant κ such that

$$|\operatorname{Im} q(x, \xi)| \leq \kappa \operatorname{Re} q(x, \xi) \quad \forall x, \xi \in \mathbb{R}^d. \tag{5.9}$$

Then it is not hard to see that $k(x, r) = k_0 := \left(\arccos \sqrt{2/3}\right)^{-1}$. This proves the following result.

Corollary 5.6. *Let $(X_t)_{t \geq 0}$ be a Feller process as in Theorem 5.5. If the symbol $q(x, \xi)$ satisfies the sector condition (5.9), then we have for all $x \in \mathbb{R}^d$ and $r, t > 0$*

$$\mathbb{P}^x(\tau_r^x \geq t) \leq c \left(t \sup_{|\xi| \leq 1/(k_0 r)} \inf_{|y-x| \leq r} \operatorname{Re} q(y, \xi) \right)^{-1} \tag{5.10}$$

and

$$\mathbb{P}^x \left(\sup_{s \leq t} |X_s - x| \leq r \right) \leq c \left(t \sup_{|\xi| \leq 1/(k_0 r)} \inf_{|y-x| \leq r} \operatorname{Re} q(y, \xi) \right)^{-1} \tag{5.11}$$

with $k_0 := \left(\arccos \sqrt{2/3}\right)^{-1}$ and $c = 4/\cos \sqrt{2/3}$.

Example 5.7. a) Let $q(x, \xi)$ be the symbol from Example 5.4. Then, for all $x \in \mathbb{R}^d$ and $r, t > 0$,

$$\mathbb{P}^x(\tau_r^x \geq t) \leq ct^{-1} r^\alpha \sup_{|y-x| \leq r} (1 + |y|^\beta)^{-1} \leq ct^{-1} r^\alpha$$

with $c = 4/\cos \sqrt{2/3}$.

b) Let $q(x, \xi) = a(x)|\xi|^\alpha + ib(x) \cdot \xi$, where $a : \mathbb{R}^d \to \mathbb{R}$ and $b : \mathbb{R}^d \to \mathbb{R}^d$ are bounded and Lipschitz continuous functions such that $a(x) > 0$ for all $x \in \mathbb{R}^d$, and $\alpha \in (0, 2)$. Example 3.9 shows that there is a corresponding Feller process. Note that the sector condition (5.9) fails for this symbol. By Theorem 5.5, we have for all $x \in \mathbb{R}^d$ and $r, t > 0$,

$$\mathbb{P}^x(\tau_r^x \geq t) \leq \frac{4}{\cos \sqrt{2/3} \arccos^\alpha \sqrt{2/3}} \frac{r^\alpha}{t \inf_{|y-x| \leq r} a(y)}.$$

The proof of Theorem 5.5 needs one further ingredient. Recall that $\hat{A}_b$ denotes the full generator, see Definition 1.35. By definition, if $(u, Au) \in \hat{A}_b$, then the process defined by (5.2) is a *local* martingale. From Theorem 2.37(i) we know that

$$C_c^\infty(\mathbb{R}^d) \subset \mathcal{D}(A) \implies C_b^2(\mathbb{R}^d) \subset \hat{\mathcal{D}}(A) := \{u : \exists! w \in B(E), (u, w) \in \hat{A}\}. \tag{5.12}$$

Here $\hat{\mathcal{D}}(A)$ is the extended generator (1.51). The necessary uniqueness is guaranteed by Theorem 2.37(a).

Proof (of Theorem 5.5). The first part is a modification of the proof of [277, Lemma 6.3]. Therefore, we only sketch the differences. For simplicity, we write k instead of $k(x, r)$. Let $\epsilon \in \mathbb{R}^d$ with $|\epsilon| \leqslant 1/k$, then

$$\mathbb{P}^x(\tau_r^x > t) = \mathbb{P}^x(|X_t - x| \leqslant r, \; \tau_r^x > t)$$

$$\leqslant \mathbb{P}^x\left(\cos \frac{(X_t - x) \cdot \epsilon}{r} \geqslant \cos \frac{1}{k}, \; \tau_r^x > t\right) \tag{5.13}$$

$$\leqslant \mathbb{P}^x\left(\cos \frac{(X_{t \wedge \tau_r^x} - x) \cdot \epsilon}{r} \geqslant \cos \frac{1}{k}\right).$$

In the first inequality we have used the fact that on the set $\{\tau_r^x > t\}$,

$$\frac{(X_t - x) \cdot \epsilon}{r} \leqslant \frac{|X_t - x| \cdot |\epsilon|}{r} \leqslant |\epsilon| \leqslant \frac{1}{k} \leqslant \arccos \sqrt{2/3} \leqslant \frac{\pi}{4},$$

and that the function $x \mapsto \cos x$ is decreasing in $[0, \pi/4]$.

Now we use the extended generator (5.12): For any $x, \epsilon \in \mathbb{R}^d$ and $r > 0$, we have

$$\cos \frac{(\cdot - x) \cdot \epsilon}{r} \in C_b^\infty(\mathbb{R}^d) \subset \hat{\mathcal{D}}(A).$$

By Proposition 2.27(c) with $e_\xi(x) := e^{ix \cdot \xi}$

$$e^{-i(z-x) \cdot \xi} A e^{i(\cdot - x) \cdot \xi}(z) = e_{-\xi}(z) A e_\xi(z) = -q(z, \xi) \quad \forall x, z, \xi \in \mathbb{R}^d$$

and thus

$$A\left(\cos \frac{(\cdot - x) \cdot \epsilon}{r}\right)(z) = A\left(\operatorname{Re}\left[\exp \frac{i(\cdot - x) \cdot \epsilon}{r}\right]\right)(z)$$

$$= \operatorname{Re}\left[A\left(\exp \frac{i(\cdot - x) \cdot \epsilon}{r}\right)(z)\right]$$

$$= -\operatorname{Re}\left[\exp \frac{i(z - x) \cdot \epsilon}{r} q(z, \epsilon/r)\right]$$

yields, together with (5.13),

$$\mathbb{P}^x(\tau_r^x > t)$$

$$\leqslant \frac{1}{\cos \frac{1}{k}} \mathbb{E}^x\left(\cos \frac{(X_{t \wedge \tau_r^x} - x) \cdot \epsilon}{r}\right)$$

$$= \frac{1}{\cos \frac{1}{k}} \left[1 - \mathbb{E}^x \int_0^{t \wedge \tau_r^x} \cos \frac{(X_s - x) \cdot \epsilon}{r} \, \text{Re} \, q(X_s, \epsilon/r) \, ds \right.$$

$$\left. + \mathbb{E}^x \int_0^{t \wedge \tau_r^x} \sin \frac{(X_s - x) \cdot \epsilon}{r} \, \text{Im} \, q(X_s, \epsilon/r) \, ds \right]$$

$$= \frac{1}{\cos \frac{1}{k}} \left\{ 1 - \mathbb{E}^x \int_0^{t \wedge \tau_r^x} \left[\cos \frac{(X_s - x) \cdot \epsilon}{r} \right. \right.$$

$$\left. \left. \times \left(\text{Re} \, q(X_s, \epsilon/r) - \tan \frac{(X_s - x) \cdot \epsilon}{r} \, \text{Im} \, q(X_s, \epsilon/r) \right) \right] ds \right\}.$$

On the set $\{\tau_r^x > s\}$ we have

$$\frac{(X_s - x) \cdot \epsilon}{r} \leq |\epsilon| \leq \frac{1}{k} \leq \arccos \sqrt{2/3}$$

and some elementary calculations yield

$$\tan \frac{(X_s - x) \cdot \epsilon}{r} \leq \frac{3}{2} \cdot \frac{|X_s - x| \cdot |\epsilon|}{r} \leq \frac{3|\epsilon|}{2}.$$

The inequality above combined with the remarks on the estimate (5.8) directly below Theorem 5.5 give

$$\mathbb{P}^x(\tau_r^x > t)$$

$$\leq \frac{1}{\cos \frac{1}{k}} \left[1 - \mathbb{E}^x \int_0^{t \wedge \tau_r^x} \left[\cos \frac{(X_s - x) \cdot \epsilon}{r} \right. \right.$$

$$\left. \left. \times \left(\text{Re} \, q(X_s, \epsilon/r) - \frac{3|\epsilon|}{2} | \text{Im} \, q(X_s, \epsilon/r) | \right) \right] ds \right]$$

$$\leq \frac{1}{\cos \frac{1}{k}} \left[1 - \frac{1}{4} \mathbb{E}^x \int_0^{t \wedge \tau_r^x} \cos \frac{(X_s - x) \cdot \epsilon}{r} \, \text{Re} \, q(X_s, \epsilon/r) \, ds \right]$$

$$\leq \frac{1}{\cos \frac{1}{k}} \left[1 - \frac{\cos \frac{1}{k}}{4} \left(\inf_{|y-x| \leq r} \text{Re} \, q(y, \epsilon/r) \right) \mathbb{E}^x(t \wedge \tau_r^x) \right] \qquad (5.14)$$

$$\leq \frac{1}{\cos \frac{1}{k}} \left[1 - \frac{t \cos \frac{1}{k}}{4} \left(\inf_{|y-x| \leq r} \text{Re} \, q(y, \epsilon/r) \right) \mathbb{P}^x(\tau_r^x > t) \right].$$

That is,

$$
\mathbb{P}^x(\tau_r^x > t) \leq \frac{4}{\cos \frac{1}{k}} \left(4 + t \inf_{|y-x| \leq r} \operatorname{Re} q(y, \epsilon/r) \right)^{-1}
$$

$$
\leq \frac{4}{\cos \frac{1}{k}} \left(t \inf_{|y-x| \leq r} \operatorname{Re} q(y, \epsilon/r) \right)^{-1}
$$

$$
\leq \frac{4}{\cos \sqrt{2/3}} \left(t \inf_{|y-x| \leq r} \operatorname{Re} q(y, \epsilon/r) \right)^{-1}.
$$

Taking the infimum with respect to $|\epsilon| \leq 1/k$, we obtain

$$
\mathbb{P}^x(\tau_r^x > t) \leq \frac{4}{\cos \sqrt{2/3}} \left(t \sup_{|\xi| \leq 1/(kr)} \inf_{|y-x| \leq r} \operatorname{Re} q(y, \xi) \right)^{-1}.
$$

Finally, since $\{\tau_r^x \geq t\} = \bigcup_{n \geq 1} \{\tau_r^x > t - \frac{1}{n}\}$, the proof is completed with the usual *continuity of measures*-argument. □

Similar to Corollary 5.3, the following conclusion is deduced from the proof above.

Corollary 5.8. *Let* $(X_t)_{t \geq 0}$ *be a d-dimensional Feller process with symbol* $q(x, \xi)$ *and with* $k(x, r)$ *as in Theorem 5.5. Then, for any* $x \in \mathbb{R}^d$ *and* $r > 0$,

$$
\mathbb{E}^x(\tau_r^x) \leq \frac{4}{\left(\cos \sqrt{2/3} \right) \sup_{|\xi| \leq 1/(rk(x,r))} \inf_{|y-x| \leq r} \operatorname{Re} q(y, \xi)}
$$

and if $q(x, \xi)$ *satisfies the sector condition* (5.9), *we can replace the constant* $k(x, r)$ *by* $k_0 = \left(\arccos \sqrt{2/3} \right)^{-1}$.

Proof. Write $k = k(x, r)$ for simplicity. From (5.14) we get for any $\epsilon \in \mathbb{R}^d$ with $|\epsilon| \leq 1/k$

$$
0 \leq \mathbb{P}^x(\tau_r^x > t) \leq \frac{1}{\cos \frac{1}{k}} \left[1 - \frac{\cos \frac{1}{k}}{4} \left(\inf_{|y-x| \leq r} \operatorname{Re} q(y, \epsilon/r) \right) \mathbb{E}^x(t \wedge \tau_r^x) \right].
$$

Thus, for any $t > 0$,

$$
\cos \frac{1}{k} \left(\inf_{|y-x| \leq r} \operatorname{Re} q(y, \epsilon/r) \right) \mathbb{E}^x(t \wedge \tau_r^x) \leq 4.
$$

Letting $t \to \infty$ and taking the infimum w.r.t. $|\epsilon| \leq 1/k$ finishes the proof. □

If we combine the estimate from Theorem 5.5 with the strong Markov property and an iteration procedure, we can improve the tail estimate for τ_r^x.

Theorem 5.9. *Let $(X_t)_{t \geqslant 0}$ be a Feller process with symbol $q(x, \xi)$ and $k(x, r)$ as in Theorem 5.5. For all $x \in \mathbb{R}^d$ and $r, t > 0$,*

$$\mathbb{P}^x(\tau_r^x > t) \leqslant \exp\left[-c_0 t \sup_{|\xi| \leqslant 1/(2rk^*(x,r))} \inf_{|y-x| \leqslant 3r} \operatorname{Re} q(y, \xi) + 1\right], \qquad (5.15)$$

where $c_0 = \cos\sqrt{2/3}/(4e)$ and

$$k^*(x, r) = \inf\left\{k \geqslant \left(\arccos\sqrt{2/3}\right)^{-1} : \sup_{|\xi| \leqslant 1/(2kr)} \sup_{|y-x| \leqslant r} \frac{\operatorname{Re} q(y, \xi)}{|\xi| |\operatorname{Im} q(y, \xi)|} \geqslant 4r\right\}.$$

In particular, it holds that

$$\mathbb{E}^x(\tau_r^x) \leqslant \frac{e}{c_0 \displaystyle\sup_{|\xi| \leqslant 1/(2rk^*(x,r))} \inf_{|y-x| \leqslant 3r} \operatorname{Re} q(y, \xi)}.$$

If $q(x, \xi)$ satisfies the sector condition (5.9), then we may replace $k^*(x, r)$ in the above estimates by $k_0 = \left(\arccos\sqrt{2/3}\right)^{-1}$.

Proof. We use the notation from Theorem 5.5. For any $x \in \mathbb{R}^d$ and $r > 0$, let

$$h(x, r) := \sup_{|y-x| \leqslant r}\left[\sup_{|\xi| \leqslant 1/(2rk(y,2r))} \inf_{|z-y| \leqslant 2r} \operatorname{Re} q(z, \xi)\right]^{-1}.$$

Then, according to Theorem 5.5,

$$\sup_{|y-x| \leqslant r} \mathbb{P}^y\left(\tau_{2r}^y > t\right) \leqslant c \sup_{|y-x| \leqslant r}\left[t \sup_{|\xi| \leqslant 1/(2rk(y,2r))} \inf_{|z-y| \leqslant 2r} \operatorname{Re} q(z, \xi)\right]^{-1} = c\, t^{-1} h(x, r).$$

In particular,

$$\sup_{|y-x| \leqslant r} \mathbb{P}^y\left(\tau_{2r}^y > ceh(x, r)\right) \leqslant e^{-1}.$$

For $t > 0$, set $n := \lfloor t/(ceh(x, r)) \rfloor$ and $t_j = ceh(x, r)j$ for $j = 0, 1, \ldots, n$. Define $(X_t - x)_t^* := \sup_{s \leqslant t} |X_s - x|$. Then an application of the strong Markov property yields

$$\mathbb{P}^x(\tau_r^x > t) \leq \mathbb{P}^x\left(\sup_{s \leq t} |X_s - x| < r\right)$$

$$\leq \mathbb{P}^x\left(|X_{t_j} - x| < r, \ \sup_{s \in (t_j, t_{j+1}]} |X_s - X_{t_j}| < 2r \ \forall j = 0, 1, \ldots, n-1\right)$$

$$= \mathbb{E}^x\left[\mathbb{1}_{\left\{(X.-x)_{t_1}^* \leq r\right\}} \prod_{j=1}^{n-1} \mathbb{E}^{X_{t_j}}\left(\mathbb{1}_{\left\{(X.-X_0)_{t_1}^* \leq 2r, |X_0-x| \leq r\right\}}\right)\right]$$

$$\leq \left[\sup_{|y-x| \leq r} \mathbb{E}^y\left(\mathbb{1}_{\left\{(X.-y)_{t_1}^* \leq 2r\right\}}\right)\right]^n$$

$$\leq e^{-n}$$

$$\leq \exp\left(-t/(ceh(x,r)) + 1\right).$$

For any $y \in \overline{\mathbb{B}}(x, r)$ we see that $|z - y| \leq 2r$ implies $|x - z| \leq 3r$; on the other hand, if $|z - x| \leq r$, then $|z - y| \leq 2r$, and so $\sup_{|y-x| \leq r} k(y, 2r) \leq k^*(x, r)$. Thus, we gather that

$$h(x,r) \leq \left[\sup_{|\xi| \leq 1/(2rk^*(x,r))} \inf_{|z-x| \leq 3r} \operatorname{Re} q(z, \xi)\right]^{-1}.$$

This proves the first estimate. The second conclusion follows from the integral identity

$$\mathbb{E}^x \tau_r^x = \int_0^\infty \mathbb{P}^x(\tau_r^x > t) \, dt$$

and (5.15). Finally, if we assume the sector condition, we get $k_0 = k^*(x, r)$. □

We close this part by re-examining Example 5.7(a).

Example 5.10. Consider $q(x, \xi) = (1 + |x|^\beta)|\xi|^\alpha$ with $\alpha \in (0, 2)$ and $\beta \leq \alpha$ as in Example 5.7(a). Then, for any $x \in \mathbb{R}^d$ and $r, t > 0$,

$$\mathbb{P}^x(\tau_r^x > t) \leq \exp\left(-ctr^\alpha \inf_{|y-x| \leq r} (1 + |y|^\beta)\right) \leq \exp\left(-ctr^\alpha\right). □$$

Existence of Exponential Moments. For a Lévy process $(X_t)_{t \geq 0}$ with symbol ψ and Lévy triplet (l, Q, ν) it is well-known, cf. Sato [267, Theorem 25.3, p. 159], that for a submultiplicative function[3] $g : \mathbb{R}^d \to \mathbb{R}$

$$\mathbb{E}\, g(X_t) < \infty \iff \int_{|y| \geq 1} g(y) \, \nu(dy) < \infty.$$

---

[3] That is, $g(x + y) \leq cg(x)g(y)$ for all $x, y \in \mathbb{R}^d$ and some constant $c > 0$.

In [183, Lemma 12] it was observed that we have a similar result for Feller processes. The following theorem contains a slightly more general result.

Theorem 5.11. *Let $(X_t)_{t\geqslant 0}$ be a Feller process in $\mathbb{R}^d$ with generator $(A, \mathcal{D}(A))$, $C_c^\infty(\mathbb{R}^d) \subset \mathcal{D}(A)$, symbol $q(x, \xi)$ satisfying $q(x, 0) = 0$ and with x-dependent Lévy triplet $(l(x), Q(x), N(x, dy))$. If the symbol has bounded coefficients, i.e.*

$$\sup_{x\in\mathbb{R}^d} |q(x,\xi)| \leqslant c(1 + |\xi|^2) \quad \forall \xi \in \mathbb{R}^d$$

cf. (2.34) in Theorem 2.31, and satisfies

$$\sup_{z\in\mathbb{R}^d} \int_{|y|\geqslant 1} e^{\zeta\cdot y} N(z, dy) < \infty \quad \forall \zeta \in \mathbb{R}^d \tag{5.16}$$

then $\mathbb{E}^x e^{\zeta\cdot X_t} < \infty$ for all $\zeta, x \in \mathbb{R}^d$.

Proof. We prove the assertion only for $d = 1$, the case $d > 1$ increases only the complexity of *notation*. Pick $\theta_m \in C_c^\infty(\mathbb{R})$, $m \geqslant 1$, with $\mathbb{1}_{B(0,m)} \leqslant \theta_m \leqslant \mathbb{1}_{B(0,2m)}$ and set

$$u_m(z) := \theta_m(z - x)e_{-i\xi}(z - x) = \theta_m(z - x)e^{(z-x)\cdot\xi} \quad \forall z, x, \xi \in \mathbb{R}, \ m \geqslant 1.$$

Recall that $\tau_k = \tau_k^x = \inf\{t > 0 : X_t \notin \overline{\mathbb{B}}(x, k)\}$, $k \geqslant 1$, is the first exit time from the closed ball $\overline{\mathbb{B}}(x, k)$, and denote by $X_t^{\tau_k} = X_{t\wedge\tau_k}$ the stopped process.

By Theorem 2.37(f) we know that for any $u \in C_c^\infty(\mathbb{R})$ the pair (u, Au) is in the extended generator $\hat{A}$; in particular $\left(u(X_t) - \int_0^t Au(X_s)\,ds\right)_{t\geqslant 0}$ is a local martingale. Consequently, as $u_m \in C_c^\infty(\mathbb{R}) \subset \mathcal{D}(A)$, we see

$$\mathbb{E}^x u_m(X_t^{\tau_k}) - 1 = \mathbb{E}^x \left(\int_0^{\tau_k \wedge t} Au_m(X_s)\,ds \right)$$

$$= \mathbb{E}^x \left(\int_{[0,\tau_k \wedge t)} Au_m(X_s^{\tau_k})\,ds \right) \quad \forall m \geqslant 1, \ t > 0.$$

Using the integro-differential representation (2.25) of the generator we get for all $z \in \mathbb{B}(x, k)$ where $m \geqslant k + 1$

$$Au_m(z) = l(z)u_m'(z) + \frac{1}{2}Q(z)u_m''(z)$$

$$+ \int_{\mathbb{R}\setminus\{0\}} \left(u_m(z + y) - u_m(z) - u_m'(z)y\,\chi(|y|) \right) N(z, dy)$$

$$= e_{-i\xi}(z - x)\left[l(z) \cdot \xi + \frac{1}{2}Q(z)\xi^2 \right.$$

$$+ \int_{\mathbb{R}\backslash\{0\}} \left(\theta_m(z - x + y)e_{-i\xi}(y) - 1 - \xi \cdot y\chi(|y|) \right) N(z, dy) \Bigg]$$

$$= e_{-i\xi}(z - x)\left[l(z) \cdot \xi + \frac{1}{2}Q(z)\xi^2 \right.$$

$$+ \int_{\mathbb{R}\backslash\{0\}} \left(e_{-i\xi}(y) - 1 - \xi \cdot y\chi(|y|) \right) N(z, dy)$$

$$+ \int_{\mathbb{R}\backslash\{0\}} \left(\theta_m(z - x + y) - 1 \right) e_{-i\xi}(y) \, N(z, dy) \Bigg]$$

$$= e_{-i\xi}(z - x)\left[-q(z, -i\xi) - \int_{|y| \geqslant 1} \left(1 - \theta_m(z - x + y) \right) e_{-i\xi}(y) \, N(z, dy) \right].$$

In the last step we used that $\theta_m(z - x + y) = 1$ for $|y| < 1$ and $z \in \mathbb{B}(x, k)$ with $m \geqslant k + 1$ along with the Lévy–Khintchine representation (2.32) for the symbol. Note that the boundedness of the coefficients and the integrability condition (5.16) of the Lévy kernel $N(x, dy)$ enables us to extend $q(x, \xi)$ to the complex plane. This proves

$$\left| A[\theta_m e_{-i\xi}](z - x) \right| \leqslant e_{-i\xi}(z - x)\kappa(\xi) \quad \forall z \in \mathbb{B}(x, k), \ m \geqslant k + 1$$

where the constant $\kappa(\xi) := \sup_{z \in \mathbb{R}} \left(|q(z, -i\xi)| + \int_{|y| \geqslant 1} e^{y \cdot \xi} N(z, dy) \right)$ appearing on the right-hand side is finite because of the integrability condition (5.16).

By construction, we have $X_s^{\tau_k} \in \mathbb{B}(x, k)$ under $\mathbb{P}^x$ for all $s < \tau_k$, and we get for $m \geqslant k + 1$ and the definition of the function $u_m = \theta_m e_{-i\xi}$

$$\left| \mathbb{E}^x \left[e_{-i\xi}(X_t^{\tau_k} - x) \right] - 1 \right| \leqslant \left| \mathbb{E}^x \left[\int_{[0, \tau_k \wedge t)} A[\theta_m e_{-i\xi}](X_s^{\tau_k} - x) \, ds \right] \right|$$

$$\leqslant \kappa(\xi) \int_{[0, t)} \mathbb{E}^x \left[e_{-i\xi}(X_s^{\tau_k} - x) \right] ds.$$

Now we can use Gronwall's lemma, see e.g. [284, Theorem A.43, p. 360], to get

$$\mathbb{E}^x \left[e^{\xi \cdot (X_t^{\tau_k} - x)} \right] \leqslant e^{t\kappa(\xi)} \quad \forall k \geqslant 1, \ x, \xi \in \mathbb{R}.$$

Since the symbol has bounded coefficients and satisfies $q(x, 0) = 0$, Theorem 5.1 applies and tells us that $\lim_{k \to \infty} \tau_k = \infty$ in probability. By Fatou's lemma (and taking a suitable subsequence) we finally get

$$\mathbb{E}^x \left[e^{\xi \cdot (X_t - x)} \right] \leqslant \varliminf_{j \to \infty} \mathbb{E}^x \left[\exp\left(\xi \cdot (X_{t \wedge \tau_{k(j)}} - x) \right) \right] \leqslant e^{t\kappa(\xi)},$$

and the proof is complete. $\qquad\qquad\qquad\qquad\qquad\qquad\qquad\qquad\qquad\qquad\square$

5.2   Hausdorff Dimension and Indices

The notion of Hausdorff dimension is very useful in order to describe the roughness or irregularity of the paths of a stochastic process. Recall that the **Hausdorff dimension** of a set $A \subset \mathbb{R}^d$, denoted by $\dim_H A$, is the unique number $\lambda \in [0, d]$ where the λ-dimensional **Hausdorff measure** $\mathcal{H}^\lambda(A)$, defined by

$$\mathcal{H}^\lambda(A) := \sup_{\epsilon > 0} \mathcal{H}^\lambda_\epsilon(A),$$

$$\mathcal{H}^\lambda_\epsilon(A) := \inf \left\{ \sum_{n=1}^\infty (\operatorname{diam} B_n)^\lambda \; : \; B_n \text{ Borel}, \; \bigcup_{n=1}^\infty B_n \supset A \text{ and } \operatorname{diam} B_n \leqslant \epsilon \right\},$$

changes from $+\infty$ to a finite value. Among the standard references for Hausdorff measure and Hausdorff dimension are Rogers [253] and Falconer [103].

Let $(Y_t)_{t \geqslant 0}$ be a d-dimensional *Lévy process* with the characteristic exponent (i.e. symbol) ψ. For the study of the Hausdorff dimension for the sample paths of Lévy processes, various indices were introduced by Blumenthal–Getoor [32, Sects. 2, 3 and 5]:

$$\beta'' = \sup \left\{ \delta > 0 : \lim_{|\xi| \to \infty} \frac{\operatorname{Re} \psi(\xi)}{|\xi|^\delta} = \infty \right\},$$

$$\beta' = \sup \left\{ \delta > 0 : \int |\xi|^{\delta - d} \frac{1 - e^{-\operatorname{Re} \psi(\xi)}}{\operatorname{Re} \psi(\xi)} \, d\xi < \infty \right\},$$

$$\beta = \inf \left\{ \delta > 0 : \lim_{|\xi| \to \infty} \frac{|\psi(\xi)|}{|\xi|^\delta} = 0 \right\}.$$

It is not hard to see that $0 \leqslant \beta'' \leqslant \beta' \leqslant \beta \leqslant 2$; if $(Y_t)_{t \geqslant 0}$ is a symmetric α-stable Lévy process, all indices coincide; on the other hand there are examples such that the inequalities between the various indices are strict. The results for the Hausdorff dimension of the image sets of a Lévy process can be summarized as follows, see Blumenthal–Getoor [32], Pruitt [244] and Millar [228].

Theorem 5.12. *Let $(Y_t)_{t \geqslant 0}$ be a d-dimensional Lévy process with indices β'', β' and β given above. For every analytic set $E \subset [0, 1]$, we have*

$$\min\{d, \beta \dim_H E\} \geqslant \dim_H Y(E) \geqslant \begin{cases} \beta' \dim_H E, & \text{if } \beta' \leqslant d \\ \min\{1, \beta'' \dim_H E\}, & \text{if } \beta' > d = 1 \end{cases}$$

almost surely.

If $E = [0, 1]$ there are sharp results due to Pruitt [244] and Khoshnevisan et al. [180]. If $\dim_H E < 1$ and if $\beta' < \beta$, one cannot expect that $\dim_H Y(E)$ is a simple function of $\dim_H E$. Nevertheless, the Hausdorff dimension can be characterized, cf. Khoshnevisan–Xiao [179]; the proof of this result involves additive (multiparameter) Lévy processes and capacities. We refer to Xiao [353] for an up-to-date survey.

It is possible to generalize the results of Theorem 5.12. To do so we need a substitute for the Blumenthal–Getoor–Pruitt indices. The following definition is essentially taken from [277].[4]

Definition 5.13. Let $q(x, \xi)$ be a negative definite symbol. For every compact set $K \subset \mathbb{R}^d$ the **generalized Blumenthal–Getoor–Pruitt indices** (at infinity) are the numbers

$$\underline{\delta}_\infty^K := \sup \left\{ \lambda > 0 : \varliminf_{|\xi| \to \infty} \frac{\inf_{x \in K} \inf_{|\eta| \leqslant |\xi|} \inf_{|z-x| \leqslant 1/|\xi|} \operatorname{Re} q(z, \eta)}{|\xi|^\lambda} = \infty \right\},$$

$$\bar{\delta}_\infty^K := \sup \left\{ \lambda > 0 : \varlimsup_{|\xi| \to \infty} \frac{\inf_{x \in K} \inf_{|\eta| \leqslant |\xi|} \inf_{|z-x| \leqslant 1/|\xi|} \operatorname{Re} q(z, \eta)}{|\xi|^\lambda} = \infty \right\},$$

$$\underline{\beta}_\infty^K := \inf \left\{ \lambda > 0 : \varliminf_{|\xi| \to \infty} \frac{\sup_{x \in K} \sup_{|\eta| \leqslant |\xi|} \sup_{|z-x| \leqslant 1/|\xi|} |q(z, \eta)|}{|\xi|^\lambda} = 0 \right\},$$

$$\beta_\infty^K := \inf \left\{ \lambda > 0 : \varlimsup_{|\xi| \to \infty} \frac{\sup_{x \in K} \sup_{|\eta| \leqslant |\xi|} \sup_{|z-x| \leqslant 1/|\xi|} |q(z, \eta)|}{|\xi|^\lambda} = 0 \right\}.$$

If $K = \{x\}$ we use β_∞^x etc. as shorthand for $\beta_\infty^{\{x\}}$ etc.

For Lévy processes and random walks the indices of type β are due to Blumenthal–Getoor [32], while the δ-indices were defined by Pruitt [245] for random walks and Lévy processes. A different generalization of the Blumenthal–Getoor index to Itô semimartingales has been suggested by Aït-Sahalia–Jacod [3, Sect. 2.1, Equation (5)]; note, however, that their index is itself a random process.

Remark 5.14. a) It is easy to see that for all compact sets $K \subset \mathbb{R}^d$

$$0 \leqslant \underline{\delta}_\infty^K \leqslant \underline{\beta}_\infty^K \leqslant \beta_\infty^K \leqslant 2,$$

$$0 \leqslant \underline{\delta}_\infty^K \leqslant \bar{\delta}_\infty^K \leqslant \beta_\infty^K \leqslant 2.$$

The relations between $\underline{\beta}_\infty^K$ and $\bar{\delta}_\infty^K$ depend, in general, on the symbol $q(x, \xi)$.
b) One can show that

$$\underline{\delta}_\infty^K = \sup \left\{ \lambda > 0 : \varliminf_{|\xi| \to \infty} \frac{\inf_{x \in K} \inf_{|z-x| \leqslant 1/|\xi|} \operatorname{Re} q(z, \xi)}{|\xi|^\lambda} = \infty \right\},$$

---

[4]Proposition 5.2 in [277] assumes that the symbol has bounded coefficients, cf. Theorem 2.31, but this is actually not needed. Some of the definitions are streamlined and adapted to the improved probability estimates from Sect. 5.1.

$$\beta_\infty^K = \inf\left\{\lambda > 0 : \lim_{|\xi|\to\infty} \frac{\sup_{x\in K}\sup_{|z-x|\leq 1/|\xi|}|q(z,\xi)|}{|\xi|^\lambda} = 0\right\},$$

whereas similar relations for the indices $\underline{\beta}_\infty^K$ and $\bar{\delta}_\infty^K$ cannot be expected since their definition involves lower and upper limits, see [277, Lemma 5.1 and Proposition 5.2].

c) In [277] only the indices for $K = \{x\}$ have been considered. Depending on the smoothness of the function $x \mapsto q(x,\xi)$ there are simple relations between the above indices and the indices where $K = \{x\}$.

If $q(x,\xi)$ satisfies the sector condition (5.9), we can interchange q and $\mathrm{Re}\,q$ in the definition of the indices.

If there is no dominating diffusion (i.e. $Q(x) \equiv 0$) and drift (i.e. the sector condition (5.9) holds), then it is possible to characterize the generalized Blumenthal–Getoor indices by integration properties of the Lévy kernel $N(x, dy)$. For details, cf. [277, Proposition 5.4].

d) If $q(x,\xi) = |\xi|^{\alpha(x)}$ is the symbol of a stable-like process where $x \mapsto \alpha(x)$ is Lipschitz continuous, then $\delta_\infty^x = \bar{\delta}_\infty^x = \underline{\beta}_\infty^x = \beta_\infty^x = \alpha(x)$. Moreover, $\beta_\infty^K = \sup_{x\in K}\beta_\infty^x$.

More generally, we have the following formula for β_∞^x, cf. [285, Proposition 5.6]:

$$\beta_\infty^x = \varlimsup_{|\eta|\to\infty} \sup_{|y-x|\leq 2/|\eta|} \frac{\log|q(y,\eta)|}{\log|\eta|}. \qquad \square$$

The following result presents upper estimates for the Hausdorff dimension for the image of a Feller process, and it partly extends Theorem 5.12.

Theorem 5.15. *Let $(X_t)_{t\geq 0}$ be a Feller process with the generator $(A, \mathcal{D}(A))$ such that $C_c^\infty(\mathbb{R}^d) \subset \mathcal{D}(A)$ and symbol $q(x,\xi)$. For every bounded analytic time-set $E \subset [0,\infty)$,*

$$\dim_H X(E) \leq \min\left\{d,\ \sup_K \beta_\infty^K \dim_H E\right\} \tag{5.17}$$

where $\sup_K$ denotes the supremum taken for all compact sets $K \subset \mathbb{R}^d$.

Proof. Without loss of generality we consider only X_t for $t \in [0,1]$. Fix $x \in \mathbb{R}^d$, $R > |x| \geq 0$ and set $\Omega_R := \{\sup_{s\leq 1}|X_s| < R\}$. Since $(X_t)_{t\geq 0}$ has càdlàg paths, $\lim_{R\to\infty}\mathbb{P}^x(\Omega_R) = 1$.

Step 1. We claim that for every $p > \sup_K \beta_\infty^K$ and $t \in [0,1]$

$$\mathbb{P}^x\left(\left\{\sup_{|s-t|\leq h}|X_s - X_t| > u\right\} \cap \Omega_R\right) \leq c\,hu^{-p} \quad \forall x \in \mathbb{R}^d,\ u \in (0, u_0] \tag{5.18}$$

where u_0 and c are two positive constants independent of h. By the Markov property, it is enough to verify that for sufficiently small $u > 0$

$$\sup_{|y| \leqslant R} \mathbb{P}^y \left(\sup_{0 \leqslant r \leqslant h} |X_r - y| > u \right) \leqslant c \, h \, u^{-p}.$$

According to Corollary 5.2,

$$\sup_{|y| \leqslant R} \mathbb{P}^y \left(\sup_{0 \leqslant r \leqslant h} |X_r - y| > u \right) \leqslant c \, h \sup_{|y| \leqslant R} \sup_{|z-y| \leqslant u} \sup_{|\eta| \leqslant 1/u} |q(z, \eta)|.$$

This, along with the very definition of β_∞^K, proves (5.18).

Step 2. Now we follow the proofs of [276, Theoerm 4] and Shieh–Xiao [299, Lemma 4.7 and Proposition 4.8] (with some significant modifications) to get the desired assertion.

Case 1: Suppose that $\dim_H E < 1$. For every constant $\gamma \in (\dim_H E, 1)$, there exists a sequence of balls $\{\mathbb{B}(t_{j,k}, h_{j,k})\}_{j,k \geqslant 1}$ such that

$$E \subset \bigcup_{k=1}^{\infty} \mathbb{B}(t_{j,k}, h_{j,k}) \quad \forall j \geqslant 1, \qquad \lim_{j \to \infty} \sup_{k \geqslant 1} h_{j,k} = 0 \quad \text{and} \quad \sup_{j \geqslant 1} \sum_{k=1}^{\infty} h_{j,k}^{\gamma} < \infty.$$

Without loss of generality, we may assume that $h_{j,k} \leqslant 1/j$ for all $j, k \geqslant 1$. For $j \geqslant 1$, let

$$\Omega_j^R := \left\{ \omega : \sup_{|s-t| \leqslant 1/j} |X_s(\omega) - X_t(\omega)| \leqslant u_0 \right\} \cap \Omega_R,$$

where u_0 is the constant appearing in (5.18). Note that

$$X(E) \subset \bigcup_{k=1}^{\infty} \mathbb{B}(X_{t_{j,k}}, D(t_{j,k}), h_{j,k}) \quad \text{with} \quad D(t, h) = \sup_{|s-t| \leqslant h} |X_s - X_t|.$$

Then, for any $j \geqslant j_0$ and $p > \sup_K \beta_\infty^K$,

$$\sum_{k=1}^{\infty} \mathbb{E}^x \left((D(t_{j,k}, h_{j,k}))^{\gamma p} \mathbb{1}_{\Omega_{j_0}^R} \right)$$

$$\leqslant \gamma p \sum_{k=1}^{\infty} \int_0^{u_0} u^{\gamma p-1} \mathbb{P}^x \left(\{(t_{j,k}, h_{j,k}) > u\} \cap \Omega_R \right) du$$

$$= \gamma p \sum_{k=1}^{\infty} \int_{0}^{u_0} u^{\gamma p - 1} \, \mathbb{P}^x \left(\left\{ \sup_{|s - t_{j,k}| \leqslant h_{j,k}} |X_s - X_{t_{j,k}}| > u \right\} \cap \Omega_R \right) du$$

$$\leqslant c \gamma p \sum_{k=1}^{\infty} \int_{0}^{u_0} u^{\gamma p - 1} \left(1 \wedge (h_{j,k} u^{-p}) \right) du,$$

where the last estimate follows from (5.18). It is elementary to verify that, up to a constant, the integral in the last term is bounded by $c_1 h_{j,k}^{\gamma}$ with some constant c_1 which does not depend on j_0 and $h_{j,k}$. Therefore,

$$\sup_{j \geqslant j_0} \sum_{k=1}^{\infty} \mathbb{E}^x \left(D(t_{j,k}, h_{j,k})^{\gamma p} \, \mathbb{1}_{\Omega_{j_0}^R} \right) \leqslant c_2 \sup_{j \geqslant j_0} \sum_{k=1}^{\infty} h_{j,k}^{\gamma} < \infty,$$

which yields $\sup_{j \geqslant j_0} \mathbb{E}^x \left(\mathcal{H}_{1/j}^{\gamma p}(X(E) \cap \Omega_{j_0}^R) \right) < \infty$ for every $x \in \mathbb{R}^d$. Using monotone convergence we get that

$$\mathbb{E}^x \left(\mathcal{H}_{1/j_0}^{\gamma p}(X(E) \cap \Omega_{j_0}^R) \right) < \infty \quad \text{and} \quad \mathcal{H}_{1/j_0}^{\gamma p}(X(E) \cap \Omega_{j_0}^R) < \infty \text{ a.s.}$$

This implies that $\dim_{\mathrm{H}} X(E) \leqslant \gamma p$ almost surely on $\Omega_{j_0}^R$. From (5.18) we infer $\lim_{j \to \infty} \mathbb{P}^x(\Omega_j^R) = \mathbb{P}(\Omega_R)$. Letting first $j_0 \to \infty$ and $R \to \infty$, then $\gamma \to \dim_{\mathrm{H}} E$, and finally $p \to \sup_K \beta_{\infty}^K$ along countable sequences proves the first inequality in (5.17).

Case 2: Suppose that $\dim_{\mathrm{H}} E = 1$. We may assume that $E = [0, 1]$. For every $j \geqslant 1$ we cover $[0, 1]$ by finitely many closed intervals $E_{j,k} := [(k-1)/j, k/j]$ for $k = 1, 2, \dots, j$. As in the first case, we see that for any $j \geqslant j_0$, $R > 0$ and $p > \sup_K \beta_{\infty}^K$,

$$\sup_{j \geqslant j_0} \sum_{k=1}^{j} \mathbb{E} \left(D\left(\tfrac{k-1}{j}, \tfrac{1}{j}\right)^p \mathbb{1}_{\Omega_{j_0}^R} \right) \leqslant c p \sup_{j \geqslant j_0} \sum_{k=1}^{j} \frac{1}{j} = c p$$

holds with some constant c which is independent of j_0. Thus, $\dim_{\mathrm{H}} X(E) \leqslant p$ almost surely on $\Omega_{j_0}^R$, and we obtain the first inequality as before, letting first $j_0 \to \infty$, then $R \to \infty$, and finally $p \to \sup_K \beta_{\infty}^K$ along countable sequences. $\qquad\square$

5.3 Asymptotic Behaviour of the Sample Paths

We can use the symbol to obtain information on the local and global growth of the sample paths of a Feller process. For Lévy processes this is a classic topic and very precise LIL-type results and integral criteria are available. One of the first (and still unsurpassed) work in this direction is Khintchine [178] (the essence of it is contained in Skorokhod's monograph [305, Appendix]), many details can be found in Fristedt [112] and Sato [267]. In particular for the LIL we mention the work by Dupuis [95], Savov [269] and Aurzada–Döring–Savov [7].

Although such refined results are not available for general Feller processes, we can still give some polynomial asymptotics as $t \to 0$ and $t \to \infty$. As for Lévy processes, the proofs for both regimes are similar, but the variable coefficients of a Feller generator require stronger uniform control for global (i.e. $t \to \infty$) assertions than for local results (i.e. $t \to 0$). Let us illustrate this for a one-dimensional stable-like process $(X_t)_{t \geq 0}$ with symbol $q(x, \xi) = |\xi|^{\alpha(x)}$, $0 < \alpha(x) < 2$, cf. Theorem 3.31; without loss of generality we assume that α is smooth. Set $\alpha_0 := \alpha(x_0)$ and denote by $(Y_t)_{t \geq 0}$ the symmetric α_0-stable Lévy process with symbol $|\xi|^{\alpha_0}$. Intuitively, we have

$$\varlimsup_{t \to 0} \frac{|X_t|}{t^{1/\lambda}} = \varlimsup_{t \to 0} \frac{|Y_t|}{t^{1/\lambda}} = \begin{cases} 0, & \text{if } \lambda > \alpha_0 = \alpha(x_0) \\ \infty, & \text{if } \lambda < \alpha_0 = \alpha(x_0) \end{cases} \qquad (\text{a.s. } \mathbb{P}^{x_0})$$

cf. Blumenthal–Getoor [32]. While for the Lévy process the behaviour as $t \to \infty$ can also be expressed in terms of α_0, the behaviour of the Feller process will certainly depend on $\alpha(x)$ as $|x| \to \infty$. This explains why the short-time results for Feller processes are sharper than the long-time results. Let us, therefore, concentrate on the short-time regime.

Using the probability estimates from Sect. 5.1 and the generalized Blumenthal–Getoor indices from Definition 5.13 we can generalize [277, Theorem 4.3].

Theorem 5.16. *Let $(X_t)_{t \geq 0}$ be a d-dimensional Feller process with symbol $q(x, \xi)$. Then, $\mathbb{P}^x$-a.s.*

$$\varlimsup_{t \to 0} \frac{\sup_{0 \leq s \leq t} |X_s - x|}{t^{1/\lambda}} = 0 \qquad \forall \lambda > \beta_\infty^x \tag{5.19}$$

$$\lim_{t \to 0} \frac{\sup_{0 \leq s \leq t} |X_s - x|}{t^{1/\lambda}} = 0 \qquad \forall \lambda > \underline{\beta}_\infty^x. \tag{5.20}$$

and, if the sector condition (5.9) is satisfied,

$$\varlimsup_{t \to 0} \frac{\sup_{0 \leq s \leq t} |X_s - x|}{t^{1/\lambda}} = \infty \qquad \forall \lambda < \bar{\delta}_\infty^x \tag{5.21}$$

$$\lim_{t \to 0} \frac{\sup_{0 \leq s \leq t} |X_s - x|}{t^{1/\lambda}} = \infty \qquad \forall \lambda < \delta_\infty^x \tag{5.22}$$

where β_∞^x, $\underline{\beta}_\infty^x$, $\bar{\delta}_\infty^x$ and δ_∞^x are the generalized Blumenthal–Getoor–Pruitt indices.

Clearly, the upper and lower limits in (5.19) and (5.22) are actually limits.

Proof (of Theorem 5.16). Write $(X. - x)_t^* := \sup_{s \leq t} |X_s - x|$.

Proof of (5.19). Assume that $\lambda > \gamma > \alpha > \beta_\infty^x$. From Corollary 5.2 and the definition of the index β_∞^x we know that

$$\mathbb{P}^x\left((X.-x)_t^* > t^{1/\gamma}\right) \le ct \sup_{|x-y|\le t^{1/\gamma}} \sup_{|\eta|\le t^{-1/\gamma}} |q(y,\eta)| \le c't^{1-\frac{\alpha}{\gamma}}.$$

Now we take $t = t_k = 2^{-k}$, $k \ge 1$, and sum up the corresponding probabilities to get

$$\sum_{k=1}^\infty \mathbb{P}^x\left((X.-x)_{t_k}^* > t_k^{1/\gamma}\right) \le c' \sum_{k=1}^\infty 2^{-k(1-\frac{\alpha}{\gamma})} < \infty$$

since $\alpha < \gamma$. With the usual Borel–Cantelli argument we deduce that

$$\mathbb{P}^x\left(\varlimsup_{k\to\infty}\left\{(X.-x)_{t_k}^* > t_k^{1/\gamma}\right\}\right) = 0$$

or $(X.(\omega)-x)_{t_k}^* \le t_k^{1/\gamma}$ for all $k \ge k(\omega)$ for $\mathbb{P}_x$-a.e. ω. Fix ω and $k(\omega)$, pick $t \in (t_{k+1}, t_k]$ and observe that

$$(X.(\omega)-x)_t^* \le (X.(\omega)-x)_{t_k}^* \le t_k^{1/\gamma} \le 2^{1/\gamma}t^{1/\gamma}.$$

Consequently, we find for all $\lambda > \gamma$

$$t^{-1/\lambda}(X.-x)_t^* \le 2^{1/\gamma}t^{1/\gamma-1/\lambda} \xrightarrow[t\to 0]{} 0 \qquad (\mathbb{P}^x\text{-a.s.})$$

Proof of (5.20). Assume that $\lambda > \gamma > \beta_{-\infty}^x$. By definition of the index $\beta_{-\infty}^x$, there is a sequence $(t_k)_{k\ge 1} \subset (0,\infty)$ such that $t_k \to \infty$ as $k \to \infty$ and

$$\lim_{k\to\infty} t_k^{-\gamma} \sup_{|\eta|\le t_k} \sup_{|x-y|<1/t_k} |q(x,\eta)| = 0.$$

Again using Corollary 5.2 we get

$$\mathbb{P}^x\left((X.-x)_{t_k}^* > t_k^{-1}\right) \le ct_k^{-\gamma} \sup_{|\eta|\le t_k} \sup_{|x-y|<1/t_k} |q(x,\eta)| \xrightarrow[k\to\infty]{} 0.$$

Fatou's Lemma implies

$$0 = \lim_{k\to\infty} \mathbb{P}^x\left((X.-x)_{t_k}^* > t_k^{-1}\right) = 1 - \varliminf_{k\to\infty} \mathbb{P}^x\left((X.-x)_{t_k}^* \le t_k^{-1}\right)$$

$$\ge 1 - \mathbb{P}^x\left(\varliminf_{k\to\infty}\left\{(X.-x)_{t_k}^* \le t_k^{-1}\right\}\right).$$

This shows

$$\mathbb{P}^x \left((X. - x)^{*-\gamma}_{t_k} \leqslant t_k^{-1} \text{ for infinitely many } k \right) = 1$$

or

$$\lim_{s \to 0} \frac{(X. - x)^*_s}{s^{1/\lambda}} = \lim_{s \to 0} \frac{(X. - x)^*_s}{s^{1/\gamma}} \lim_{s \to 0} \frac{s^{1/\gamma}}{s^{1/\lambda}} = 1 \cdot 0 = 0.$$

The proofs of (5.21) and (5.22) are similar to (5.20) and (5.19), respectively. All we have to do is to use δ^x_∞, δ^x_∞ and Corollary 5.6 instead of β^x_∞, β^x_∞ and Corollary 5.2 with some obvious further changes. □

In order to describe the long-time behaviour, we need a corresponding set of generalized Blumenthal–Getoor indices *at zero*. Again we follow [277].

Definition 5.17. Let $q(x, \xi)$ be a negative definite symbol with bounded coefficients, i.e. $\sup_{x \in \mathbb{R}^d} |q(x, \xi)| \leqslant c(1 + |\xi|^2)$. The **generalized Blumenthal–Getoor–Pruitt indices** (at zero) are the numbers

$$\delta_0 := \inf \left\{ \delta > 0 : \lim_{|\xi| \to 0} \frac{\inf_{z \in \mathbb{R}^d} \inf_{|\eta| \leqslant |\xi|} \operatorname{Re} q(z, \eta)}{|\xi|^\delta} = \infty \right\},$$

$$\bar{\delta}_0 := \inf \left\{ \delta > 0 : \overline{\lim_{|\xi| \to 0}} \frac{\inf_{z \in \mathbb{R}^d} \inf_{|\eta| \leqslant |\xi|} \operatorname{Re} q(z, \eta)}{|\xi|^\delta} = \infty \right\},$$

$$\underline{\beta}_0 := \sup \left\{ \delta > 0 : \lim_{|\xi| \to 0} \frac{\sup_{z \in \mathbb{R}^d} \sup_{|\eta| \leqslant |\xi|} |q(z, \eta)|}{|\xi|^\delta} = 0 \right\},$$

$$\beta_0 := \sup \left\{ \delta > 0 : \lim_{|\xi| \to 0} \frac{\sup_{z \in \mathbb{R}^d} \sup_{|\eta| \leqslant |\xi|} |q(z, \eta)|}{|\xi|^\delta} = 0 \right\}.$$

Clearly, $\beta_0 \leqslant \underline{\beta}_0 \leqslant \delta_0$ and $\beta_0 \leqslant \bar{\delta}_0 \leqslant \delta_0$.

The long-time counterpart of Theorem 5.16 now reads as follows.

Theorem 5.18. *Let $(X_t)_{t \geqslant 0}$ be a d-dimensional Feller process with symbol $q(x, \xi)$. Then, $\mathbb{P}^x$-a.s.*

$$\overline{\lim_{t \to \infty}} \frac{\sup_{0 \leqslant s \leqslant t} |X_s - x|}{t^{1/\lambda}} = 0 \qquad \forall \lambda < \beta_0 \tag{5.23}$$

$$\lim_{t \to \infty} \frac{\sup_{0 \leqslant s \leqslant t} |X_s - x|}{t^{1/\lambda}} = 0 \qquad \forall \lambda < \underline{\beta}_0 \tag{5.24}$$

and, if the sector condition (5.9) is satisfied,

$$\overline{\lim_{t \to \infty}} \frac{\sup_{0 \leqslant s \leqslant t} |X_s - x|}{t^{1/\lambda}} = \infty \qquad \forall \lambda > \bar{\delta}_0 \tag{5.25}$$

$$\lim_{t \to \infty} \frac{\sup_{0 \leqslant s \leqslant t} |X_s - x|}{t^{1/\lambda}} = \infty \qquad \forall \lambda > \delta_0 \qquad (5.26)$$

where β_0, $\underline{\beta}_0$, $\bar{\delta}_0$ and δ_0 are the generalized Blumenthal–Getoor–Pruitt indices.

5.4   The Strong Variation of the Sample Paths

Let $p \in (0, \infty)$ and $f : [0, \infty) \to \mathbb{R}^d$ be a (non-random) càdlàg function. The **(strong) p-variation** is defined as

$$V_p(f, [0, t]) = \sup \sum_{j=0}^{m-1} |f(t_{j+1}) - f(t_j)|^p$$

where the sup is taken over all finite partitions $0 = t_0 < t_1 < \cdots < t_{m-1} < t_m = t$, $m \geqslant 1$, of the interval $[0, t]$. There are other notions of variation which are often used in connection with stochastic processes, notably the quadratic variation which is defined as a limit along a given sequence of partitions whose mesh tends to zero. By definition, the strong p-variation is the largest of all these quantities,[5] and we restrict our attention to this quantity. A thorough discussion on p-variation can be found in Dudley–Norvaiša [88, 89, 90].

Let us mention, in passing, the quadratic variation of a Feller process $(X_t)_{t \geqslant 0}$ is

$$[X, X]_t = \lim_{|\Pi| \to 0} \sum_{t_{j-1}, t_j \in \Pi} |X_{t_j} - X_{t_{j-1}}|^2 = \langle X^c, X^c \rangle_t + \sum_{s \leqslant t} |\Delta X_s|^2. \qquad (5.27)$$

Here $\Pi = \{0 \leqslant t_1 \leqslant \cdots \leqslant t_m = t\}$ is a generic partition of $[0, T]$ with mesh $|\Pi| := \max_{1 \leqslant j \leqslant m}(t_j - t_{j-1})$, the limit in (5.27) denotes convergence in probability (uniformly in compact t-sets) and along any sequence of partitions; $\langle X^c, X^c \rangle$ is the square bracket for the continuous part of the Feller process. This follows from the fact that a Feller process is a semimartingale and standard results from stochastic calculus for jump processes, e.g. Protter [243, Chap. II.6, pp. 66–67].

We will use the following result on the strong variation of Markov processes due to Manstavičius [217, Theorem 1.3] and [218, Theorem 3].

Theorem 5.19. *Let $(X_t)_{t \geqslant 0}$ be a strong Markov process with values in $\mathbb{R}^d$ and assume that there are constants $\alpha > 0$, $\beta > (3 - e)/(e - 1) \approx 0.16395$ and $C, r_0 > 0$ such that*

---

[5]By a classical result of Lévy, Brownian motion has a.s. finite quadratic variation while the strong 2-variation is a.s. infinite, see e.g. [284, Theorem 9.8, p. 143] for an elementary proof.

$$a(t,r) := \sup_{x \in \mathbb{R}^d} \sup_{0 < s \leq t} \mathbb{P}^x \left(|X_s - x| \geq r \right) \leq C t^\beta r^{-\alpha} \quad \forall t > 0, \, r \in [0, r_0). \quad (5.28)$$

Then $\mathbb{P}^x(V_p(X, [0, t]) < \infty) = 1$ for all $p > \alpha/\beta$ and $x \in \mathbb{R}^d$.

Since a Feller process $(X_t)_{t \geq 0}$ has càdlàg paths, there exists for every $T > 0$ a random variable $M_T(\omega)$ such that $\sup_{t \in [0,T]} |X_t(\omega) - X_0(\omega)| \leq M_T(\omega) < \infty$ a.s. This observation allows us to localize Manstavičius' condition.

Corollary 5.20. *Let $(X_t)_{t \geq 0}$ be a strong Markov process with values in $\mathbb{R}^d$ and assume that there are constants $\alpha, \beta, C, r_0 > 0$ such that for every compact set $K \subset \mathbb{R}^d$*

$$a(t, r, K) := \sup_{x \in K} \sup_{s \leq t} \mathbb{P}^x \left(|X_s - x| \geq r \right) \leq C_K t^\beta r^{-\alpha} \quad \forall t > 0, \, r \in [0, r_0).$$
$$(5.29)$$

Then $\mathbb{P}^x(V_p(X, [0, t]) < \infty) = 1$ for all $p > \alpha/\beta$ and $x \in \mathbb{R}^d$.

Using Theorem 5.1 and the generalized Blumenthal–Getoor–Pruitt indices as in Definition 5.13, we get the following result on the p-variation of a Feller process.

Proposition 5.21. *Let $(X_t)_{t \geq 0}$ be a Feller process with symbol $q(x, \xi)$ and denote by $\beta^* := \sup_K \beta_\infty^K$ where the supremum ranges over all compact sets $K \subset \mathbb{R}^d$. Then*

$$\mathbb{P}^x \left(V_p(X, [0, t]) < \infty \right) = 1 \quad \forall p > \beta^*, \, x \in \mathbb{R}^d, \, t > 0.$$

Proof. We want to use Corollary 5.20.

It is clear that for all $t, r > 0$ and any compact set $K \subset \mathbb{R}^d$,

$$a(t, r, K) \leq \sup_{x \in K} \mathbb{P}^x \left(\sup_{s \leq t} |X_s - x| \geq r \right).$$

Applying Corollary 5.2 yields

$$a(t, r, K) \leq c t \sup_{x \in K} \sup_{|\eta| \leq 1/r} \sup_{|y - x| \leq r} |q(y, \eta)|.$$

By the very definition of β_∞^K we see that for $p > \beta^* \geq \beta_\infty^K$ there exists some (small) $r_{0,K} > 0$ such that

$$\sup_{x \in K} \sup_{|\eta| \leq 1/r} \sup_{|y - x| \leq r} |q(y, \eta)| \leq c' r^{-p} \quad \forall r \in (0, r_{0,K}).$$

Now we can apply Corollary 5.20. $\qquad \square$

Example 5.22 (Continuation of Example 5.4). Let $(X_t)_{t \geq 0}$ be a Feller process, whose symbol is given by $q(x, \xi) = (1 + |x|^\beta)|\xi|^\alpha$ with $\alpha \in (0, 2)$ and $\beta \leq \alpha$.

Then, for any $p > \alpha$, the strong p-variation of the sample function $(X_t)_{t\geq 0}$ is almost surely finite.   □

Example 5.23. Let $(X_t)_{t\geq 0}$ be a Lévy-driven Ornstein–Uhlenbeck process, i.e. the strong solution of the following linear stochastic differential equation

$$dX_t = AX_t\, dt + dL_t, \quad X_0 = x \in \mathbb{R}^d,$$

where $A \in \mathbb{R}^{d\times d}$ is a non-trivial $d \times d$ matrix and $(L_t)_{t\geq 0}$ is a Lévy process in $\mathbb{R}^d$. The corresponding symbol is given by $q(x,\xi) = -iAx \cdot \xi + \psi(\xi)$, where $\psi(\xi)$ is the symbol of the Lévy process. Let β_∞^ψ be the upper Blumenthal–Getoor index of the Lévy process, i.e.

$$\beta_\infty^\psi = \inf\left\{\lambda > 0 : \lim_{r\to\infty} \frac{\sup_{|\xi|\leq r} |\psi(\xi)|}{r^\lambda} = 0\right\} = \inf\left\{\lambda > 0 : \lim_{|\xi|\to\infty} \frac{\psi(\xi)}{|\xi|^\lambda} = 0\right\}.$$

Then, for any $p > \beta_\infty^\psi \vee 1$, the p-variation of the process $(X_t)_{t\geq 0}$ is finite almost surely.   □

Example 5.24. Let $(L_t)_{t\geq 0}$ be an n-dimensional Lévy process with the characteristic exponent $\psi : \mathbb{R}^n \to \mathbb{C}$, and consider the following SDE

$$dX_t = \Phi(X_{t-})\, dL_t, \quad X_0 = x \in \mathbb{R}^d$$

where $\Phi : \mathbb{R}^d \to \mathbb{R}^{d\times n}$ is locally Lipschitz continuous and satisfies the linear growth condition. According to the comment following Theorem 3.8 the unique strong solution $(X_t)_{t\geq 0}$ of the SDE has the symbol $q(x,\xi) = \psi(\Phi^\top(x)\xi)$.

Let β_∞^ψ be the index from the previous Example 5.23. Then, for any $p > \beta_\infty^\psi$, the p-variation of the process $(X_t)_{t\geq 0}$ is finite almost surely.

Proof. Note that for $x, y \in \mathbb{R}^d$, $\sqrt{\psi(\xi + \eta)} \leq \sqrt{\psi(\xi)} + \sqrt{\psi(\eta)}$, cf. Proposition 2.17(c). Thus, for any compact set $K \subset \mathbb{R}^d$,

$$\sup_{x\in K} \sup_{|\xi|\leq 1/r} \sup_{|y-x|\leq r} |q(y,\xi)| = \sup_{x\in K} \sup_{|\xi|\leq 1/r} \sup_{|y-x|\leq r} |\psi(\Phi^\top(y)\xi)|$$

$$\leq 2\sup_{x\in K} \sup_{|\xi|\leq 1/r} \sup_{|y-x|\leq r} |\psi(\Phi^\top(x)\xi)|$$

$$+ 2\sup_{x\in K} \sup_{|\xi|\leq 1/r} \sup_{|y-x|\leq r} |\psi([\Phi^\top(y) - \Phi^\top(x)]\xi)|.$$

Since $\Phi(x)$ is locally Lipschitz continuous we find for every compact set K some constant $L_K > 0$ such that for all

$$|\Phi(x) - \Phi(y)| \leq L_K|x - y| \quad \forall x \in K,\ y \in \mathbb{R}^d,\ |x - y| \leq r < 1.$$

Thus, for all $|\xi| \leqslant 1/r$ and $x \in K$, $y \in \mathbb{R}^d$ with $|y - x| \leqslant r$,

$$|[\Phi^\top(y) - \Phi^\top(x)]\xi| \leqslant |\Phi(y) - \Phi(x)| \cdot |\xi| \leqslant L_K |y - x| \cdot |\xi| \leqslant L_K,$$

and so

$$\sup_{x \in K} \sup_{|\xi| \leqslant 1/r} \sup_{|y-x| \leqslant r} |\psi\left([\Phi^\top(y) - \Phi^\top(x)]\xi\right)| \leqslant \sup_{|\eta| \leqslant L_K} |\psi(\eta)|.$$

Since Φ is locally bounded, $\sup_{x \in K} |\Phi(x)| = M_K < \infty$ which shows that for all $x \in K$ and $|\xi| \leqslant 1/r$ we have $|\Phi^\top(x)\xi| \leqslant M_K r^{-1}$. Hence,

$$\sup_{x \in K} \sup_{|\xi| \leqslant 1/r} \sup_{|y-x| \leqslant r} |\psi(\Phi^\top(x)\xi)| \leqslant \sup_{|\eta| \leqslant M_K/r} |\psi(\eta)|.$$

Combining all estimates, we arrive at

$$\sup_{x \in K} \sup_{|\xi| \leqslant 1/r} \sup_{|y-x| \leqslant r} |q(y, \xi)| \leqslant 2 \sup_{|\eta| \leqslant M_K/r} |\psi(\eta)| + 2 \sup_{|\eta| \leqslant L_K} |\psi(\eta)|.$$

The conclusion follows from the definition of β_∞^ψ and Proposition 5.21. $\qquad\square$

5.5 Besov Regularity of Feller Processes

Describing the smoothness of functions is a central problem in analysis. Typically, one measures smoothness in terms of integrability and differentiability properties of functions. Take, as an example, the classical Sobolev spaces. By definition, $u \in W^{k,p}(\mathbb{R}^d)$ if, and only if, $u \in \mathcal{S}'(\mathbb{R}^d)$ such that the distributional derivatives $\nabla^\alpha u \in L^p(\mathbb{R}^d)$ for all multiindices $0 \leqslant |\alpha| \leqslant k$. By Sobolev's embedding theorem $W^{k,p}(\mathbb{R}^d) \hookrightarrow C_\infty^m(\mathbb{R}^d)$ whenever $k > d/p + m$, i.e. we get the existence of classical derivatives; if we *fill the gaps* between the integers $k = 0, 1, 2, \ldots$, we get *fractional* smoothness in the Hölder scale. Technically, this amounts to interpolation between $L^p(\mathbb{R}^d)$ and $W^{k,p}(\mathbb{R}^d)$, and this opens up the possibility to measure the smoothness of functions which have jumps, e.g. càdlàg functions which appear as trajectories of stochastic processes. The interpolation spaces which appear here are the Besov $B_{p,q}^s(\mathbb{R}^d)$ and Triebel–Lizorkin $F_{p,q}^s(\mathbb{R}^d)$ scales.

There is a huge literature on function spaces and we refer summarily to the work of Triebel [321, 322, 323, 324] for a general overview. Function spaces on domains are especially treated in Jonsson–Wallin [171], function with extreme indices are covered by Runst–Sickel [260]. For a non-technical survey we recommend the chapter *How to measure smoothness* in Triebel [322, Chap. 1, pp. 1–86] and [324, Chap. 1, pp. 1–125] for recent developments.

It is interesting to note that function spaces and embeddings are at the very origin of the theory of stochastic processes: Wiener constructed "his" Brownian motion on the space $\Omega = \{\omega : [0,\infty) \to \mathbb{R}^d : \omega(0) = 0,\ \omega \text{ continuous}\}$ and established, later on, the embedding of the Brownian paths into the Hölder continuous functions. The Kolmogoroff–Chentsov theorem is a Sobolev embedding theorem in disguise, see [280], and the paper Garsia–Rodemich–Rumsey [121] contains a sleek improvement of many Sobolev inequalities but seems to be only known by Gauss process aficionados.

The first papers which treat Besov regularity of stochastic processes were Roynette [257] (Brownian Motion), Ciesielski–Kerkyacharian–Roynette [69] (Gaussian processes and symmetric α-stable Lévy processes), Herren [133] (Lévy processes) and, for Feller processes, [273].

As usual, we write

$$\sigma_p := d(p^{-1} - 1)^+ \quad \text{and} \quad \sigma_{p,q} := d \max\{p^{-1} - 1,\ q^{-1} - 1,\ 0\}$$

and denote by Δ_h^M the M-fold iterated difference of step h where

$$\Delta_h^1 u(x) = \Delta_h u(x) := u(x + h) - u(x) \quad \text{and} \quad \Delta_h^0 u = u.$$

Definition 5.25. Let $1 < p, q \leq \infty$ and set for $u \in L^{p \vee 1}(\mathbb{R}^d)$

$$\|u\|_{B_{p,q}^s} := \|u\|_{L^p} + \left(\int_{|h|<\epsilon} |h|^{-sq} \left\| \Delta_h^M u \right\|_{L^p}^q \frac{dh}{|h|^d} \right)^{\frac{1}{q}} \quad \forall \epsilon > 0 \tag{5.30}$$

if $\sigma_p < s < M$, and

$$\|u\|_{F_{p,q}^s} := \|u\|_{L^p} + \left\| \left[\int_0^1 t^{-sq} \left(\frac{1}{t^d} \int_{|h|<t} |\Delta_h^M u(x)|\, dh \right)^q \frac{dt}{t} \right]^{\frac{1}{q}} \right\|_{L^p} \tag{5.31}$$

if $\sigma_{p,q} < s < M$; the usual modifications apply if $p = \infty$ or $q = \infty$. The **Besov space** $B_{p,q}^s$ and **Triebel–Lizorkin space** $F_{p,q}^s$ are defined as

$$B_{p,q}^s(\mathbb{R}^d) = \left\{ u \in L^{p \vee 1}(\mathbb{R}^d) : \|u\|_{B_{p,q}^s} < \infty \right\},$$

$$F_{p,q}^s(\mathbb{R}^d) = \left\{ u \in L^{p \vee 1}(\mathbb{R}^d) : \|u\|_{F_{p,q}^s} < \infty \right\}.$$

Note that (5.30) gives a family of equivalent (quasi-)norms,[6] i.e. the definition of $B_{p,q}^s(\mathbb{R}^d)$ does not depend on the parameter $\epsilon > 0$. The spaces $B_{p,q}^s(\mathbb{R}^d)$ and

---

[6]*Quasi* indicates that the triangle inequality only holds with a constant: $\|u + v\| \leq c(\|u\| + \|v\|)$. A **quasi-Banach space** is a complete quasi-normed vector space. Typical examples are the spaces L^p with $0 < p < 1$.

$F^s_{p,q}(\mathbb{R}^d)$ are Banach spaces if $p, q \in [1, \infty]$ and we have quasi-Banach spaces if p or q is less than 1.

There are various characterizations of the B- and F-scales which work for *all* indices $p, q \in (0, \infty]$ and $s \in \mathbb{R}$; for our purposes the range covered by Definition 5.25 is more than enough.

Remark 5.26. All spaces appearing below are spaces of functions on $\mathbb{R}^d$ and, for brevity, we will suppress the $\mathbb{R}^d$ in the notation.

a) Besov- and Triebel–Lizorkin scales accommodate many of the classical function spaces, cf. Runst–Sickel [260, Sect. 2.1.2, pp. 11–14]. For example for $s > 0$, $m = 1, 2, \ldots$ and $p, q \in (0, \infty)$ we have

$$B^s_{p,p} = F^s_{p,p} = W^{s,p}, \qquad F^m_{p,2} = W^{m,p} = H^{m,p},$$

$$B^s_{\infty,\infty} = \mathscr{C}^s, \qquad F^0_{p,2} = L^p, \qquad F^s_{p,2} = H^{s,p}$$

where W are the Sobolev(–Slobodeckij) spaces, H are the Bessel potential spaces and $\mathscr{C}^s$ are the Zygmund spaces (which coincide with the Hölder spaces C^s whenever s is not an integer).

b) Let $\epsilon > 0$, $s > 0$, $p, q \in (0, \infty]$. We have the following continuous embeddings, cf. Runst–Sickel [260, Sect. 2.2, pp. 29–32].

$$A^s_{p,q_0} \hookrightarrow A^s_{p,q_1} \quad (\forall q_0 \leqslant q_1) \quad \text{and} \quad A^{s+\epsilon}_{p,\infty} \hookrightarrow A^s_{p,q} \qquad (5.32)$$

where A stands for B or F, respectively. Moreover,

$$B^s_{p,u} \hookrightarrow F^s_{p,q} \hookrightarrow B^s_{p,w} \iff 0 < u \leqslant p \wedge q \leqslant p \vee q \leqslant w \leqslant \infty, \qquad (5.33)$$

and for $0 < p_0 < p < p_1 \leqslant \infty$, $u, w \geqslant 0$,

$$B^{s_0}_{p_0,u} \hookrightarrow F^s_{p,q} \hookrightarrow B^{s_1}_{p_1,w} \quad \text{if} \quad s_0 - \frac{d}{p_0} = s - \frac{d}{p} = s_1 - \frac{d}{p_1}. \qquad (5.34)$$

c) We have the following Sobolev-type embeddings, cf. Runst–Sickel [260, Sect. 2.2.4, pp. 32–33],

$$B^s_{p,q} \hookrightarrow C \iff \begin{cases} sp > d, \\ sp = d \text{ and } 0 < q \leqslant 1, \end{cases} \qquad (5.35)$$

$$F^s_{p,q} \hookrightarrow C \iff \begin{cases} sp > d, \\ sp = d \text{ and } 0 < p \leqslant 1. \end{cases} \qquad (5.36)$$

If we combine the definition of the Besov- and Triebel–Lizorkin spaces with our maximal estimates from Sect. 5.1 and the Blumenthal–Getoor–Pruitt indices, cf.

Definition 5.13, we can show the following generalization of [279, Theorem 4.2 and Theorem 5.2]. Note that we do not assume that the generator has bounded coefficients, cf. Theorem 2.31.[7]

Theorem 5.27. *Let $(X_t)_{t \geqslant 0}$ be a Feller process with the negative definite symbol $q(x, \xi)$ such that $q(x, 0) = 0$. For every $x \in \mathbb{R}^d$*

$$[t \mapsto X_{t \vee 0}] \in B^{s,\mathrm{loc}}_{p,q}(\mathbb{R}) \quad \mathbb{P}^x\text{-}a.s.[8]$$

whenever

$$q \in (0, \infty), \quad s > \max\{0, p^{-1} - 1\} \quad and \quad s \cdot \sup_{K \subset \mathbb{R}^d, \, cpt.} \{p, q, \beta^K_\infty\} < 1$$

$$or \quad q = \infty, \quad s > \max\{0, p^{-1} - 1\} \quad and \quad s \cdot \sup_{K \subset \mathbb{R}^d, \, cpt.} \{p, \beta^K_\infty\} < 1.$$

Similarly,

$$[t \mapsto X_{t \vee 0}] \in F^{s,\mathrm{loc}}_{p,q}(\mathbb{R}) \quad \mathbb{P}^x\text{-}a.s.[8]$$

whenever

$$q \in (0, \infty), \quad s > \max\{0, p^{-1} - 1, q^{-1} - 1\}) \quad and \quad s \cdot \sup_{K \subset \mathbb{R}^d, \, cpt.} \{p, q, \beta^K_\infty\} < 1$$

$$or \quad q = \infty, \quad s > \max\{0, p^{-1} - 1\} \quad and \quad s \cdot \sup_{K \subset \mathbb{R}^d, \, cpt.} \{p, \beta^K_\infty\} < 1.$$

We can use the Sobolev embedding Remark 5.26(c) to show that certain trajectories do not appear in spaces of the B- and F-scales. For this it is important to make sure that the Feller process does have jumps. For this we need that the jump-part of the symbol is uniformly (in x) non-degenerate. If $(X_t)_{t \geqslant 0}$ is a Feller process with symbol $q(x, \xi)$, then write $p(x, \xi)$ for the jump part,

$$p(x, \xi) = \int_{\mathbb{R}^d \setminus \{0\}} \left(1 - e^{iy \cdot \xi} + i\xi \cdot y \chi(|y|)\right) N(x, dy),$$

cf. (2.32). If $\inf_x \operatorname{Re} p(x, \xi_0) > 0$ for some $\xi_0 \in \mathbb{R}^d$, then the Feller process has a.s. jumps, cf. [279, Corollary 6.4]. Thus, cf. [279, Theorem 6.5], we arrive at:

---

[7]In [279] weighted Besov- and Triebel–Lizorkin spaces are considered which allow to treat the *whole* trajectory on $[0, \infty)$. Here we confine ourselves to local assertions and for $t \in [0, 1]$, say. Since we do not have bounded coefficients, we have to use the indices β^K_∞ instead of β^x_∞.

[8]As usual, the exponent "loc" indicates that the function is **locally**, that is when multiplied with a test function $\chi \in C^\infty_c(\mathbb{R}^d)$, in the function space.

Corollary 5.28. *Let $(X_t)_{t \geqslant 0}$ be a Feller process with the negative definite symbol $q(x, \xi)$ such that $q(x, 0) = 0$ and with uniformly non-degenerate jump part $p(x, \xi)$. Then we have for all $x \in \mathbb{R}^d$ and $s > 0$, $p \in (0, \infty)$, $q \in (0, \infty]$ and any test function $\phi \in C_c^\infty(\mathbb{R})$*

$$\mathbb{P}^x \left([t \mapsto \phi(t) X_{t \vee 0}] \notin B_{p,q}^s(\mathbb{R}) \right) = 1 \quad \text{if } sp > 1 \text{ or } sp = 1, \ q \in (0, 1],$$

$$\mathbb{P}^x \left([t \mapsto \phi(t) X_{t \vee 0}] \notin F_{p,q}^s(\mathbb{R}) \right) = 1 \quad \text{if } sp > 1 \text{ or } sp = 1, \ p \in (0, 1].$$

Remark 5.29. Theorem 5.27 covers all earlier results on Lévy and Feller processes from Ciesielski–Kerkyacharian–Roynette [69], Herren [133] and Schilling [273]. These papers rely on the characterization of Besov spaces by atoms and wavelets. Since this method might be interesting for other processes, let us briefly explain this characterization by wavelets (the characterization by atoms is very similar, cf. [273, Theorem 3.3] and the references given there). For details on the following result we refer to Triebel [324, Sect. 1.7.3, pp. 30–33]. To keep notation simple, we use $a_{p,q}^s$, $A_{p,q}^s$ to denote either the Besov $b_{p,q}^s$, $B_{p,q}^s$ or the Triebel–Lizorkin $f_{p,q}^s$, $F_{p,q}^s$ scales. For a double sequence $\lambda = (\lambda_{j,m})_{j \geqslant 0, m \in \mathbb{Z}}$ we set

$$\| \lambda \|_{b_{p,q}^s}^q := \sum_{j=0}^\infty 2^{j(s - \frac{1}{p})q} \left(\sum_{m=-\infty}^\infty |\lambda_{j,m}|^p \right)^{q/p},$$

$$\| \lambda \|_{f_{p,q}^s} := \left\| \left(\sum_{j \geqslant 0, \, m \in \mathbb{Z}} 2^{jsq} |\lambda_{j,m} \mathbb{1}_{[(m-1)2^{-j}, \, (m+1)2^{-j}]}(\cdot)|^q \right)^{1/q} \right\|_{L^p(\mathbb{R})}$$

(if $p = \infty$ or $q = \infty$ we replace, as usual, the ℓ^p, L^p etc. norm(s) by the (essential) supremum norm for ℓ^∞, L^∞ etc.) Let $(\Psi_{j,m})_{j \geqslant 0, m \in \mathbb{Z}}$ be the one-dimensional Daubechies wavelets of smoothness $k \in \mathbb{N}$ where

$$k > \max \left\{ s, \ \tfrac{2}{p} + \tfrac{1}{2} - s \right\} \qquad \text{for the } B\text{-scale}$$

$$k > \max \left\{ s, \ \tfrac{2}{p \wedge q} + \tfrac{1}{2} - s \right\} \qquad \text{for the } F\text{-scale.}$$

Then a tempered distribution $u \in \mathcal{S}'(\mathbb{R})$ is in the space $A_{p,q}^s$ with $0 < p, q \leqslant \infty$ and $s \in \mathbb{R}$ if, and only if,

$$u = \sum_{j \geqslant 0, m \in \mathbb{Z}} \lambda_{j,m} 2^{-j/2} \Psi_{j,m} \quad \text{and} \quad \| \lambda \|_{a_{p,q}^s} < \infty.$$

This representation is unique and we have $\lambda_{j,m} = 2^{j/2} \langle u, \Psi_{j,m} \rangle$ where $\langle \cdot, \cdot \rangle$ means the application of the distribution to the wavelet.

When dealing with paths of stochastic processes, $u(t) = X_{t \vee 0}(\omega)$ will be a (random) càdlàg *function*, and all we have to do is to check the finiteness of the expression $\|\lambda\|_{a_{p,q}^s}$. This can be done by (uniform) probability estimates since we know the structure of the coefficients $\lambda_{j,m}$. This approach (for atoms rather than for wavelets) can be found in [273], modelled on the presentation in Ciesielski–Kerkyacharian–Roynette [69].   □

We close this section with a novel characterization of the Besov spaces $B_{p,q}^s([0,1])$ in terms of dyadic p-variation sums. Let

$$V_p^j(u,[0,1]) := \sum_{k=1}^{2^j} \left| u(k2^{-j}) - u((k-1)2^{-j}) \right|^p.$$

Then we have the following result by Rosenbaum [256].

Theorem 5.30. *Let* $1/p < s < 1$ *and* $1 \leq p, q \leq \infty$. *Then* $u \in B_{p,q}^s([0,1])$ *if, and only if,*

$$\|u\|^q := |u(0)| \vee \sum_{j=0}^{\infty} 2^{-jq(\frac{1}{p}-s)|V_p^j(u,[0,1])|^{q/p}} < \infty.$$

Moreover, if $u : [0,1] \to \mathbb{R}$ *is a Borel function and* $0 < p < \infty$, *then* $u \in B_{p,\infty}^{1/p}([0,1])$, *if the strong* p-*variation* $V_p(u,[0,1]) < \infty$ *is finite.*

Combining Proposition 5.21 with Theorem 5.30, we get the following partial improvement of Theorem 5.27 for the Besov scale.

Proposition 5.31. *Let* $\beta^* := \sup_K \beta_\infty^K$ *be the supremum (taken over all compact sets* $K \subset \mathbb{R}^d$) *of the generalized Blumenthal–Getoor–Pruitt index* β_∞^K, *cf. Definition 5.13 and Proposition 5.21. Then for all* $p > \beta^*$

$$t \mapsto X(t,\omega)\big|_{[0,1]} \in B_{p,\infty}^{1/p}([0,1]) \qquad \mathbb{P}^x\text{-}a.s.$$

Chapter 6
Global Properties

The sample path properties in Chap. 5 describe the behaviour of the stochastic process *locally*. We will now turn our attention to *global* properties of the process and its semigroup; following Fukushima–Oshima–Takeda [118] we mean by *global* behaviour of a process questions like transience, recurrence, irreducibility, functional inequalities and so on. When studying such properties, one often needs not only Feller but also L^2-semigroups. We refer, in particular, to Sect. 1.5 for situations when a Feller semigroup is also an L^p-semigroup. Unless otherwise stated, we work again with $E = \mathbb{R}^d$.

6.1 Functional Inequalities

Functional inequalities are powerful and efficient tools to analyse Markov semigroups and their generators, see e.g. Wang [340] for the general theory of functional inequalities and their applications or Saloff-Coste [262]. The more general group setting is treated in Varopoulos–Saloff-Coste–Coulhon [332]. Let us briefly mention some well-known facts.

(a) The Nash and Sobolev inequalities correspond to uniform heat kernel upper bounds of the semigroup, see Davies [79].
(b) The log-Sobolev inequality is equivalent to Nelson's hypercontractivity (Nelson [233]) of the semigroup, see Gross [127].
(c) The super log-Sobolev inequality (also known as the log-Sobolev inequality with parameter) is equivalent to the supercontractivity, and in some cases it implies the ultracontractivity of the semigroup, see Davies–Simon [81].
(d) The Poincaré inequality is equivalent to the exponential convergence of the semigroup, see Chen [59].
(e) Slower than exponential convergence rates are characterized by the weak Poincaré inequality, see Röckner–Wang [251] and Wang [338].

B. Böttcher et al., *Lévy Matters III*, Lecture Notes
in Mathematics 2099, DOI 10.1007/978-3-319-02684-8_6,
© Springer International Publishing Switzerland 2013

(f) The super Poincaré inequality is equivalent to the uniform integrability of the
    semigroup, as well as to the absence of the essential spectrum of the generator
    if the semigroup has an asymptotic density, see Wang and co-authors [336, 337,
    125, 339].

In order to establish functional inequalities, many explicit criteria are known for
diffusion processes and Markov chains, but only few results are available for jump
processes such as Lévy- and Lévy-type processes. Using Bochner's subordination,
cf. Sect. 4.2, we may deduce functional inequalities for certain jump processes from
those for diffusion processes, see Bendikov–Maheux [19], Wang [341], Gentil–
Maheux [122] and [289], [293, Chap. 13.3, pp. 233–242]. The following result from
[289, Theorem 1], see also [293, Theorem 13.41, p. 236] for an abstract version,
shows that certain functional inequalities are preserved under subordination. Let
$(E, \mathscr{B}(E), m)$ be a measure space with a σ-finite measure m with full topological
support. We write $\langle \cdot, \cdot \rangle_{L^2}$ and $\| \cdot \|_{L^2}$ for the scalar product and norm in $L^2(E, m)$,
respectively; $\| \cdot \|_{L^1}$ denotes the norm in $L^1(E, m)$.

Theorem 6.1. *Let $(T_t)_{t \geq 0}$ be a strongly continuous contraction semigroup of sym-*
metric operators on $L^2(E, m)$ and assume that for each $t \geq 0$, $T_t|_{L^2(E,m) \cap L^1(E,m)}$
has an extension which is a contraction on $L^1(E, m)$, i.e. $\|T_t u\|_{L^1} \leq \|u\|_{L^1}$ for all
$u \in L^1(E, m) \cap L^2(E, m)$. Suppose that the generator $(A, \mathcal{D}(A))$ of $(T_t)_{t \geq 0}$ satisfies
the following Nash-type inequality

$$\|u\|_{L^2}^2 \, B \left(\|u\|_{L^2}^2 \right) \leq \langle -Au, u \rangle_{L^2}, \quad \forall u \in \mathcal{D}(A), \; \|u\|_{L^1} = 1,$$

where $B : (0, \infty) \to (0, \infty)$ is some measurable, increasing function. Then, for any
Bernstein function f, the generator $f(-A)$ of the subordinate semigroup satisfies

$$\tfrac{1}{2} \|u\|_{L^2}^2 \, f \left(B \left(\tfrac{1}{2} \|u\|_{L^2}^2 \right) \right) \leq \langle f(-A)u, u \rangle_{L^2}, \quad \forall u \in \mathcal{D}(f(A)), \; \|u\|_{L^1} = 1.$$

For the Bernstein functions $f(\lambda) = \lambda^\alpha$, $0 < \alpha < 1$, and the fractional powers
$(-A)^\alpha$ the result of Theorem 6.1 is due to Bendikov–Maheux [19, Theorem 1.3].
Our result is valid for *all* Bernstein functions, hence, for *all subordinate generators*
$f(-A)$. Theorem 6.1 has several applications, e.g. the super Poincaré inequality
and the weak Poincaré inequality for subordinate semigroups and the hyper-, super-
and ultracontractivity of subordinate semigroups, see [289, Sect. 4]. Let us also
mention that there is a corresponding version for non-symmetric semigroups, cf.
[289, Theorem 2] and [293, Theorem 13.41, p. 236].

In general it is wrong, and at best extremely difficult to verify, that a Lévy-
type or Feller process is subordinate to some diffusion process; take, for instance,
an Ornstein–Uhlenbeck process driven by a symmetric α-stable Lévy process.
Therefore, it is necessary to provide general criteria for functional inequalities of
Lévy-type processes. Let us explain the method with a basic example and consider
an Ornstein–Uhlenbeck process driven by an symmetric α-stable Lévy process. Let
Δ be the Laplacian on $\mathbb{R}^d$ and set

$$L_\alpha u(x) := -(-\Delta)^{\alpha/2} u(x) - x \cdot \nabla u(x), \quad x \in \mathbb{R}^d,$$

with $0 < \alpha < 2$. The associated sub-Markov semigroup has a unique invariant (but not reversible) probability measure μ_α, cf. Albeverio–Rüdiger–Wu [4], which can be identified by its inverse Fourier transformation

$$\tilde{\mu}_\alpha(\xi) := \int_{\mathbb{R}^d} e^{i\langle x,\xi\rangle} \mu_\alpha(dx) = e^{-\frac{1}{\alpha}|\xi|^\alpha}, \quad \xi \in \mathbb{R}^d.$$

For any $u \in C_c^\infty(\mathbb{R}^d)$, see Lescot–Röckner [202, Proposition 4.1] or Röckner–Wang [252, (1.9)],

$$D_\alpha(u,u) := - \int_{\mathbb{R}^d} u L_\alpha u \, d\mu_\alpha = \frac{1}{2} \iint_{\mathbb{R}^d \times \mathbb{R}^d} \frac{|u(x) - u(y)|^2}{|x - y|^{d+\alpha}} \, dy \, \mu_\alpha(dx),$$

$$\mathcal{D}(D_\alpha) := \{ u \in L^2(\mu_\alpha) : D_\alpha(u,u) < \infty \}. \tag{6.1}$$

According to Röckner–Wang [252, Example 3.2(2)], the semigroup $(P_t^\alpha)_{t\geq 0}$ generated by L_α is not hypercontractive, i.e.

$$\| P_t^\alpha \|_{L^p(\mu_\alpha) \to L^q(\mu_\alpha)} = \infty \quad \forall t > 0, \, q > p \geq 1.$$

Therefore, the log-Sobolev inequality does not hold for D_α. In fact, since we have $\mu_\alpha(dx) = m_\alpha(x) \, dx$ with

$$\frac{1}{c(1 + |x|^2)^{(d+\alpha)/2}} \leq m_\alpha(x) \leq \frac{c}{(1 + |x|^2)^{(d+\alpha)/2}} \quad \forall x \in \mathbb{R}^d$$

for some constant $c > 1$, see e.g. Blumenthal–Getoor [31, Theorem 2.1] or Chen [60, (1.5)], Corollary 6.3 below provides an even stronger statement: The super Poincaré inequality is not available either. Recall that the log-Sobolev inequality, cf. Gross [127],

$$\int u^2 \log u^2 \, d\mu_\alpha \leq C D_\alpha(u,u), \quad \forall u \in \mathcal{D}(D_\alpha), \int u^2 \, d\mu_\alpha = 1 \tag{6.2}$$

holds for some constant $C > 0$ if, and only if, the super Poincaré inequality

$$\int u^2 \, d\mu_\alpha \leq r D_\alpha(u,u) + e^{c(1+r^{-1})} \left(\int |u| \, d\mu_\alpha \right)^2 \quad \forall u \in \mathcal{D}(D_\alpha) \tag{6.3}$$

holds for all $r > 0$ and some constant $c > 0$, see Wang [340, Theorem 3.3.13]. On the other hand, Corollary 6.3 below also implies that the Poincaré inequality

$$\int u^2 \, d\mu_\alpha \leq C D_\alpha(u,u) \quad \forall u \in \mathcal{D}(D_\alpha), \int u \, d\mu_\alpha = 0 \tag{6.4}$$

holds for some constant $C > 0$.

This example is typical for jump processes: Poincaré's inequality is, in general, the best we can hope for.

Let us now generalize (6.1) and consider

$$D_{\rho,V}(u,w) := \iint_{\mathbb{R}^d \times \mathbb{R}^d} (u(x) - u(y))(w(x) - w(y))\rho(|x - y|)\, dy\, \mu_V(dx), \tag{6.5}$$

$$\mathcal{D}(D_{\rho,V}) := \{u \in L^2(\mu_V) : D_{\rho,V}(u,u) < \infty\},$$

where $\rho : (0, \infty) \to (0, \infty)$ is measurable with $\int_{(0,\infty)} (1 \wedge r^2)\rho(r)r^{d-1}\, dr < \infty$, and $V : \mathbb{R}^d \to \mathbb{R}$ is a measurable function such that

$$\mu_V(dx) := \frac{1}{\int_{\mathbb{R}^d} e^{-V(x)}\, dx}\, e^{-V(x)}\, dx$$

is a probability measure. Then $(D_{\rho,V}, \mathcal{D}(D_{\rho,V}))$ is a symmetric Dirichlet form on $L^2(\mu_V)$, see e.g. Fukushima–Oshima–Takeda [118, Example 1.2.4, pp. 13–15]. The following result is taken from [348, Theorem 1].

Theorem 6.2.   a) *If there exists a constant $c > 0$ such that for any $x, y \in \mathbb{R}^d$ with $x \neq y$,*

$$\left(e^{V(x)} + e^{V(y)}\right)\rho(|x - y|) \geq c, \tag{6.6}$$

then the following Poincaré inequality

$$\int \left(u - \int u\, d\mu_V\right)^2 d\mu_V \leq c^{-1} D_{\rho,V}(u,u), \quad u \in \mathcal{D}(D_{\rho,V}), \tag{6.7}$$

holds.

b) *For all probability measures μ_V, the following weak Poincaré inequality*

$$\int \left(u - \int u\, d\mu_V\right)^2 d\mu_V \leq \alpha(r) D_{\rho,V}(u,u) + r\|u\|_\infty^2, \quad u \in \mathcal{D}(D_{\rho,V}), \tag{6.8}$$

holds for all $r > 0$ with $\alpha(r)$ given by

$$\inf\left\{ \frac{1}{\inf_{0<|x-y|\leq s}\left[(e^{V(x)} + e^{V(y)})\rho(|x - y|)\right]} \right| \iint_{|x-y|>s} \mu_V(dy)\,\mu_V(dx) \leq \frac{r}{2},\ s > 0 \right\}.$$

c) *Suppose that there is a locally bounded measurable function $w : \mathbb{R}^d \to [0, \infty)$ such that*

$$\lim_{|x| \to \infty} w(x) = \infty,$$

and for any $x, y \in \mathbb{R}^d$ with $x \neq y$,

$$e^{V(x)} + e^{V(y)} \geq \frac{w(x) + w(y)}{\rho(|x - y|)}.$$

Then the following super Poincaré inequality

$$\int u^2 \, d\mu_V \leq r D_{\rho,V}(u, u) + \beta(r) \mu_V(|u|)^2, \quad u \in \mathcal{D}(D_{\rho,V}), \qquad (6.9)$$

holds for all $r > 0$ with

$$\beta(r) = \inf \left\{ \frac{2 \int w \, d\mu_V}{\inf\limits_{|x| \geq t} w(x)} + \beta_t(t \wedge s) \; \middle| \; \frac{2}{\inf_{|x| \geq t} w(x)} + s \leq r \text{ and } t, s > 0 \right\},$$

where for every $t > 0$ the function $\beta_t(s)$ is of the form

$$\inf \left\{ \frac{2 \sup_{|z| \leq 2t} e^{2V(z)}}{\operatorname{Leb}(\mathbb{B}(0, u)) \inf\limits_{|z| \leq t} e^{V(z)}} \; \middle| \; \frac{2 \sup_{\epsilon \leq u} \frac{1}{\rho(\epsilon)} \sup_{|z| \leq 2t} e^{V(z)}}{\operatorname{Leb}(\mathbb{B}(0, u)) \inf\limits_{|z| \leq t} e^{V(z)}} \leq s \text{ and } u > 0 \right\}.$$

To get an idea of the proof of Theorem 6.2, we sketch the argument for part (a).

Proof (of Theorem 6.2(a)). For $u \in \mathcal{D}(D_{\rho,V})$ we have

$$\frac{1}{2} \iint (u(x) - u(y))^2 \, \mu_V(dy) \, \mu_V(dx)$$

$$= \frac{1}{2} \iint \left(u^2(x) + u^2(y) - 2u(x)u(y) \right) \mu_V(dy) \, \mu_V(dx)$$

$$= \int u^2 \, d\mu_V - \left(\int u \, d\mu_V \right)^2$$

$$= \int \left(u - \int u \, d\mu_V \right)^2 \, d\mu_V.$$

Furthermore, we find that

$$\frac{1}{2} \iint (u(x) - u(y))^2 \, \mu_V(dy) \, \mu_V(dx)$$

$$= \frac{1}{2} \iint_{x \neq y} (u(x) - u(y))^2 \, \rho(|x-y|) \rho(|x-y|)^{-1} \, \mu_V(dy) \, \mu_V(dx)$$

$$\leqslant c^{-1} \iint_{x \neq y} (u(x) - u(y))^2 \, \rho(|x-y|) \, \frac{e^{-V(x)} + e^{-V(y)}}{2} \, dy \, dx$$

$$= c^{-1} D_{\rho, V}(u, u),$$

where we have used (6.6)

$$\frac{e^{-V(x) - V(y)}}{(e^{-V(x)} + e^{-V(y)}) \rho(|x-y|)} \leqslant c^{-1}. \qquad \qquad \Box$$

Using Theorem 6.2, we can get the following statement for the Dirichlet form $(D_\alpha, \mathcal{D}(D_\alpha))$.

Corollary 6.3. *Let* $\mu_V(dx) = e^{-V(x)} dx := C_\epsilon (1 + |x|)^{-(d+\epsilon)} dx$ *with* $\epsilon > 0$, *and* $\rho(r) = r^{-(d+\alpha)}$ *with* $\alpha \in (0, 2)$. *Then the following assertions hold:*

a) *The Poincaré inequality (6.7) holds if, and only if,* $\epsilon \geqslant \alpha$.
b) *If* $0 < \epsilon < \alpha$, *then the weak Poincaré inequality (6.8) holds with*

$$\alpha(r) = c_1 \left(1 + r^{-(\alpha-\epsilon)/\epsilon}\right)$$

for some constant $c_1 > 0$.
c) *The super Poincaré inequality (6.9) holds for some* $\beta : (0, \infty) \to (0, \infty)$ *if, and only if,* $\epsilon > \alpha$; *in this case*

$$\beta(r) = c_2 \left(1 + r^{-\frac{d}{\alpha} - \frac{(d+\epsilon)(d+2\alpha)}{\alpha(\epsilon-\alpha)}}\right)$$

for some constant $c_2 > 0$.

Using techniques from harmonic analysis, a sufficient condition for the Poincaré inequality for $(D_{\alpha,\delta}, \mathcal{D}(D_{\alpha,\delta}))$, cf. (6.1) with $\rho(r) = r^{-d-\alpha} e^{-\delta r}$, $\alpha \in (0, 2)$ and $\delta > 0$, has been proved by Mouhot–Russ–Sire [229]. Using the generator of the Dirichlet form $(D_{\rho,V}, \mathcal{D}(D_{\rho,V}))$ and a proper Lyapunov function explicit criteria for functional inequalities of the non-local Dirichlet form $(D_{\rho,V}, \mathcal{D}(D_{\rho,V}))$ have been obtained in [351, 63, 64, 348].

6.2 Coupling Methods

In this section we discuss recent results on the coupling of Lévy and Ornstein–Uhlenbeck processes, cf. [48, 292, 287, 289]. As an important tool in modern probability theory, the coupling method is primarily used in order to estimate total variation distances. Recently it became useful to prove ergodicity of stochastic processes, to get the rate of convergence towards the invariant measure, and to obtain the regularity properties for the transition semigroups, e.g. gradient estimates and Harnack inequalities. It has also turned out to be highly successful in connection with renewal processes and particle systems. Results on coupling for continuous-time jump processes are quite recent and, to the best of our knowledge, there are only theorems for Lévy processes and Lévy-driven SDEs.

There are three different ways to study the coupling property of Lévy processes, and we will present them separately. Then we discuss the coupling property for Ornstein–Uhlenbeck processes and stochastic differential equations driven a Lévy process. Our standard references for various types of coupling are Lindvall [204] and Thorisson [320] for the classical theory, and Chen [59] for applications to Markov processes.

Let us briefly recall the definition of the coupling property for Markov processes.

Definition 6.4. Let $(X_t)_{t \geq 0}$ be a Markov process on $\mathbb{R}^d$ with transition function $p_t(x, \cdot)$ and transition semigroup $T_t u(x) = \int_{\mathbb{R}^d} u(y)\, p_t(x, dy)$.

a) A **coupling of the Markov process** $(X_t)_{t \geq 0}$ is an $\mathbb{R}^{2d}$-valued process $(X_t', X_t'')_{t \geq 0}$ where $(X_t')_{t \geq 0}$ and $(X_t'')_{t \geq 0}$ are Markov processes which have the same transition function $p_t(x, \cdot)$ as $(X_t)_{t \geq 0}$, but possibly different initial distributions.
b) $(X_t')_{t \geq 0}$ and $(X_t'')_{t \geq 0}$ are called the **marginal processes** of the coupling process.
c) The random time $T := \inf\{t \geq 0 : X_t' = X_t''\} \in [0, \infty]$ is called the **coupling time**.
d) The coupling is said to be **successful** if T is a.s. finite.
e) A Markov process $(X_t)_{t \geq 0}$ admits a **successful coupling** (also: enjoys the **coupling property**) if for any two initial distributions $\mu_1, \mu_2 \in \mathcal{M}^1(\mathbb{R}^d)$, there exists a successful coupling with marginal processes starting from μ_1 and μ_2, respectively.

It is known, see Lindvall [204] or Thorisson [319], that the coupling property is equivalent to the statement that

$$\lim_{t \to \infty} \|\mu_1 T_t - \mu_2 T_t\|_{TV} = 0 \quad \forall \mu_1, \mu_2 \in \mathcal{M}^1(\mathbb{R}^d). \tag{6.10}$$

As usual, $\mathcal{M}^1(\mathbb{R}^d)$ is the set of probability measures on $\mathbb{R}^d$, $\|\cdot\|_{TV}$ stands for the total variation norm, and $\mu T_t(A) = \int_{\mathbb{R}^d} p_t(x, A)\, \mu(dx)$ is the left action of T_t.

Successful Couplings for Lévy Processes. The motivation behind this approach is the paper by Wang [342], which uses the conditional Girsanov theorem on Poisson space as its main ingredient, and assumes that the Lévy measure has a non-trivial absolutely continuous part. The following result is taken from [287, Theorem 1.1] where we use an explicit expression of the compound Poisson semigroup and combine this with the Mineka and Lindvall–Rogers couplings for random walks, see [206].

Theorem 6.5. *Let* $(X_t)_{t \geq 0}$ *be a Lévy process in* $\mathbb{R}^d$ *with Lévy triplet* (l, Q, v) *and transition function* $p_t(x, dy) = \mathbb{P}^x(X_t \in dy)$.[1] *For every* $\epsilon > 0$, *define* v_ϵ *by*

$$
v_\epsilon(B) := \begin{cases} v(B), & v(\mathbb{R}^d) < \infty; \\ v\{z \in B : |z| \geq \epsilon\}, & v(\mathbb{R}^d) = \infty. \end{cases} \tag{6.11}
$$

Assume that there exist $\epsilon, \delta > 0$ *such that*

$$
\inf_{x \in \mathbb{R}^d, |x| \leq \delta} v_\epsilon \wedge (\delta_x * v_\epsilon)(\mathbb{R}^d) > 0.[2] \tag{6.12}
$$

Then, there is a constant $C = C(\epsilon, \delta, v) > 0$ *such that for all* $x, y \in \mathbb{R}^d$ *and* $t > 0$,

$$
\| p_t(x, \cdot) - p_t(y, \cdot) \|_{TV} \leq \frac{C(1 + |x - y|)}{\sqrt{t}} \wedge 2. \tag{6.13}
$$

In particular, the Lévy process $(X_t)_{t \geq 0}$ *admits a successful coupling.*

Remark 6.6. Let us briefly indicate the strategy of the proof of Theorem 6.5. Since $X = (X_t)_{t \geq 0}$ is a Lévy process, we can decompose it into two *independent* Lévy processes $X' = (X_t')_{t \geq 0}$ and $X'' = (X_t'')_{t \geq 0}$. Denote by $p_t(x, \cdot), p_t'(x, \cdot), p_t''(x, \cdot)$ and T_t, T_t', T_t'' the transition functions and transition operators of X, X' and X'', respectively. By independence, $T_t = T_t' \circ T_t''$, and we see that

$$
\begin{aligned}
\| p_t(x, \cdot) - p_t(y, \cdot) \|_{TV} &= \sup_{\|u\|_\infty \leq 1} \left| T_t u(x) - T_t u(y) \right| \\
&= \sup_{\|u\|_\infty \leq 1} \left| T_t' T_t'' u(x) - T_t' T_t'' u(y) \right| \\
&\leq \sup_{\|w\|_\infty \leq 1} \left| T_t' w(x) - T_t' w(y) \right| \\
&= \| p_t'(x, \cdot) - p_t'(y, \cdot) \|_{TV}.
\end{aligned} \tag{6.14}
$$

---

[1] In fact, $p_t(x, B) = \mathbb{P}^x(X_t \in B) = \mathbb{P}^0(X_t \in B - x) = p_t(0, B - x)$ since a Lévy process is a translation invariant Feller process, cf. Sect. 2.1.

[2] For $\mu, v \in \mathcal{M}_b^+(\mathbb{R}^d)$ we define $\mu \wedge v := \mu - (\mu - v)^+$, where $(\mu - v)^\pm$ is the Hahn–Jordan decomposition of the signed measure $\mu - v$.

Since (6.10) is equivalent to the existence of a successful coupling, we see that X has the coupling property if one of its components, say X', has the coupling property.

If we split X in such a way that X' has only large jumps and $X'' := X - X'$, we can assume that X' is a compound Poisson process; thus, $X'_t = S_{N_t}$ for some Poisson process $(N_t)_{t \geq 0}$ and the random walk $S_n = X_1 + \ldots + X_n$ where the X_j are independent, identically distributed random variables which correspond to the jump heights of the process X', i.e. the large jumps of X.

From this it becomes clear that X' admits a successful coupling whenever the random walk $(S_n)_{n \geq 0}$ has a successful (maximal) coupling—and the latter can be ensured by a hands-on construction using the Mineka coupling; condition (6.12) is need for the construction of this particular coupling. $\qquad \square$

Formula (6.13) from Theorem 6.5 shows that

$$\| p_t(x, \cdot) - p_t(y, \cdot) \|_{TV} = O(t^{-1/2}) \quad \text{as } t \to \infty$$

holds locally uniformly for all $x, y \in \mathbb{R}^d$. This order of convergence is known to be optimal for compound Poisson processes, see Wang [342, Remark 3.1].

Roughly speaking, a pure jump Lévy process admits a successful coupling only if there is enough jump activity, and a Lévy measure with discrete support is usually not good enough for this. The condition (6.12) is one possibility to guarantee sufficient jump activity; intuitively, it is satisfied, if for sufficiently small values of $\epsilon, \delta > 0$ and all $x \in \mathbb{B}(0, \delta)$ we have $(x + \text{supp}(\nu_\epsilon)) \cap \text{supp}(\nu_\epsilon) \neq \emptyset$.

Example 6.7. In order to see that (6.12) is sharp, we consider a one-dimensional compound Poisson process with Lévy measure ν supported on $\mathbb{Z}$. Clearly,

$$\forall \delta \in (0, 1) \ \forall x \in \mathbb{B}(0, \delta) \ : \ \nu \wedge (\delta_x * \nu)(\mathbb{Z}) = 0.$$

On the other hand, all functions with $f(x + n) = f(x)$ for $x \in \mathbb{R}^d$ and $n \in \mathbb{Z}$ are harmonic. By Cranston–Greven [75] and Cranston–Wang [76], this process cannot have the coupling property. $\qquad \square$

Wang [342, Theorem 3.1] establishes the coupling property for Lévy processes whose jump measure ν has a non-trivial absolutely continuous part. The following consideration, cf. [289, Proposition 1.5], shows that this condition is more restrictive than (6.12).

Proposition 6.8. *Let $(X_t)_{t \geq 0}$ be a d-dimensional Lévy process and assume that the Lévy measure satisfies $\nu(dz) \geq \rho_0(z) \, dz$ where*

$$\int_{\overline{\mathbb{B}}(z_0, \epsilon)} \frac{dz}{\rho_0(z)} < \infty$$

for some $\rho_0 \in L^1_{\text{loc}}(\mathbb{R}^d \setminus \{0\})$, $z_0 \in \mathbb{R}^d$ and $\epsilon > 0$. Then, there exists a closed subset $F \subset \overline{\mathbb{B}}(z_0, \epsilon)$ and a constant $\delta > 0$ such that

$$\inf_{x\in\mathbb{R}^d,\,|x|\leqslant\delta}\int_F \big(\rho_0(z)\wedge\rho_0(z-x)\big)\,dz > 0.$$

Proposition 6.8 shows that Theorem 6.5 improves Wang's [342, Theorem 3.1], even if the Lévy measure v of $(X_t)_{t\geqslant0}$ has an absolutely continuous component, see [289, Example 1.6]. In fact, based on the Lindvall–Rogers zero–two law for random walks, cf. Lindvall–Rogers [206, Proposition 1], we can give a necessary and sufficient condition guaranteeing that a Lévy process has the coupling property, see [287, Theorem 4.1].

Theorem 6.9. *The following statements are equivalent.*

a) *The Lévy process* $(X_t)_{t\geqslant0}$ *has the coupling property.*
b) *There exists some* $t_0 > 0$ *such that for any* $t \geqslant t_0$, *the transition probability* $p_t(x,\cdot)$ *has (with respect to Lebesgue measure) an absolutely continuous component.*

If the Lévy process has the strong Feller property, i.e. if the corresponding semigroup maps $B_b(\mathbb{R}^d)$ into $C_b(\mathbb{R}^d)$, Theorem 6.9 becomes particularly simple.[3]

Corollary 6.10. *Suppose that there exists some* $t_0 > 0$ *such that the semigroup* T_{t_0} *maps* $B_b(\mathbb{R}^d)$ *into* $C_b(\mathbb{R}^d)$. *Then, the Lévy process* $(X_t)_{t\geqslant0}$ *has the coupling property.*

In particular, every Lévy process which enjoys the strong Feller property has the coupling property.

Based on Theorem 6.9, we also can find the connection between (6.12) and the existence of a non-trivial absolutely continuous component of the Lévy measure which extends Theorem 6.5 and Wang's [342, Theorem 3.1], cf. [287, Theorem 4.3],

Theorem 6.11. *Let* $(X_t)_{t\geqslant0}$ *be a Lévy process on* $\mathbb{R}^d$ *with Lévy measure* $v \not\equiv 0$, *and for any* $\epsilon > 0$ *let* v_ϵ *be the finite measure defined in (6.11). If there exists some* $\epsilon > 0$ *such that one of the following conditions is satisfied*

a) *the* k-*fold convolution* v_ϵ^{*k}, $k \geqslant 1$, *has an absolutely continuous component,*
b) *there exist* $\delta > 0$ *and* $k \geqslant 1$ *such that* $\displaystyle\inf_{x\in\mathbb{R}^d,\,|x|\leqslant\delta} v_\epsilon^{*k} \wedge (\delta_x * v_\epsilon^{*k})(\mathbb{R}^d) > 0$,

then the process $(X_t)_{t\geqslant0}$ *has the coupling property.*

Conversely, assume that $(X_t)_{t\geqslant0}$ *is a compound Poisson process with a finite Lévy measure* $v(\mathbb{R}^d) < \infty$. *If* $(X_t)_{t\geqslant0}$ *has the coupling property, then there is some* $k \geqslant 1$ *such that* v^{*k} *has an absolutely continuous component.*

Successful Couplings of Subordinate Brownian Motions. Let $(B_t)_{t\geqslant0}$ be a d-dimensional Brownian motion with initial value B_0. By definition,

---

[3]A Lévy process is a strong Feller process if, and only if, the transition function $p_t(0, dy)$ is absolutely continuous with respect to Lebesgue measure, cf. Example 1.13(b).

$$\mathbb{E}\left(e^{i\xi\cdot(B_t-B_0)}\right) = e^{-t|\xi|^2}, \quad \xi \in \mathbb{R}^d, \; t > 0.$$

Recall from Sect. 4.2 that a subordinator $(S_t)_{t\geq 0}$ is a nonnegative increasing Lévy process with $S_0 = 0$. Denote by $(\mu_t)_{t\geq 0}$ the transition probabilities of the subordinator $(S_t)_{t\geq 0}$, i.e. $\mu_t(ds) = \mathbb{P}(S_t \in ds)$; these are uniquely characterized by the Laplace transform,

$$\mathcal{L}\mu_t(\lambda) = \int_0^\infty e^{-\lambda s} \, \mu_t(ds) = e^{-tf(\lambda)}, \quad \lambda > 0,$$

where $f(\lambda)$ is a Bernstein function.

We assume that $(S_t)_{t\geq 0}$ and $(B_t)_{t\geq 0}$ are independent. The subordinate Brownian motion $X_t = B_t^f$ defined by $X_t = B_{S_t}$ is a symmetric Lévy process such that

$$\mathbb{E}\left(e^{i\xi\cdot(X_t-X_0)}\right) = e^{-tf(|\xi|^2)}, \quad \xi \in \mathbb{R}^d, \; t > 0,$$

i.e. the symbol of a subordinate Brownian motion is $f(|\xi|^2)$.

Before we construct a (maximal Markovian) coupling for the subordinate Brownian motion $(X_t)_{t\geq 0}$, let us review some facts of the so-called reflection coupling of a d-dimensional Brownian motion, cf. Lindvall–Rogers [205] or Chen–Li [61]. Fix $x, y \in \mathbb{R}^d$ with $x \neq y$, and denote by $(B_t^x)_{t\geq 0}$ a Brownian motion on $\mathbb{R}^d$ starting from $B_0^x = x \in \mathbb{R}^d$. By $\mathbb{H}_{x,y} \subset \mathbb{R}^d$ we denote the hyperplane such that the vector $x - y$ is normal with respect to $\mathbb{H}_{x,y}$ and such that $\frac{1}{2}(x + y) \in \mathbb{H}_{x,y}$, i.e.

$$\mathbb{H}_{x,y} = \left\{ u \in \mathbb{R}^d \; : \; \left\langle u - \tfrac{1}{2}(x + y), x - y \right\rangle = 0 \right\}.$$

Denote by $R_{x,y} : \mathbb{R}^d \to \mathbb{R}^d$ the reflection with respect to the hyperplane $\mathbb{H}_{x,y}$; for every $z \in \mathbb{R}^d$

$$R_{x,y}z = z - 2\left\langle z - \tfrac{1}{2}(x + y), x - y \right\rangle \frac{x - y}{|x - y|^2}.$$

Define

$$\tau_{x,y} = \inf\left\{ t > 0 \; : \; B_t^x \in \mathbb{H}_{x,y} \right\}$$

and

$$\hat{B}_t^y := \begin{cases} R_{x,y}B_t^x, & t \leq \tau_{x,y}; \\ B_t^x, & t > \tau_{x,y}; \end{cases}$$

that is, up to the random time $\tau_{x,y}$, $(\hat{B}_t^y)_{t\leq\tau_{x,y}}$ is the reflection of $(B_t^x)_{t\leq\tau_{x,y}}$ with respect to the hyperplane $\mathbb{H}_{x,y}$, and from $\tau_{x,y}$ onwards, $(\hat{B}_t^y)_{t\geq\tau_{x,y}} = (B_t^x)_{t\geq\tau_{x,y}}$.

Because of the strong Markov property, $(\hat{B}_t^y)_{t\geq 0}$ is a Brownian motion starting from y.

Set $\tilde{B}_t^{x,y} := (B_t^x, \hat{B}_t^y)$. Then $(\tilde{B}_t^{x,y})_{t\geq 0}$ is a coupling of two Brownian motions starting from x and $y \in \mathbb{R}^d$, respectively. The coupling time

$$T_{x,y}^B := \inf\{t > 0 : B_t^x = \hat{B}_t^y\}$$

coincides with the stopping time $\tau_{x,y}$.

Assume that $(S_t)_{t\geq 0}$ is a subordinator which is independent of $(\tilde{B}_t^{x,y})_{t\geq 0}$. Set

$$\tilde{X}_t^{x,y} := \tilde{B}_{S_t}^{x,y} = (B_{S_t}^x, \hat{B}_{S_t}^y).$$

Since $S_0 = 0$, we see that $(\tilde{X}_t^{x,y})_{t\geq 0}$ is a coupling process of $(X_t)_{t\geq 0}$ starting from (x,y). For simplicity, let $\tilde{X}_t^{x,y} := (X_t^x, \hat{X}_t^y)$, and call $(\tilde{X}_t^{x,y})_{t\geq 0}$ the *reflection-sub-ordinate coupling* of $(X_t)_{t\geq 0}$. Then the coupling time of $(\tilde{X}_t^{x,y})_{t\geq 0}$ is

$$T_{x,y}^X := \inf\left\{t \geq 0 : X_t^x = \hat{X}_t^y\right\}.$$

We claim that for any $x, y \in \mathbb{R}^d$ we have $T_{x,y}^X < \infty$ almost surely, cf. [48, Theorem 2.1].

Theorem 6.12. *Let $(X_t)_{t\geq 0}$ be a subordinate Brownian motion on $\mathbb{R}^d$ corresponding to the Bernstein function f, and denote by $p_t(x,\cdot)$ be the transition function.[4] Then $(X_t)_{t\geq 0}$ has the coupling property; moreover, for any $t > 0$ and $x, y \in \mathbb{R}^d$,*

$$\|p_t(x,\cdot) - p_t(y,\cdot)\|_{TV} \leq 2\,\mathbb{P}\left(T_{x,y}^X > t\right) \leq \frac{|x-y|}{\sqrt{2\pi}} \int_0^\infty e^{-tf(r)} \frac{dr}{\sqrt{r}}.$$

If, in addition,

$$\lim_{s\to 0} \frac{f^{-1}(\lambda s)}{f^{-1}(s)} \in (0, \infty) \quad \forall \lambda > 0,$$

then there exists a constant $C > 0$ such that for sufficiently large values of $t > 0$

$$\|p_t(x,\cdot) - p_t(y,\cdot)\|_{TV} \leq C|x-y|\sqrt{f^{-1}\left(\tfrac{1}{t}\right)}.$$

The key ingredient in the proof of Theorem 6.12 is the observation that

$$T_{x,y}^X = \inf\left\{t \geq 0 : S_t \geq T_{x,y}^B\right\}$$

where $T_{x,y}^B$ is the coupling time of the $\mathbb{H}_{x,y}$-reflected Brownian motion.

---

[4]See the footnote to Theorem 6.5 on p. 148.

Theorem 6.12 can be used to study the coupling of symmetric α-stable processes.

Example 6.13. Let $f(\lambda) = \lambda^{\alpha/2}$ for some $\alpha \in (0,2)$. The corresponding subordinate Brownian motion $(X_t)_{t \geq 0}$ is the rotationally symmetric α-stable Lévy process. By Theorem 6.12 we find for sufficiently large values of $t > 0$

$$\|p_t(x,\cdot) - p_t(y,\cdot)\|_{TV} \leq \frac{C|x-y|}{t^{1/\alpha}}.$$

We also have a lower estimate. Fix $x, y \in \mathbb{R}^d$ and denote by $\boldsymbol{Q}_{xy}(z,r)$ the d-dimensional cube with centre z and side-length r such that $x - y$ is normal to one of the cube's faces. Then let $r > 2|x - y|$, assume that $t \geq 1$ and note that, by scaling,

$$\|p_t(x,\cdot) - p_t(y,\cdot)\|_{TV}$$
$$\geq \left| \mathbb{P}\left(X_t + x \in \boldsymbol{Q}_{xy}(x, rt^{1/\alpha})\right) - \mathbb{P}\left(X_t + y \in \boldsymbol{Q}_{xy}(x, rt^{1/\alpha})\right) \right|$$
$$= \left| \mathbb{P}\left(X_t \in \boldsymbol{Q}_{xy}(0, rt^{1/\alpha})\right) - \mathbb{P}\left(X_t \in \boldsymbol{Q}_{xy}(x - y, rt^{1/\alpha})\right) \right|$$
$$= \left| \mathbb{P}\left(X_1 \in \boldsymbol{Q}_{xy}(0, r)\right) - \mathbb{P}\left(X_1 \in \boldsymbol{Q}_{xy}((x-y)t^{-1/\alpha}, r)\right) \right|$$
$$= \int_{\boldsymbol{Q}_{xy}(0,r) \setminus \boldsymbol{Q}_{xy}((x-y)t^{-1/\alpha},r)} p_1(0,z)\,dz + \int_{\boldsymbol{Q}_{xy}((x-y)t^{-1/\alpha},r) \setminus \boldsymbol{Q}_{xy}(0,r)} p_1(0,z)\,dz$$
$$\geq 2\frac{|x-y|}{t^{1/\alpha}}r^{d-1}\inf_{z \in B(0,2r)} p_1(0,z)$$

($p_1(0,z)$ denotes the α-stable density at $t = 1$; it is continuous and strictly positive; see e.g. Sharpe [297] or Blumenthal–Getoor [31, Theorem 2.1] which implies that $p_t(0,z) \asymp t^{-d/\alpha} \wedge (t\,|z|^{-d-\alpha})$). Consequently, we find for rotationally symmetric α-stable processes

$$\|p_t(x,\cdot) - p_t(y,\cdot)\|_{TV} \asymp \frac{|x-y|}{t^{1/\alpha}}, \quad t \to \infty. \qquad \square$$

Theorem 6.12 can be easily generalized to prove the coupling property of Lévy processes which can be decomposed into two independent parts, one of which is a subordinate Brownian motion, cf. [48, Proposition 2.8].

Proposition 6.14. *Let $(X_t)_{t \geq 0}$ be a d-dimensional Lévy process which can be decomposed in the following way*

$$X_t = B_t^f + X_t'',$$

*where $(B_t^f)_{t\geq 0}$ is a Brownian motion subordinated w.r.t. the Bernstein function f,
and $(X_t'')_{t\geq 0}$ is some Lévy process. Let $p_t(x,\cdot)$ be the transition function of $(X_t)_{t\geq 0}$.*[5]
Then there exists a constant $C > 0$ such that for $t > 0$ and $x, y \in \mathbb{R}^d$,

$$\|p_t(x,\cdot) - p_t(y,\cdot)\|_{TV} \leq \left(\frac{|x-y|}{\sqrt{2\pi}}\int_0^\infty e^{-tf(r)}\frac{dr}{\sqrt{r}}\right) \wedge \frac{C(1+|x-y|)}{\sqrt{t}} \wedge 2.$$

Proof. Let T_t' and T_t'' be the transition operators of $(B_t^f)_{t\geq 0}$ and $(X_t'')_{t\geq 0}$, respectively. Then the assertion follows from the calculation (6.14) and Theorem 6.12. $\square$

The Rate of Successful Couplings for Lévy Processes. We know from Theorem 6.9 that a Lévy process has the coupling property if the transition function $p_t(\cdot)$ is absolutely continuous with respect to Lebesgue measure for sufficiently large times $t > 0$. In this case, the transition function $p_t(x, dy) = \mathbb{P}^x(X_t \in dy)$ satisfies

$$\|p_t(x,\cdot) - p_t(y,\cdot)\|_{TV} \leq \frac{C(1+|x-y|)}{\sqrt{t}} \wedge 2 \quad \forall t > 0, \; x, y \in \mathbb{R}^d. \qquad (6.15)$$

It is clear that $\|p_t(x,\cdot) - p_t(y,\cdot)\|_{TV} \leq 2$ for any $x, y \in \mathbb{R}^d$ and $t \geq 0$, while

$$t \mapsto \|p_t(x,\cdot) - p_t(y,\cdot)\|_{TV}$$

is decreasing. This shows that it is enough to estimate $\|p_t(x,\cdot) - p_t(y,\cdot)\|_{TV}$ for large values of t.

We call any estimate for $\|p_t(x,\cdot) - p_t(y,\cdot)\|_{TV}$ an **estimate of the coupling time**. For a general Lévy process admitting a successful coupling the rate $1/\sqrt{t}$ from (6.15) is not optimal. For example, for symmetric α-stable Lévy processes we can prove, see Example 6.13, that

$$\|p_t(x,\cdot) - p_t(y,\cdot)\|_{TV} \asymp \frac{1}{t^{1/\alpha}} \quad \text{as } t \to \infty. \qquad (6.16)$$

This indicates that one should derive explicit estimates for the coupling property of Lévy processes.

Let $p_t(x,\cdot)$ and T_t be the transition function and the semigroup of the Lévy process $(X_t)_{t\geq 0}$ and denote by $\psi : \mathbb{R}^d \to \mathbb{C}$ the characteristic exponent. Assume that ψ satisfies the following **Hartman–Wintner condition** for some $t_0 > 0$:

$$\lim_{|\xi|\to\infty} \frac{\operatorname{Re}\psi(\xi)}{\log(1+|\xi|)} > \frac{d}{t_0}; \qquad (6.17)$$

---

[5]See the footnote to Theorem 6.5 on p. 148.

this condition ensures that the transition function of the Lévy process $(X_t)_{t \geq 0}$ has for all $t > t_0$ a (smooth) density with respect to Lebesgue measure, see e.g. Hartman–Wintner [130] or [184]. The following coupling estimate is from [292, Theorem 1.1].

Theorem 6.15. *Let $(X_t)_{t \geq 0}$ be a Lévy process with transition function $p_t(x, \cdot)$[6] and characteristic exponent $\psi : \mathbb{R}^d \to \mathbb{C}$. Suppose that (6.17) holds and*

$$\operatorname{Re}\psi(\xi) \asymp f(|\xi|) \quad as \quad |\xi| \to 0,$$

where $f : [0, \infty) \to \mathbb{R}$ is a strictly increasing function which is differentiable near zero and satisfies

$$\lim_{r \to 0} f(r)|\log r| < \infty \quad and \quad \overline{\lim_{s \to 0}} \frac{f^{-1}(2s)}{f^{-1}(s)} < \infty.$$

Then the Lévy process $(X_t)_{t \geq 0}$ has the coupling property, and there exist two constants $c, t_1 > 0$ such that for any $x, y \in \mathbb{R}^d$ and $t \geq t_1$,

$$\|p_t(x, \cdot) - p_t(y, \cdot)\|_{TV} \leq c \, f^{-1}(1/t).$$

We see from (6.16), which holds for symmetric α-stable Lévy processes, that the estimate in Theorem 6.15 is sharp. In [48, Theorem 1.1 and (1.3)] we showed that the following condition on the Lévy measure ν ensures that a (pure jump) Lévy process admits a successful coupling:

$$\nu(dz) \geq |z|^{-d} \, g(|z|^{-2}) \, dz \tag{6.18}$$

where g is some Bernstein function. Theorem 6.15 uses a different condition in terms of the characteristic exponent $\psi(\xi)$. Let us briefly compare [48, Theorem 1.1] and Theorem 6.15. If (6.18) holds, then we know that

$$\psi(\xi) = \psi_\rho(\xi) + \psi_\mu(\xi),$$

where ψ_ρ and ψ_μ denote the pure-jump characteristic exponents given by the Lévy measures $\rho(dz) = |z|^{-d} \, g(|z|^{-2}) \, dz$ and $\mu = \nu - \rho$, respectively. Observe that (6.18) guarantees that μ is a *nonnegative* measure. By [164, Lemma 2.1] and some tedious, but otherwise routine, calculations one can see that $\psi_\rho(\xi) \asymp g(|\xi|^2)$ as $|\xi| \to 0$. If g satisfies [48, (2.10) and (2.11)]—these conditions coincide with the asymptotic properties required of f in Theorem 6.15—we can apply Theorem 6.15 to the symbol $\psi_\rho(\xi)$ with $f(s) = g(s^2)$, and follow the argument of [48, Proposition 2.9 and Remark 2.10] to get a new proof of [48, Theorem 1.1]. This argument relies

---

[6]See the footnote to Theorem 6.5 on p. 148.

on the fact that we can decompose the Lévy process with exponent $\psi(\xi)$ into two independent Lévy processes with characteristic exponents $\psi_\rho(\xi)$ and $\psi_\mu(\xi)$.

The considerations above can be adapted to show that we may replace the two-sided estimate $\mathrm{Re}\,\psi(\xi) \asymp f(|\xi|)$ in Theorem 6.15 by $\mathrm{Re}\psi(\xi) \geqslant c\,f(|\xi|)$; this, however, requires that we know in advance that $\psi(\xi) - cf(|\xi|)$ is a negative definite function, i.e. the characteristic exponent of some Lévy process. While this was obvious under (6.18) and for the difference of two Lévy measures being again a non-negative measure, there are no good conditions, in general, when the difference of two characteristic exponents is again a characteristic exponent of some Lévy process.

Theorem 6.15 trivially applies to most subordinate stable Lévy processes: Here the characteristic exponent is of the form $f(|\xi|^\alpha)$, $0 < \alpha \leqslant 2$, but the corresponding Lévy measures cannot be given in closed form. Nevertheless the methods of [48] are only applicable in *one* particular case, while Theorem 6.15 applies to *all* non-degenerate settings.

The Coupling Property of Ornstein–Uhlenbeck Processes. Let $(X_t^x)_{t\geqslant 0}$ be an n-dimensional Ornstein–Uhlenbeck process, which is defined as the unique strong solution of the following stochastic differential equation

$$dX_t = AX_t\,dt + B\,dZ_t, \qquad X_0 = x \in \mathbb{R}^n. \tag{6.19}$$

Here $A \in \mathbb{R}^{n\times n}$, $B \in \mathbb{R}^{n\times d}$, and $(Z_t)_{t\geqslant 0}$ is a d-dimensional Lévy process; note that we allow $(Z_t)_{t\geqslant 0}$ to take values in any proper subspace of $\mathbb{R}^d$. It is well known that

$$X_t^x = e^{tA}x + \int_0^t e^{(t-s)A} B\,dZ_s.$$

Let A be an $n \times n$ matrix. We say that an eigenvalue λ of A is **semisimple** if the dimension of the corresponding eigenspace is equal to the algebraic multiplicity of λ as a root of the characteristic polynomial of A; for symmetric matrices A all eigenvalues are real and semisimple.

The following result provides a sufficient condition for the coupling property of an Ornstein–Uhlenbeck process, cf. [289, Theorem 1.1].

Theorem 6.16. *Let $p_t(x,\cdot)$ be the transition probability of the Ornstein–Uhlenbeck process $(X_t^x)_{t\geqslant 0}$ given by* (6.19). *Assume that* $\mathrm{Rank}(B) = n$ *(which implies $n \leqslant d$), and that* (6.12) *holds for some $\epsilon, \delta > 0$ for the Lévy measure of Z_t.*

If the real parts of all eigenvalues of A are negative, and if all purely imaginary eigenvalues are semisimple, then there is a constant $C = C(\epsilon, \delta, \nu, A, B) > 0$ such that for all $x, y \in \mathbb{R}^n$ and $t > 0$,

$$\|p_t(x,\cdot) - p_t(y,\cdot)\|_{TV} \leqslant \frac{C(1 + |x - y|)}{\sqrt{t}} \wedge 2.$$

As a consequence of Theorem 6.16, we immediately obtain the following result which partly answers a question about Liouville theorems for non-local operators:

> A challenging task would be to apply other probabilistic techniques, based on Malliavin calculus [...] or on coupling [...] to non-local operators.
>
> (Priola–Zabczyk [241, p. 458])

In fact, under the conditions of Theorem 6.16, the Ornstein–Uhlenbeck process $(X_t^x)_{t \geq 0}$ admits a successful coupling and has the Liouville property, i.e. every bounded harmonic function is constant.

Recently, the coupling property for Ornstein–Uhlenbeck processes with values in a Banach space was studied in [350]. Unlike in the finite-dimensional case where the coupling property also holds for Lévy processes without drift, see Theorem 6.16, the infinite-dimensional setting requires a drift term to ensure the quasi-invariance of the process under shifts in the initial data; the latter is a key ingredient for the existence of a successful coupling of an infinite-dimensional Ornstein–Uhlenbeck process, cf. [350, Theorem 1.2].

Example 6.17. a) Let $(Z_t)_{t \geq 0}$ be a rotationally symmetric α-stable Lévy process $(0 < \alpha < 2)$ in $\mathbb{R}^n$, and denote by $(X_t^x)_{t \geq 0}$ the n-dimensional Ornstein–Uhlenbeck process driven by Z_t, i.e.

$$dX_t = AX_t \, dt + dZ_t, \quad X_0 = x \in \mathbb{R}^n.$$

If the real part of at least one eigenvalue of A is strictly positive, then $(X_t^x)_{t \geq 0}$ does not have the coupling property. Indeed, according to Priola–Zabczyk [241, Example 3.4 and Theorem 3.5], we know that $(X_t^x)_{t \geq 0}$ does not have the Liouville property, i.e. there exists a bounded harmonic function which is not constant.

By Lindvall [204, Theorem 21.12] or Cranston–Greven [75, the second remark following Theorem 1], $(X_t^x)_{t \geq 0}$ does not have the coupling property. This example indicates that the non-positivity of the real parts of the eigenvalues of A is also necessary.

b) By Theorem 6.9, any Lévy process with the strong Feller property has the coupling property. This is no longer true for an Ornstein–Uhlenbeck process. Consider, for instance, the one-dimensional Ornstein–Uhlenbeck process given by

$$dX_t = X_t \, dt + dZ_t, \quad X_0 = x \in \mathbb{R},$$

where $(Z_t)_{t \geq 0}$ is an α-stable Lévy process on $\mathbb{R}$. According to Priola–Zabczyk [242, Theorem 1.1 and Proposition 2.1], we know that $(X_t)_{t \geq 0}$ has the strong Feller property. From the first part of this example, we see that this process does not have the coupling property. $\qquad\Box$

We will now estimate $\|p_t(x, \cdot) - p_t(y, \cdot)\|_{TV}$ for large values of t with the help of the characteristic exponent $\psi(\xi)$ of the driving Lévy process $(Z_t)_{t \geq 0}$. We restrict ourselves to the case that the Lévy process $(Z_t)_{t \geq 0}$ has no Gaussian part. Define

$$\phi_t(\rho) := \sup_{|\xi| \leq \rho} \int_0^t \mathrm{Re}\psi\left(B^\top e^{sA^\top}\xi\right) ds \quad \forall t, \rho > 0$$

where $M^\top$ denotes the transpose of the matrix M. Then we have, cf. [289, Theorem 1.7]:

Theorem 6.18. *Let $p_t(x, \cdot)$ be the transition function of the Ornstein–Uhlenbeck process $(X_t^x)_{t \geq 0}$ on $\mathbb{R}^n$ given by (6.19). Assume that there exists some $t_0 > 0$ such that*

$$\lim_{|\xi| \to \infty} \frac{\int_0^{t_0} \mathrm{Re}\psi\left(B^\top e^{sA^\top}\xi\right) ds}{\log(1 + |\xi|)} > 2n + 2. \tag{6.20}$$

If

$$\int_{\mathbb{R}^n} \exp\left(-\int_0^t \mathrm{Re}\psi\left(B^\top e^{sA^\top}\xi\right) ds\right) |\xi|^{n+2}\, d\xi = O\left(\phi_t^{-1}(1)^{2n+2}\right) \quad \text{as } t \to \infty,$$

then there exist $t_1, C > 0$ such that for all $x, y \in \mathbb{R}^n$ and $t \geq t_1$,

$$\|p_t(x, \cdot) - p_t(y, \cdot)\|_{TV} \leq C |e^{tA}(x - y)|\, \phi_t^{-1}(1). \tag{6.21}$$

In particular, if

$$\xi \mapsto \int_0^\infty \mathrm{Re}\psi\left(B^\top e^{sA^\top}\xi\right) ds \quad \text{is locally bounded,} \tag{6.22}$$

it is enough to assume (6.20) in order to get (6.21).

Note that (6.22) is satisfied, if the real parts of all eigenvalues of A are negative, and

$$\overline{\lim_{|\xi| \to 0}} \frac{\mathrm{Re}\psi\left(B^\top \xi\right)}{|\xi|^\kappa} < \infty$$

for some constant $\kappa > 0$.

The Coupling Property of Stochastic Differential Equations with Additive Noise. In this part we study the coupling property for the following d-dimensional stochastic differential equation (SDE) with jumps

$$X_t = x + \int_0^t b(X_s)\, ds + Z_t, \quad x \in \mathbb{R}^d, \ t \geq 0, \tag{6.23}$$

where $b : \mathbb{R}^d \to \mathbb{R}^d$ is a continuous function, and $(Z_t)_{t \geq 0}$ is a pure jump Lévy process on $\mathbb{R}^d$.

For such processes we do not have explicit formulae for the transition functions. This means that the approaches used in the previous parts will not work in the present setting. Based on the construction of an efficient coupling operator for the Markov generator corresponding to the solution of the SDE (6.23), the following result from [346, Theorem 1.1], which improves [203, Theorem 1.1], gives conditions for the coupling property of stochastic differential equations with additive noise.

Theorem 6.19. *Let $(Z_t)_{t \geq 0}$ be a d-dimensional Lévy process with the Lévy measure ν satisfying the condition (6.12) for some constants $\epsilon, \delta > 0$. If*

$$\langle b(x) - b(y), x - y \rangle \leq 0 \quad \forall x, y \in \mathbb{R}^d \tag{6.24}$$

then for every $\beta \in (0, 1)$ there exists a constant $C := C(\beta) > 0$ such that for $x, y \in \mathbb{R}^d$ and $t > 0$,

$$\|p_t(x, \cdot) - p_t(y, \cdot)\|_{TV} \leq \frac{C(1 + |x - y|)}{t^{1/(3-\beta)}}.$$

In particular, the solution $(X_t)_{t \geq 0}$ of the SDE (6.23) has the coupling property.

Example 6.17(a) shows that we need the condition (6.24) for the drift term b in order to get the coupling property of Ornstein–Uhlenbeck processes driven by an α-stable Lévy process. If the driving noise is a subordinate Brownian motion, we can derive explicit estimates for the coupling property.

Consider the following stochastic differential equation

$$X_t = x + \int_0^t b(X_s) \, ds + W_{S_t}, \tag{6.25}$$

where $(W_{S_t})_{t \geq 0}$ is a subordinate Brownian motion on $\mathbb{R}^d$ and $(S_t)_{t \geq 0}$ is a subordinator which is independent of $(W_t)_{t \geq 0}$; we assume that the drift term $b : \mathbb{R}^d \to \mathbb{R}^d$ is continuous. The following theorem is from [346, Theorem 1.4].

Theorem 6.20. *Let $(W_{S_t})_{t \geq 0}$ be a subordinate Brownian motion on $\mathbb{R}^d$ where $(S_t)_{t \geq 0}$ is an independent subordinator with Bernstein function f. Suppose that $b : \mathbb{R}^d \to \mathbb{R}^d$ is a continuous function satisfying (6.24), and there are two constants $c > 0$ and $m \geq 1$ such that $|b(x)| \leq c(1 + |x|^m)$. Let $(X_t)_{t \geq 0}$ be the unique solution to the SDE (6.25). Then, $(X_t)_{t \geq 0}$ has the coupling property, and there exists a constant $C > 0$ such that for any $t > 0$ and $x, y \in \mathbb{R}^d$,*

$$\|p_t(x, \cdot) - p_t(y, \cdot)\|_{TV} \leq C|x - y| \int_0^\infty e^{-tf(r)} \frac{dr}{\sqrt{r}}.$$

If, in addition,

$$\varliminf_{r \to \infty} \frac{f(r)}{\log r} > 0, \quad \varliminf_{r \to 0} f(r)|\log r| < \infty \quad \text{and} \quad \varlimsup_{s \to 0} \frac{f^{-1}(2s)}{f^{-1}(s)} < \infty,$$

then, there exists a constant $C_1 > 0$ such that for $t > 0$ sufficiently large

$$\|p_t(x,\cdot) - p_t(y,\cdot)\|_{TV} \leqslant C_1 |x - y| \sqrt{f^{-1}\left(\tfrac{1}{t}\right)}.$$

The proof of Theorem 6.20 is based on the time-change argument developed in Zhang [355], and it is easy to see that Theorem 6.20 is optimal for symmetric α-stable processes.

6.3 Transience and Recurrence

For Feller processes there have been several approaches to classify their recurrence and transience behaviour based on the generator or the associated Dirichlet form. For a survey of the potential theoretic approach, see Getoor [123] for unified criteria, and the related techniques we refer to Jacob [159, Chap. 6] and the references given therein. For non-symmetric Dirichlet forms we use Oshima's Erlangen lectures [234, Chap. 1.3].

In this section we will focus on the Markov chain approach using Foster–Lyapunov criteria which was pioneered by Meyn–Tweedie [223, 224, 225, 226, 227]. Our exposition follows closely [43].

In the sequel, Leb denotes Lebesgue measure in $\mathbb{R}$ or $\mathbb{R}^d$, and B is always a Borel measurable set.

Definition 6.21. A process $(X_t)_{t \geqslant 0}$ on $\mathbb{R}^d$ is called

a) **Leb-irreducible** if

$$\mathbb{E}^x \left(\int_0^\infty \mathbb{1}_B(X_t)\, dt \right) > 0 \quad \forall B \in \mathscr{B}(\mathbb{R}^d) \text{ with } \operatorname{Leb}(B) > 0, \ x \in \mathbb{R}^d;$$

b) **recurrent** with respect to Leb if

$$\mathbb{E}^x \left(\int_0^\infty \mathbb{1}_B(X_t)\, dt \right) = \infty \quad \forall B \in \mathscr{B}(\mathbb{R}^d) \text{ with } \operatorname{Leb}(B) > 0, \ x \in \mathbb{R}^d;$$

c) **Harris recurrent** with respect to Leb if

$$\mathbb{P}^x \left(\int_0^\infty \mathbb{1}_B(X_t)\, dt = \infty \right) = 1 \quad \forall B \in \mathscr{B}(\mathbb{R}^d) \text{ with } \operatorname{Leb}(B) > 0, \ x \in \mathbb{R}^d;$$

d) **transient** if there exists a countable cover of $\mathbb{R}^d$ with Borel sets B_j such that for each $j \geqslant 1$

$$\mathbb{E}^x \left(\int_0^\infty \mathbb{1}_{B_j}(X_t)\, dt \right) < \infty \quad \forall x \in \mathbb{R}^d;$$

e) a **T-model** if for some probability measure $\mu \in \mathcal{M}^1[0, \infty)$ there exists a measure kernel $T(x, B)$ on $\mathbb{R}^d \times \mathcal{B}(\mathbb{R}^d)$ with $T(x, \mathbb{R}^d) > 0$ for all x such that the function $x \mapsto T(x, B)$ is lower semi-continuous for all $B \in \mathcal{B}(\mathbb{R}^d)$ and

$$\int_0^\infty \mathbb{E}^x \mathbb{1}_B(X_t)\, \mu(dt) \geq T(x, B) \quad \forall B \in \mathcal{B}(\mathbb{R}^d),\ x \in \mathbb{R}^d.$$

Of fundamental importance is the following recurrence–transience dichotomy which we take from [43, Theorem 4.2].

Theorem 6.22 (Recurrence–Transience Dichotomy). *Let $(X_t)_{t \geq 0}$ be a Leb-irreducible T-model, then it is either Harris recurrent or transient.*

In order to use this result we need criteria which ensure that a given process $(X_t)_{t \geq 0}$ is a T-model. Following Tweedie [327, Theorems 5.1, 7.1] we have

Theorem 6.23. a) *$(X_t)_{t \geq 0}$ is a T-model, if every compact set $K \subset \mathbb{R}^d$ is* **petite**, *i.e. there exists a probability measure $\mu \in \mathcal{M}^1[0, \infty)$ and a non-trivial (positive) measure $\nu \in \mathcal{M}^+(\mathbb{R}^d)$ such that*

$$\int_0^\infty \mathbb{E}^x \mathbb{1}_B(X_t)\, \mu(dt) \geq \nu(B) \quad \forall B \in \mathcal{B}(\mathbb{R}^d),\ x \in K. \tag{6.26}$$

b) *Let $(X_t)_{t \geq 0}$ be Leb-irreducible and $x \mapsto \mathbb{E}^x(f(X_t))$ continuous for all functions $f \in C_b(\mathbb{R}^d)$. Then $(X_t)_{t \geq 0}$ is a T-model.*

Note that Theorem 6.23(b) shows, in particular, that every Leb-irreducible C_b-Feller process is a T-model. By Theorem 1.9 the condition that $T_t 1 \in C_b$ is a sufficient condition that a $(C_\infty$-)Feller process is also a C_b-Feller process. The next theorem, cf. [43, Theorem 4.5], provides direct criteria for a process to be a Leb-irreducible T-model.

Theorem 6.24. *Let $(X_t)_{t \geq 0}$ be a Markov process on $\mathbb{R}^d$ and denote its transition function by $p_t(x, B) := \mathbb{P}^x(X_t \in B)$. Then*

a) *$(X_t)_{t \geq 0}$ is Leb-irreducible if*

$$p_t(x, B) > 0 \quad \forall B \in \mathcal{B}(\mathbb{R}^d) \text{ with } \mathrm{Leb}(B) > 0,\ t > 0,\ x \in \mathbb{R}^d. \tag{6.27}$$

b) *$(X_t)_{t \geq 0}$ is a Leb-irreducible T-model if (6.27) holds and there exits a compact set $T \subset [0, \infty]$ and a non-trivial (positive) measure $\nu \in \mathcal{M}^+(\mathbb{R}^d)$ such that for all compact sets $K \subset \mathbb{R}^d$*

$$\inf_{t \in T} \inf_{x \in K} p_t(x, B) \geq \nu(B) \quad \forall B \in \mathcal{B}(\mathbb{R}^d). \tag{6.28}$$

Let us mention, in particular, two simple consequences of Theorem 6.24(b).

Corollary 6.25. *Let $(X_t)_{t \geq 0}$ be a Markov process on $\mathbb{R}^d$ with transition function $p_t(x, B) = \mathbb{P}^x(X_t \in B)$. Then $(X_t)_{t \geq 0}$ is a Leb-irreducible T-model if the transition probability $p_t(x, \cdot)$ is the sum of a discrete measure and a measure which is absolutely continuous with respect to Leb such that the density $\tilde{p}_t(x, y)$ satisfies*

$$\tilde{p}_t(x, y) > 0 \quad \forall x, y \in \mathbb{R}^d, \ t > 0, \tag{6.29}$$

$$\inf_{t \in [1,2]} \inf_{x \in K} \tilde{p}_t(x, y) > 0 \quad \forall y \in R^d, \ K \subset \mathbb{R}^d \text{ compact.} \tag{6.30}$$

Thus, we have in particular

Corollary 6.26. *A Feller process with strictly positive transition density is a Leb-irreducible T-model. Hence it is either Harris recurrent or transient.*

Having this dichotomy, we need further conditions which imply recurrence or transience. The following result from [43, Theorem 4.6] provides a necessary and sufficient characterization via the hitting times of balls.

We write $\sigma_{\overline{\mathbb{B}}(0,R)} := \inf\{t > 0 : X_t \in \overline{\mathbb{B}}(0, R)\}$ for the first entrance time of the process $(X_t)_{t \geq 0}$ into the closed ball $\overline{\mathbb{B}}(0, R)$.

Theorem 6.27. *Let $(X_t)_{t \geq 0}$ be Leb-irreducible T-model, and $R > 0$ be some positive constant, then*

a) $\forall x \in \mathbb{R}^d \ : \ \mathbb{P}^x\left(\sigma_{\overline{\mathbb{B}}(0,R)} < \infty\right) = 1 \iff (X_t)_{t \geq 0}$ *is Harris recurrent.*
b) $\exists x \in \mathbb{R}^d \ : \ \mathbb{P}^x\left(\sigma_{\overline{\mathbb{B}}(0,R)} < \infty\right) < 1 \iff (X_t)_{t \geq 0}$ *is transient.*

Note that Theorem 6.27 shows that two Leb-irreducible T-models, which coincide outside some ball, have the same recurrence and transience behaviour, respectively. For Feller processes this has the following consequence.

Corollary 6.28. *Let $(X_t)_{t \geq 0}$ and $(Y_t)_{t \geq 0}$ be Feller processes on $\mathbb{R}^d$ with symbols q^X and q^Y and strictly positive transition densities. If there exists a compact set $K \subset \mathbb{R}^d$ such that*

$$q^X(x, \xi) = q^Y(x, \xi) \quad \forall x \notin K, \ \xi \in \mathbb{R}^d \tag{6.31}$$

and if for all starting points $x \notin K$ the processes $X_t^{\sigma_K} := X_{t \wedge \sigma_K}$ and $Y_t^{\sigma_K} := Y_{t \wedge \sigma_K}$ have the same finite-dimensional distributions,

$$(X_t^{\sigma_K})_{t \geq 0} \sim (Y_t^{\sigma_K})_{t \geq 0}, \tag{6.32}$$

then $(X_t)_{t \geq 0}$ and $(Y_t)_{t \geq 0}$ have the same recurrence or transience behaviour.

Note that in Corollary 6.28 the Condition (6.32) follows from (6.31) if the processes are unique for the given symbol.

In contrast to Theorem 6.27 the following **Foster–Lyapunov criteria** provide only sufficient conditions. These where developed in Meyn–Tweedie [226], we follow the presentation of Sandrić [264, Theorem 2.3].

Theorem 6.29. *Let* $(X_t)_{t \geq 0}$ *be a non-explosive* Leb*-irreducible càdlàg Markov process on* $\mathbb{R}^d$ *with full generator* $\hat{A}$, *cf.* (1.50).

a) *If every compact set is petite, cf.* (6.26), *and if there exists a compact set* $K \subset \mathbb{R}^d$ *with* Leb$(K) > 0$ *and a norm-like function* u, *i.e.*

$$(u, Au) \in \hat{A} \quad \text{and} \quad \{x \in \mathbb{R}^d : u(x) \leq r\} \text{ relatively compact for all } r \geq 0,$$

and if there exists a constant $c > 0$ *such that*

$$Au(x) \leq c\mathbb{1}_K(x) \quad \forall x \in \mathbb{R}^d, \tag{6.33}$$

then $(X_t)_{t \geq 0}$ *is Harris recurrent.*

b) *If there exist a bounded Borel measurable function* $u : \mathbb{R}^d \to \mathbb{R}_+$ *and closed, disjoint sets* $E, F \subset \mathbb{R}^d$ *such that*

$$\text{Leb}(E) > 0, \ \text{Leb}(F) > 0 \quad \text{and} \quad \sup_{x \in E} u(x) < \inf_{x \in F} u(x) \tag{6.34}$$

and $(u, Au) \in \hat{A}$ *such that*

$$Au(x) \geq 0 \quad \forall x \notin E, \tag{6.35}$$

then $(X_t)_{t \geq 0}$ *is transient.*

The function u appearing in Theorem 6.29(a) is sometimes called a **Lyapunov function**.

The paper [192] contains the following generalization of Theorem 6.29(b) with a very simple proof based on Dynkin's formula and the paper Azéma–Kaplan-Duflo–Revuz [8].

Lemma 6.30. *Let* $(X_t)_{t \geq 0}$ *be an* $\mathbb{R}^d$*-valued, càdlàg strong Markov process with generator* $(A, \mathcal{D}(A))$ *and full generator* $\hat{A}_b$. *Let* $B \in \mathcal{B}(\mathbb{R}^d)$ *be a bounded Borel set and assume that there exists a sequence* $(v_j)_{j \geq 1} \subset C_b(\mathbb{R}^d)$ *and some function* $v \in C(\mathbb{R}^d)$, *such that the following conditions are satisfied:*

a) A *has an extension* $\tilde{A}$ *such that* $\tilde{A}v_j(x)$ *is defined for all* $x \in \mathbb{R}^d$, $(v_j, \tilde{A}v_j) \in \hat{A}$ *and the limit* $\lim_{j \to \infty}(v_j, \tilde{A}v_j) = (v, \tilde{A}v)$ *exists locally uniformly.*
b) $v \geq 0$ *and* $\inf_{x \in B} v(x) > a > 0$ *for some* $a > 0$.
c) $v(y_0) < a$ *for some* $y_0 \notin \overline{B}$.
d) $\tilde{A}v(x) \leq 0$ *for all* $x \notin B$.

Then $(X_t)_{t \geq 0}$ *is transient.*

We close this section with a few examples.

Example 6.31. a) (*Symmetric stable process*) A symmetric stable process on $\mathbb{R}$ with characteristic exponent $|\xi|^\alpha$ is Harris recurrent if, and only if, $\alpha \geq 1$.

b) (*Stable-like process—sharp switching*) Consider the following negative definite symbol

$$q(x,\xi) = \begin{cases} |\xi|^\alpha, & \text{for } x > 0, \\ |\xi|^\beta, & \text{for } x < 0, \end{cases} \quad x,\xi \in \mathbb{R},$$

where $0 < \alpha, \beta \leqslant 2$. Assume that there exists a one-dimensional Feller process with this symbol and an everywhere strictly positive transition density. Then this process is Harris recurrent if, and only if, $\alpha + \beta \geqslant 2$, cf. [43, Theorem 5.2, Corollary 5.5].

c) (*Stable-like process—periodic*) A one-dimensional Feller process with symbol $q(x,\xi) = |\xi|^{\alpha(x)}$, where $\alpha : \mathbb{R} \to [0,2]$, $x \mapsto \alpha(x)$, is a periodic function with $\mathrm{Leb}\{\alpha(\cdot) = \inf_{y\in\mathbb{R}} \alpha(y)\} > 0$ is recurrent if, and only if, $\inf_{x\in\mathbb{R}} \alpha(x) \geqslant 1$, cf. Franke [109, 110].

d) (*Symmetric Stable-like processes*) Related processes where the generator is a symmetric Dirichlet form were discussed by Uemura [328, 329]. For the discrete time analogues we refer to Sandrić [263, 265].

e) (*Stable-like processes with drift*) Sandrić [264, Theorem 1.1] proves the following results. Consider a Feller process on $\mathbb{R}$ with the symbol

$$q(x,\xi) = -i\beta(x) + \gamma(x)|\xi|^{\alpha(x)}, \quad x,\xi \in \mathbb{R},$$

and set

$$c(x) := \gamma(x)\frac{\alpha(x)2^{\alpha(x)-1}\Gamma\left(\frac{\alpha(x)+1}{2}\right)}{\pi^{\frac{1}{2}}\Gamma\left(1 - \frac{\alpha(x)}{2}\right)}.$$

If $\underline{\lim}_{|x|\to\infty} \alpha(x) \geqslant 1$ and

$$\overline{\lim}_{|x|\to\infty}\left(\mathrm{sgn}(x)\frac{\alpha(x)}{c(x)}|x|^{\alpha(x)-1}\beta(x) + \pi \cot\left(\frac{\pi\alpha(x)}{2}\right)\right) < 0,$$

then the process is recurrent.

If

$$\underline{\lim}_{|x|\to\infty}\left(\mathrm{sgn}(x)\frac{\alpha(x)}{c(x)}|x|^{\alpha(x)-1}\beta(x) + \pi \cot\left(\frac{\pi\alpha(x)}{2}\right)\right) > 0,$$

then the process is transient.

Lyapunov function methods for general one-dimensional Lévy-type processes are discussed in [343].

f) Consider the solution $(X_t, P_t)_{t \geq 0}$ of the following system

$$dX_t = P_t \, dt,$$
$$dP_t = -\nabla_x V(X_t) \, dt - \nabla_x c(X_t) \, dL_t,$$

with initial condition $(X_0, P_0) = (x_0, p_0) \in \mathbb{R}^{2d}$. The driving noise $(L_t)_{t \geq 0}$ is a d-dimensional Lévy process, $c \in C^2(\mathbb{R}^d, \mathbb{R}^d)$ and $V \in C^2(\mathbb{R}^d, [0, \infty))$.

If $\nabla_x c(x)$ is uniformly bounded, the process $(X_t, P_t)_{t \geq 0}$ does not explode in finite time, cf. [192, Theorem 3].

If $d \geq 3$ and if $(L_t)_{t \geq 0}$ is a rotationally symmetric α-stable Lévy process with $0 < \alpha < 2$, then $(X_t, P_t)_{t \geq 0}$ is transient. The proof, cf. [192], uses an explicit Lyapunov function v in the sense of Lemma 6.30: For $\gamma > 0$ set

$$v(x, p) = \left(\tfrac{1}{2} p^2 + V(x) - V_0 \right)^{-\gamma} \quad \text{where} \quad V_0 = \inf_{x \in \mathbb{R}^d} V(x) - 1,$$

and with the set

$$B = \{ (x, p) \in \mathbb{R}^{2d} \ : \ |x| + |p| \leq 1 \} \quad \text{and} \quad a = \inf_{(x,p) \in B} u_\gamma(x, p).$$

The generator of the process $(X_t, P_t)_{t \geq 0}$ is for $u \in C_c^2(\mathbb{R}^{2d})$ of the form

$$Au(x, p) = \nabla_x u(x, p) \cdot p - \nabla_p u(x, p) \cdot (\nabla_x V(x) + \nabla_x c(x) \beta)$$

$$+ \frac{1}{2} \operatorname{trace} \left(\nabla_p^2 u(x, p) \nabla_x c(x) \, Q \, \nabla_x c(x)^\top \right)$$

$$+ \int_{\mathbb{R}^d \setminus \{0\}} \left(u(x, p - \nabla_x c(x) y) - u(x, p) + \right.$$

$$\left. + \nabla_p u(x, p) \nabla_x c(x) y \, \mathbb{1}_{[0,1]}(|y|) \right) v(dy)$$

where $\beta = \mathbb{E}^0 \left[L_1 - \sum_{s \leq 1} \Delta L_s \mathbb{1}_{(1,\infty)}(|\Delta L_s|) \right] \in \mathbb{R}^d$ and $\Delta L_s = L_s - L_{s-}$ denotes the jump at time s.

Chapter 7
Approximation

We begin with a general remark on the convergence of operator semigroups based on Strang [308].

Let $(A^{(k)}, \mathcal{D}(A^{(k)}))$, $k \geqslant 1$, be infinitesimal generators of strongly continuous semigroups $(T_t^{(k)})_{t \geqslant 0}$ on a Banach space $(\mathcal{X}, \| \cdot \|)$.[1] We are interested in the convergence of the semigroups. The limit $(T_t)_{t \geqslant 0}$ of $(T_t^{(k)})_{t \geqslant 0}$ as k tends to ∞ can be considered in various topologies, a natural notion of convergence which preserves the semigroup property and the strong continuity is

$$\lim_{k \to \infty} \sup_{t \leqslant t_0} \|T_t^{(k)} u - T_t u\| = 0 \quad \forall u \in \mathcal{X}, \ t_0 > 0. \tag{7.1}$$

Since locally uniform convergence preserves continuity, it is clear that (7.1) defines a strongly continuous semigroup $(T_t)_{t \geqslant 0}$; by $(A, \mathcal{D}(A))$ we denote its generator.

Obviously, the following conditions are necessary for (7.1).

(a) **Uniform equi-boundedness** (also known as **stability**): For each $t_0 > 0$ there exists some $C = C(t_0) > 0$ such that

$$\sup_{k \geqslant 1} \sup_{t \in [0, t_0]} \sup_{\|u\| \leqslant 1} \|T_t^{(k)} u\| \leqslant C.$$

(b) **Dense subset**: $\lim_{k \to \infty} \sup_{t \leqslant t_0} \|T_t^{(k)} u - T_t u\| = 0$ holds for all u from a dense subset of $\mathcal{X}$.

For contraction semigroups, the uniform equi-boundedness condition is always satisfied with $C(t_0) = 1$. For the convergence on a dense subset the following

---

[1] Similar to the situation in $(C_\infty(\mathbb{R}^d), \| \cdot \|_\infty)$, cf. Definitions 1.1 and 1.2, we say that $(T_t)_{t \geqslant 0}$ is a **contraction semigroup** if $\|T_t u\| \leqslant \|u\|$ for all $u \in \mathcal{X}$, and a **strongly continuous semigroup** if $\lim_{t \to 0} \|T_t u - u\| = 0$ for all $u \in \mathcal{X}$. The (infinitesimal) generator $(A, \mathcal{D}(A))$, the resolvent etc. are defined analogously.

B. Böttcher et al., *Lévy Matters III*, Lecture Notes in Mathematics 2099, DOI 10.1007/978-3-319-02684-8_7, © Springer International Publishing Switzerland 2013

conditions are commonly used. We write $\mathcal{C}$ for any operator core of $(A, \mathcal{D}(A))$, i.e. $\overline{A|_{\mathcal{C}}} = A$.

$$\exists \mathcal{D} \overset{dense}{\subset} \mathcal{D}(A) \ \forall u \in \mathcal{D}, \ t_0 > 0 : \ \lim_{k \to \infty} \sup_{0 \leqslant t \leqslant t_0} \|(A^{(k)} - A)T_t u\| = 0, \qquad \text{(Lax)}$$

$$\forall u \in \mathcal{C} : \ \lim_{k \to \infty} \|A^{(k)} u - Au\| = 0, \qquad \text{(Trotter)}$$

$$\exists \lambda > 0 \ \forall u \in \mathcal{X} : \ \lim_{k \to \infty} \|R_\lambda^{(k)} u - R_\lambda u\| = 0, \qquad \text{(Kato)}$$

where $R_\lambda u = (\lambda - A)^{-1}$ and $R_\lambda^{(k)} = (\lambda - A^{(k)})^{-1}$ are the resolvent operators at $\lambda > 0$. If we assume that the semigroups $(T_t^{(k)})_{t \geqslant 0}$ are uniformly equi-bounded, then

$$\text{(Lax)} \implies \text{(Trotter)} \implies \text{(Kato)}.$$

Counterexamples for the reverse directions can be found in Strang [308]. Trotter's condition is most commonly used, and gives rise to the following equivalent statements, which we adapt from Kallenberg [172, Theorem 19.25, p. 385].

Theorem 7.1. *Let* $(X_t)_{t \geqslant 0}$, $(X_t^{(k)})_{t \geqslant 0}$, $k \geqslant 1$, *be d-dimensional Feller processes with transition semigroups* $(T_t)_{t \geqslant 0}$, $(T_t^{(k)})_{t \geqslant 0}$ *on the Banach space* $(C_\infty(\mathbb{R}^d), \|\cdot\|_\infty)$ *and generators* $(A, \mathcal{D}(A))$, $(A^{(k)}, \mathcal{D}(A^{(k)}))$, *respectively. Furthermore, let $\mathcal{C}$ be an operator core of* $(A, \mathcal{D}(A))$. *Then the following assertions are equivalent:*

a) $\forall u \in \mathcal{C} \ \exists (u_k)_{k \geqslant 1} \subset \mathcal{D}(A^{(k)}) : \lim_{k \to \infty} \left(\|u - u_k\|_\infty + \|Au - A^{(k)} u_k\|_\infty \right) = 0$.

b) $\lim_{k \to \infty} \|T_t u - T_t^{(k)} u\|_\infty = 0 \quad \forall u \in C_\infty(\mathbb{R}^d), \ t > 0$.

c) $\lim_{k \to \infty} \sup_{t \leqslant t_0} \|T_t u - T_t^{(k)} u\|_\infty = 0 \quad \forall u \in C_\infty(\mathbb{R}^d), \ t_0 > 0$.

d) *If* $X_0^{(k)} \xrightarrow[k \to \infty]{d} X_0$ *(convergence in distribution), then* $X^{(k)} \xrightarrow[k \to \infty]{d} X$ *in the space of càdlàg functions* $D_{[0,\infty)}(\mathbb{R}^d)$ *equipped with the Skorohod J_1 topology.*[2]

For the implications a)$\Rightarrow$b), c), d) in Theorem 7.1 it is important to know that $(A, \mathcal{D}(A))$ is the generator of a Feller semigroup.

Note that only the fourth statement in Theorem 7.1 is probabilistic; it is based on the work of Mackevičius [214].

---

[2]Skorokhod's J_1 topology was introduced in Skorokhod [303], it is discussed e.g. in Billingsley [28, Chap. 3, pp. 109–153], Ethier–Kurtz [100, Chaps. 3.5–3.9, pp. 116–147], Jacod–Shiryaev [170, Chap. VI] or Kallenberg [172, Chap. 16, pp. 307–326].

7.1 Constructive Approximation

So far, we have assumed that we already know that the limiting operator A is the generator of a Feller semigroup. If we drop this assumption, we find that, given the uniform boundedness, the convergence provided by Kato's condition yields that the corresponding semigroups converge uniformly on finite intervals towards a limiting semigroup. If the corresponding generators converge, we get the following theorem of Trotter [325, Theorem 5.2].

Theorem 7.2. *Let* $(T_t^{(k)})_{t \geq 0}$, $k \geq 1$, *be strongly continuous contraction semigroups on a Banach space* $(\mathfrak{X}, \| \cdot \|)$ *with generators* $(A^{(k)}, \mathcal{D}(A^{(k)}))$. *Assume that*

a) *the strong limit* $Au := \lim_{k \to \infty} A^{(k)} u$ *exists for all* $u \in \mathcal{D}$ *where* $\mathcal{D}$ *is a dense subset of* $\bigcap_{k \geq 1} \mathcal{D}(A^{(k)})$,
b) *for some* $\lambda > 1$ *the range of* $\lambda - A$ *is dense in* $C_\infty(\mathbb{R}^d)$.

Then the closure of $(A, \mathcal{D})$ *is the generator of a strongly continuous contraction semigroup* $(T_t)_{t \geq 0}$ *where* $T_t u = \lim_{k \to \infty} T_t^{(k)} u$ *strongly for all* $u \in C_\infty(\mathbb{R}^d)$.

The following result due to Hasegawa [131] allows to relax the conditions of Theorem 7.2; the price to pay is that the semigroup generator is not necessarily the smallest closed extension (i.e. the closure) of $(A, \mathcal{D})$.

Theorem 7.3. *Let* $(T_t^{(k)})_{t \geq 0}$, $k \geq 1$, *be strongly continuous contraction semigroups on a Banach space* $(\mathfrak{X}, \| \cdot \|)$ *with generators* $(A^{(k)}, \mathcal{D}(A^{(k)}))$, *and assume that there is a dense subset* $\mathcal{D} \subset \bigcup_{m \geq 1} \bigcap_{k \geq m} \mathcal{D}(A^{(k)})$ *such that*

$$\lim_{k,m \to \infty} \|A^{(k)} u - A^{(m)} u\| = 0 \quad \forall u \in \mathcal{D}.$$

Then there exists a closed extension of $(A, \mathcal{D})$ *which generates a strongly continuous contraction semigroup* $(T_t)_{t \geq 0}$ *given by the strong limit* $T_t u = \lim_{k \to \infty} T_t^{(k)} u$ *for all* $u \in C_\infty(\mathbb{R}^d)$ *if, and only if, one of the following equivalent conditions holds:*

a) $\lim_{n,m \to \infty} \|T_s^{(n)} T_t^{(m)} u - T_t^{(m)} T_s^{(n)} u\| = 0 \quad \forall s, t > 0, \ u \in \mathfrak{X}$;
b) $\lim_{n,m \to \infty} \|R^{(n)} R^{(m)} u - R^{(m)} R^{(n)} u\| = 0 \quad \forall u \in \mathfrak{X}$;
c) $\lim_{n,m \to \infty} \|T_t^{(n)} u - T_t^{(m)} u\| = 0 \quad \forall t > 0, \ u \in \mathfrak{X}$;
d) $\lim_{n,m \to \infty} \|R^{(n)} u - R^{(m)} u\| = 0 \quad \forall u \in \mathfrak{X}$;

where $R^{(n)} u = \int_0^\infty e^{-t} T_t^{(n)} u \, dt$ *is the 1-potential or resolvent operator (at* $\lambda = 1$) *for the semigroup* $(T_t^{(n)})_{t \geq 0}$.

Furthermore, if there exists a dense set $\hat{\mathcal{D}} \subset \mathfrak{X}$ such that for all $u \in \hat{\mathcal{D}}$ the limit $Ru := \lim_{n \to \infty} R^{(n)} u$ exists strongly such that $Ru \in \mathcal{D}$, then $(T_t)_{t \geq 0}$ is generated by the smallest closed extension (i.e. the closure) of $(A, \mathcal{D})$.

Note the last statement of Theorem 7.3 just recovers Theorem 7.2.

In order to apply the previous two results to Feller semigroups on the Banach space $(C_\infty(\mathbb{R}^d), \|\cdot\|_\infty)$, one can use the following result which is implicitly used in all standard proofs of the Hille–Yosida–Ray theorem, Theorem 1.30.

Theorem 7.4. *Let* $(T_t^{(k)})_{t\geqslant 0}$, $k \geqslant 1$, *be Feller semigroups on* $(C_\infty(\mathbb{R}^d), \|\cdot\|_\infty)$ *such that*

$$\lim_{k,m\to\infty} \|T_t^{(k)}u - T_t^{(m)}u\|_\infty = 0 \quad \forall u \in C_\infty(\mathbb{R}^d). \tag{7.2}$$

Then the limit $T_t u = \lim_{k\to\infty} T_t^{(k)} u$ *exists strongly for all* $u \in C_\infty(\mathbb{R}^d)$ *and defines a sub-Markovian semigroup which satisfies the Feller property.*

If, in addition, the limit (7.2) *is uniform for bounded time intervals, i.e.*

$$\lim_{k,m\to\infty} \sup_{t\leqslant t_0} \|T_t^{(k)}u - T_t^{(m)}u\|_\infty = 0 \quad \forall u \in C_\infty(\mathbb{R}^d),\ t_0 > 0, \tag{7.3}$$

then $(T_t)_{t\geqslant 0}$ *is a Feller semigroup.*

Proof. Since $C_\infty(\mathbb{R}^d)$ is complete, the Cauchy condition (7.2) defines a family of linear operators $T_t : C_\infty(\mathbb{R}^d) \to C_\infty(\mathbb{R}^d)$. Since $\lim_{k\to\infty} T_t^{(k)}u(x) = T_t u(x)$, we see that the limit inherits the semigroup, positivity and sub-Markov properties of the families $(T_t^{(k)})_{t\geqslant 0}$.

For every $\epsilon > 0$ there is, because of (7.3), some $K \geqslant 1$ such that

$$\sup_{t\leqslant t_0} \|T_t^{(k)}u - T_t^{(m)}u\|_\infty \leqslant \epsilon \quad \forall k, m \geqslant K.$$

Thus, for all $k \geqslant K$ and $t \leqslant t_0$

$$\begin{aligned}
\|T_t u - u\|_\infty &\leqslant \|T_t u - T_t^{(k)}u\|_\infty + \|T_t^{(k)}u - u\|_\infty \\
&= \lim_{m\to\infty} \|T_t^{(m)}u - T_t^{(k)}u\|_\infty + \|T_t^{(k)}u - u\|_\infty \\
&\leqslant \varlimsup_{m\to\infty} \sup_{t\leqslant t_0} \|T_t^{(m)}u - T_t^{(k)}u\|_\infty + \|T_t^{(k)}u - u\|_\infty \\
&\leqslant \epsilon + \|T_t^{(k)}u - u\|_\infty.
\end{aligned}$$

Thus,

$$\varlimsup_{t\to 0} \|T_t u - u\|_\infty \leqslant \epsilon + \lim_{t\to 0} \|T_t^{(k)}u - u\|_\infty = \epsilon \xrightarrow[\epsilon\to 0]{} 0$$

which proves the strong continuity of $(T_t)_{t\geqslant 0}$. $\qquad\square$

For a **Lipschitz semigroup** $(T_t)_{t \geq 0}$, i.e. a semigroup satisfying

$$\|T_t u\|_{\text{Lip}} \leq e^{Kt} \|u\|_{\text{Lip}} \quad \forall t > 0, \ u \in \text{Lip}(\mathbb{R}^d)$$

with the **Lipschitz constant** e^{Kt} for the operator T_t, $t \geq 0$, where

$$\text{Lip}(\mathbb{R}^d) := \left\{ u \in C(\mathbb{R}^d) : \|u\|_{\text{Lip}} := \sup_{x \neq y} \frac{|u(x) - u(y)|}{|x - y|} < \infty \right\},$$

we have the following result, cf. Bass [14, Theorem 2.6].

Theorem 7.5. *Let $(T_t^{(k)})_{t \geq 0}$, $k \geq 1$, be Lipschitz semigroups with infinitesimal generators $(A^{(k)}, \mathcal{D}(A^{(k)}))$. Suppose that the Lipschitz constants $e^{K_k t}$ of the semigroups and the constants $K_{A^{(k)}}$ given by*

$$\|A^{(k)} u\|_\infty \leq K_{A^{(k)}} \|u\|_{(2)} \quad \forall u \in C_c^2(\mathbb{R}^d), \quad \|u\|_{(2)} = \sum_{0 \leq |\alpha| \leq 2} \|\nabla^\alpha u\|_\infty$$

are uniformly bounded in k.[3] Then there exists a subsequence $(T_t^{(k_j)})_{t \geq 0}$, $j \geq 1$, and a Lipschitz semigroup $(T_t)_{t \geq 0}$ such that

$$\lim_{j \to \infty} \|T_t^{(k_j)} u - T_t u\|_\infty = 0 \quad \forall u \in C_\infty(\mathbb{R}^d).$$

Moreover, if $\lim_{k \to \infty} \|A^{(k)} u - Au\|_\infty = 0$ for all $u \in C_c^2(\mathbb{R}^d)$ for some linear operator A, then $A|_{C_c^2(\mathbb{R}^d)}$ has an extension which generates $(T_t)_{t \geq 0}$.

7.2 Simulation

We will now focus on an approximation of Feller processes by Markov chains which allows us to simulate such processes. This is important since, in general, the transition probabilities of a Feller process are not known in a form which lends itself to simulations. There are various Markov chain approximations of Markov processes, in general, and to Feller processes, in particular, see for instance Ma–Röckner–Zhang [212] and Ma–Röckner–Sun [213]. Most of them, however, are not helpful for simulations since the increments of the chains are not given in a form which can be used for simulations form: For instance, they are given in terms of the resolvent of the process.

---

[3] Bass assumes the stronger estimate $\|A^{(k)} u\|_\infty \leq K_{A^{(k)}} \sum_{1 \leq |\alpha| \leq 2} \|\nabla^\alpha u\|_\infty$, but his proof works with $\|u\|_{(2)}$ on the right-hand side.

By the following result, we are able to approximate a Feller process through a Markov chain which has Lévy increments. Thus, once we know how to simulate Lévy processes, we can use those to simulate an approximation of a Feller process; for a survey on simulation of Lévy processes we refer to Cont–Tankov [72]. In [45] an approximation procedure was proved for processes with symbols with bounded coefficients, i.e. symbols satisfying $|q(x,\xi)| \leq c(1 + |\xi|^2)$, cf. Theorem 2.31. Here we extend this result to unbounded symbols using techniques developed in [44].

Theorem 7.6. *Let $(X_t)_{t \geq 0}$ be a d-dimensional Feller process with infinitesimal generator $(A, \mathcal{D}(A))$ such that $C_c^\infty(\mathbb{R}^d) \subset \mathcal{D}(A)$ and write $q(x,\xi)$ for the symbol. Assume that*

$$C_c^\infty(\mathbb{R}^d) \quad \text{is an operator core of } A, \tag{7.4}$$

i.e. the closure of $A|_{C_c^\infty}(\mathbb{R}^d)$ is $(A, \mathcal{D}(A))$, and

$$\lim_{|x| \to \infty} \sup_{|\xi| \leq \frac{1}{|x|}} |q(x,\xi)| = 0. \tag{7.5}$$

For each $n \geq 1$ define a Markov chain $(Y^n(k))_{k \geq 1}$ with $Y^n(0) := x_0$ and transition kernel $\mu_{x,\frac{1}{n}}(dy)$ given by

$$\int_{\mathbb{R}^d} e^{iy \cdot \xi} \mu_{x,\frac{1}{n}}(dy) = e^{ix \cdot \xi - \frac{1}{n} q(x,\xi)}, \quad x, \xi \in \mathbb{R}^d, \ n \geq 1.$$

Then

$$Y^n(\lfloor \cdot n \rfloor) \xrightarrow[n \to \infty]{d} X.$$

Here $\lfloor x \rfloor = \max\{k \in \mathbb{Z} : k \leq x\}$, and $\xrightarrow{d}$ denotes convergence in distribution in the space of càdlàg functions $D_{[0,\infty)}(\mathbb{R}^d)$ equipped with the Skorohod J_1 topology.[4]

Proof. We follow the proof of the main theorem in [45]. Let $(X_t)_{t \geq 0}$, A and $q(x,\xi)$ be as in the statement of the theorem. In particular, we assume that (7.4) and (7.5) are satisfied. Denote by $(T_t)_{t \geq 0}$ the corresponding Feller semigroup. For $u \in C_\infty(\mathbb{R}^d)$ we have $\mathbb{E}(u(Y^n(k))) = W_{\frac{1}{n}}^k u(x_0)$ where

$$W_{\frac{1}{n}} u(x) = \int_{\mathbb{R}^d} u(y) \mu_{x,\frac{1}{n}}(dy),$$

---

[4]See the footnote on p. 168.

and for $u \in C_c^\infty(\mathbb{R}^d)$,

$$W_{\frac{1}{n}}u(x) = \int e^{ix\cdot\xi}e^{-\frac{1}{n}q(x,\xi)}\hat{u}(\xi)\,d\xi.$$

By Theorem 7.1, the assertion is equivalent to

$$\lim_{n\to\infty}\left\|W_{\frac{1}{n}}^{\lfloor tn\rfloor}u - T_t u\right\|_\infty = 0 \quad \forall t > 0, \ u \in C_\infty(\mathbb{R}^d). \tag{7.6}$$

Thus, by Ethier–Kurtz [100, Theorem 6.5.iii, p. 31], we are done, if we can show

$$\lim_{n\to\infty}\left\|\frac{W_{\frac{1}{n}}u - u}{\frac{1}{n}} - Au\right\|_\infty = 0 \quad \forall u \in C_c^\infty(\mathbb{R}^d).$$

To prove this convergence, we fix $\epsilon > 0$, $u \in C_c^\infty(\mathbb{R}^d)$ and $r = r(u) > 0$ such that $\operatorname{supp} u \subset \mathbb{B}(0, r)$. We also denote for each $x \in \mathbb{R}^d$ by $(L_t^{(x)})_{t\geqslant 0}$ the Lévy process with characteristic exponent $\xi \mapsto q(x,\xi)$ and starting point $L_0^{(x)} = x$; thus, the random variable $L_t^{(x)}$ has the characteristic function $e^{ix\cdot\xi - tq(x,\xi)}$. Furthermore, we define for every x and $r > 0$

$$\tau_{\mathbb{B}(x,r)} := \inf\left\{t > 0 : L_t^{(x)} \notin \mathbb{B}(0,r)\right\}.$$

For $|x| > r$ the elementary inclusion $\mathbb{B}(0,r) \subset \mathbb{B}^c(x, |x| - r)$ implies

$$
\begin{aligned}
|W_{\frac{1}{n}}u(x) - u(x)| = |W_{\frac{1}{n}}u(x)| &= \left|\mathbb{E}\,u\left(L_{\frac{1}{n}}^{(x)}\right)\right| \\
&\leqslant \|u\|_\infty\,\mathbb{P}\left(L_{\frac{1}{n}}^{(x)} \in \mathbb{B}(0,r)\right) \\
&\leqslant \|u\|_\infty\,\mathbb{P}\left(L_{\frac{1}{n}}^{(x)} \in \mathbb{B}^c(x, |x| - r)\right) \\
&\leqslant \|u\|_\infty\,\mathbb{P}\left(\tau_{\mathbb{B}(x,|x|-r)} \leqslant \tfrac{1}{n}\right) \\
&\leqslant \tfrac{c}{n}\|u\|_\infty \sup_{|\xi|\leqslant\frac{1}{|x|-r}} |q(x,\xi)|,
\end{aligned}
$$

where the last inequality is due to Theorem 5.1. For $|x| > r$ we find by (the proof of) Lemma 3.26, see also [44, Lemma 3.1], that

$$|Au(x)| \leqslant c\,\|u\|_\infty \sup_{|\xi|\leqslant\frac{1}{|x|-r}} |q(x,\xi)|.$$

Thus, we get for $|x| > r$

$$\left| \frac{W_{\frac{1}{n}} u(x) - u(x)}{\frac{1}{n}} - Au(x) \right| \le C \|u\|_\infty \sup_{|\xi| \le \frac{1}{|x|-r}} |q(x,\xi)|.$$

Using (7.5) and Lemma 3.26, there is some $K = K(\epsilon, u, q, C)$ such that

$$\left| \frac{W_{\frac{1}{n}} u(x) - u(x)}{\frac{1}{n}} - Au(x) \right| \le \frac{\epsilon}{2} \quad \forall |x| > K.$$

Since the symbol $q(x, \xi)$ is locally bounded in x, cf. Proposition 2.27(d), there exists some constant c_K such that

$$\sup_{|x| \le K} |q(x, \xi)| \le c_K (1 + |\xi|^2) \quad \forall \xi \in \mathbb{R}^d.$$

Repeated applications of the mean value theorem with suitable intermediate values $0 < h, s < n^{-1}$ yield

$$\left| \frac{e^{-\frac{1}{n} q(x,\xi)} - 1}{\frac{1}{n}} + q(x, \xi) \right| = \left| -q(x, \xi) \left(e^{-sq(x,\xi)} - 1 \right) \right| = s |q(x, \xi)|^2 e^{-h \operatorname{Re} q(x,\xi)}.$$

Thus, for $|x| \le K$, compare with [45],

$$\left| \frac{W_{\frac{1}{n}} u(x) - u(x)}{\frac{1}{n}} - Au(x) \right| = \left| \int e^{ix \cdot \xi} \left(\frac{e^{-\frac{1}{n} q(x,\xi)} - 1}{\frac{1}{n}} + q(x, \xi) \right) \hat{u}(\xi) \, d\xi \right|$$

$$\le \int s |q(x, \xi)|^2 e^{-h \operatorname{Re} q(x,\xi)} |\hat{u}(\xi)| \, d\xi$$

$$\le s \int |q(x, \xi)|^2 |\hat{u}(\xi)| \, d\xi$$

$$\le \frac{c_K^2}{n} \int (1 + |\xi|^2)^2 |\hat{u}(\xi)| \, d\xi$$

$$\le \frac{1}{n} \tilde{C},$$

where $\tilde{C} = \tilde{C}(c_K, u)$ is a constant. But now we can select $N = N(\tilde{C}, \epsilon)$ such that

$$\frac{1}{n} \tilde{C} \le \frac{\epsilon}{2} \quad \forall n \ge N,$$

and it follows that for all $n \geq N$

$$\left\| \frac{W_{\frac{1}{n}} u - u}{\frac{1}{n}} - Au \right\|_\infty \leq \sup_{|x|>K} \left| \frac{W_{\frac{1}{n}} u(x) - u(x)}{\frac{1}{n}} - Au(x) \right|$$

$$+ \sup_{|x|\leq K} \left| \frac{W_{\frac{1}{n}} u(x) - u(x)}{\frac{1}{n}} - Au(x) \right|$$

$$\leq \frac{\epsilon}{2} + \frac{\epsilon}{2} = \epsilon.$$

This finishes the proof of the theorem. $\qquad\qquad\qquad\qquad\qquad\qquad\qquad\square$

Simulations, including the source-code, which use this approximation can be found in [42]. Moreover, in [46] it is shown that the above approximation method coincides with the Euler scheme for the SDE which is solved by the Feller process, see Proposition 3.10 and (3.16). This is quite remarkable as it provides a way to prove the convergence of the Euler scheme for an SDE where the coefficients—in particular, their smoothness properties—are unknown.

An interesting connection between the above approximation scheme and a rigorous definition of Lévy-based Feynman integrals is discussed in a series of papers [47, 54, 55, 56].

Chapter 8
Open Problems

The following list contains a few important questions and open problems of varying difficulty.

(a) Let $(T_t)_{t\geq0}$ be a Feller semigroup. The set

$$\mathcal{X}_0 := \{u \in B_b(E) : t \mapsto T_t u \text{ is strongly continuous}\}$$

is called the **domain of strong continuity** (we consider the sup-norm $\|\cdot\|_\infty$). Determine the exact domain of strong continuity.

Remark: Some discussions can be found in the early texts on semigroups, in particular in Dynkin [97, Chap. II.§§1–2, pp.Ât' 47–61] and Loève [208, Sects. 45.2 and 45.3, pp. 336–381].

(b) Let $\psi_1, \psi_2 : \mathbb{R}^d \to \mathbb{C}$ be characteristic exponents of two Lévy processes. Find necessary and sufficient conditions that their difference $\psi_1 - \psi_2$ and quotient ψ_1/ψ_2 are again characteristic exponents of a Lévy process.

(c) Is the Courrège–von Waldenfels theorem (Theorem 2.21) still valid if we require that $\mathcal{D}(A)$ contains a full and complete class (Definition 2.47) instead of $C_c^\infty(\mathbb{R}^d)$?

(d) Find a necessary and sufficient condition such that the limit (2.43) from Definition 2.41 defining the symbol of a stochastic process exists.

(e) It is known that every Lévy kernel $N(x, dy)$ can be written as a pull-back of the form $\nu(k(x, y) \in dy)$ where ν is a Lévy measure with infinite total mass. Describe the smoothness properties of $x \mapsto k(x, y)$ in terms of corresponding properties of $N(x, dy)$ or the symbol $q(x, \xi)$ (with Lévy triplet $(0, 0, N(x, dy))$).

Remark: This addresses the discussion before Theorem 3.10 on p. 77. The existence of a pullback map for any two Lévy measures was established by Skorokhod [304, Chap. 3.4, Lemma 2, p. 77] and for general Lévy kernels it

B. Böttcher et al., *Lévy Matters III*, Lecture Notes
in Mathematics 2099, DOI 10.1007/978-3-319-02684-8_8,
© Springer International Publishing Switzerland 2013

is due to El Karoui–Lepeltier [99], see also Jacod [169, Lemmas (14.50) and (14.51), pp. 469–471].

(f) Improve the non-explosion result of Theorem 2.34: Is there an integral test or a Has'minskii-condition for (non-)explosion of Feller processes?

(g) Find sharp criteria (e.g. integral criteria of Chung–Fuchs type) in terms of the symbol such that a Feller process is recurrent or transient, cf. Sect. 6.3.

(h) Provide further examples (cf. Example 2.26) of negative definite symbols $q(x, \xi)$ which are jointly continuous (in x and ξ) and which do not define Feller processes.

Are there counterexamples which are real-valued or which satisfy the sector condition $|\mathrm{Im} q(x, \xi)| \leqslant c \mathrm{Re} q(x, \xi)$?

(i) Find optimal (necessary and) sufficient conditions on the symbol $q(x, \xi)$ such that $(-q(x, D), C_c^\infty(\mathbb{R}^d))$ extends to the generator of a Feller generator.

Remark: This is likely to be a very hard problem. An educated guess for minimal sufficient conditions are: joint continuity in (x, ξ), Hölder continuity in x (in view of Example 2.26) and possibly a tightness condition as in Theorem 2.30(d), see also Lemma 3.26. The assumption of smoothness in x or ξ does not seem to be natural.

(j) Find conditions in terms of the symbol $q(x, \xi)$ or the Lévy characteristics $(l(x), Q(x), N(x, dy))$ that guarantee that a Feller generator is symmetric (with respect to a measure m).

(k) Characterize those Feller processes for which the test functions $C_c^\infty(\mathbb{R}^d)$ are an operator core.

(l) Find a non-trivial example of a Feller process, which does not contain the test functions $C_c^\infty(\mathbb{R}^d)$ in its domain. (A randomly time-changed process would be "trivial" in this sense).

(m) Let $(X_t)_{t \geqslant 0}$ be a Feller process with the symbol $q(x, \xi)$. Find a reasonable lower bound for $\mathbb{P}^x(|X_t - x| \geqslant r)$.

(n) Adapt the well-developed Wentzell–Freidlin theory of large deviations to Lévy and Lévy-type processes. In particular, one should be able to express the good rate function in terms of the symbol.

Remark: A good starting point should be the monograph Freidlin–Wentzell [111] and the work by Takeda and co-authors [300, 315, 316]. For Lévy processes (with values in a Banach space) see da Acosta [83], and for semimartingales see Liptser–Pukhalskii [207].

(o) Find a lower bound for the Hausdorff dimension of a Feller process. This question generalizes the corresponding result for Lévy processes (see Theorem 5.12), and it also completes the existing result for Feller processes (see Theorem 5.15).

(p) Prove a LIL or Chung-type LIL result for a Feller process (as $t \to 0$). This will strengthen the Hölder results from Theorem 5.16 and (5.18).

Remark: As a first step, find conditions for upper and lower functions for the sample paths of a Feller process. Some partial results are contained in the recent paper [182].

(q) Find criteria for the existence of successful couplings for stochastic differential equations driven by a *multiplicative* Lévy noise.

(r) The coupling property for Lévy processes and Lévy-driven Ornstein–Uhlenbeck processes has a number of interesting applications, see e.g. the recent papers [344, 347] on the regularity of Lévy-type semigroups and the exponential ergodicity of Ornstein–Uhlenbeck processes based on coupling. Extend the range of such applications and provide a general criterion for successful couplings of Lévy-type processes.

References[1]

1. Abels, H., Husseini, R.: On hypoellipticity of generators of Lévy processes. Ark. Mat. **48**, 231–242 (2010) (p. 37)
2. Abels, H., Kassmann, M.: The Cauchy problem and the martingale problem for integro-differential operators with non-smooth kernels. Osaka J. Math. **46**, 661–683 (2009) (p. 94)
3. Aït-Sahalia, Y., Jacod, J.: Estimating the degree of activity of jumps in high frequency data. Ann. Stat. **37**, 2202–2244 (2009) (p. 125)
4. Albeverio, S., Rüdiger, B., Wu, J.-L.: Invariant measures and symmetry property of Lévy type operators. Potential Anal. **13**, 147–168 (2000) (p. 143)
5. Applebaum, D.: Lévy Processes and Stochastic Calculus, 2nd edn. Cambridge University Press, Cambridge (2009) (p. 40, 108)
6. Aronson, D.G.: Bounds on the fundamental solution of a parabolic equation. Bull. Am. Math. Soc. **73**, 890–896 (1967) (p. 5)
7. Aurzada, F., Döring, L., Savov, M.: Small time Chung type LIL for Lévy processes. Bernoulli **19**, 115–136 (2013) (p. 129)
8. Azéma, J., Kaplan-Duflo, D., Revuz, D.: Récurrence fine des processus de Markov. Ann. Inst. Henri Poincaré (Sér. B) **2**, 185–220 (1966) (p. 163)
9. Baeumer, B., Kovács, M., Meerschaert, M.M., Schilling, R.L., Straka, P.: Reflected spectrally negative stable processes and their governing equations. Preprint [arXiv: 1301.5605] (2013) (p. 24)
10. Baldus, F.: Application of the Weyl–Hörmander calculus to generators of Feller semigroups. Math. Nachr. **252**, 3–23 (2003) (p. 88)
11. Bañuelos, R., Bogdan, K.: Lévy processes and Fourier multipliers. J. Funct. Anal. **250**, 197–213 (2007) (p. 37)
12. Barlow, M.T., Bass, R.F., Chen, Z.-Q., Kaßmann, M.: Non-local Dirichlet forms and symmetric jump processes. Trans. Am. Math. Soc. **361**, 1963–1999 (2009) (p. 84, 85)
13. Barndorff-Nielsen, O.E., Levendorskĭ, S.Z.: Feller processes of normal inverse Gaussian type. Quant. Financ. **1**, 318–331 (2001) (p. 51, 74)
14. Bass, R.F.: Markov processes with Lipschitz semigroups. Trans. Am. Math. Soc. **267**, 307–320 (1981) (p. 171)
15. Bass, R.F.: Uniqueness in law for pure jump Markov processes. Probab. Theory Relat. Fields **79**, 271–287 (1988) (p. 94, 95)
16. Bass, R.F.: Regularity results for stable-like operators. J. Funct. Anal. **257**, 2693–2722 (2009) (p. 19)

---

[1]Each reference is followed, in round brackets, by a list of page numbers where we cite it.

B. Böttcher et al., *Lévy Matters III*, Lecture Notes
in Mathematics 2099, DOI 10.1007/978-3-319-02684-8,
© Springer International Publishing Switzerland 2013

17. Bass, R.F., Levin, D.A.: Harnack inequalities for jump processes. Potential Anal. **17**, 375–388 (2002) (p. 85)
18. Behme, A., Lindner, A.: On exponential functionals of Lévy processes. J. Theor. Probab. (2013). doi:10.1007/s10959-013-0507-y (p. 4, 15, 20, 97)
19. Bendikov, A., Maheux, P.: Nash type inequalities for fractional powers of non-negative self-adjoint operators. Trans. Am. Math. Soc. **359**, 3085–3098 (2007) (p. 142)
20. Bensoussan, A., Lions, J.L.: Applications des inéquations variatioinnelles en contôle stochastique. Dunod/Bordas, Paris (1978) (p. 72)
21. Bensoussan, A., Lions, J.L.: Contôle impulsionnel et inéquations quasi variationnelles. Dunod/Bordas, Paris (1982) (p. 72)
22. Benveniste, A., Jacod, J.: Systèmes de Lévy des processus de Markov. Invent. Math. **21**, 183–198 (1973) (p. 65)
23. Berg, C., Forst, G.: Non-symmetric translation invariant Dirichlet forms. Invent. Math. **21**, 199–212 (1973) (p. 82)
24. Berg, C., Forst, G.: Potential Theory on Locally Compact Abelian Groups. Springer, Berlin (1975) (p. 4, 40, 42, 43, 44, 62)
25. Berg, C., Christensen, J.P.R., Ressel, P.: Harmonic Analysis on Semigroups. Springer, New York (1984) (p. 45)
26. Berg, C., Boyadzhiev, K., deLaubenfels, R.: Generation of generators of holomorphic semigroups. J. Aust. Math. Soc. Ser. A **55**, 246–269 (1993) (p. 103)
27. Bertoin, J.: Lévy Processes. Cambridge University Press, Cambridge (1996) (p. xii, 40, 111)
28. Billingsley, P.: Convergence of Probability Measures. Wiley, New York (1968) (p. 89, 168)
29. Bingham, N.H., Goldie, C.M., Teugels, J.L.: Regular Variation. Cambridge University Press, Cambridge (1989) (p. 33)
30. Bliedtner, J., Hansen, W.: Potential Theory. An Analytic and Probabilistic Approach to Balayage. Springer, Berlin (1986) (p. 10, 11)
31. Blumenthal, R., Getoor, R.: Some theorems on stable processes. Trans. Am. Math. Soc. **95**, 263–273 (1960) (p. xii, 143, 153)
32. Blumenthal, R., Getoor, R.: Sample functions of stochastic processes with stationary independent increments. J. Math. Mech. **10**, 493–516 (1961) (p. xii, 124, 125, 129)
33. Blumenthal, R.M., Getoor, R.K.: Markov Processes and Potential Theory. Academic, New York (1968) (p. 10)
34. Bochner, S.: Diffusion equation and stochastic processes. Proc. Natl. Acad. Sci. USA **35**, 368–370 (1949) (p. 103)
35. Bogachev, V.I.: Measure Theory, vol. 2. Springer, Berlin (2007) (p. 41)
36. Bogachev, V.I., Röckner, M., Schmuland, B.: Generalized Mehler semigroups and applications. Probab. Theory Relat. Fields **105**(2), 193–225 (1996) (p. 4)
37. Bogdan, K., Burdzy, K., Chen, Z.-Q.: Censored stable processes. Probab. Theory Relat. Fields **127**, 89–152 (2003) (p. 99)
38. Bony, J.-M., Courrège, P., Priouret, P.: Semi–groupes de Feller sur une variété à bord compacte et problème aux limites intégro-differentiels du second ordre donnant lieu au principe du maximum. Ann. Inst. Fourier **18**.2, 369–521 (1968) (p. 47)
39. Böttcher, B.: Some investigations on Feller processes generated by pseudo-differential operators. Ph.D. Thesis, University of Wales, Swansea (2004) (p. 72)
40. Böttcher, B.: A parametrix construction for the fundamental solution of the evolution equation associated with a pseudo-differential operator generating a Markov process. Math. Nachr. **278**, 1235–1241 (2005) (p. 88)
41. Böttcher, B.: Construction of time inhomogeneous Markov processes via evolution equations using pseudo-differential operators. J. Lond. Math. Soc. **78**, 605–621 (2008) (p. 88)
42. Böttcher, B.: Feller processes: The next generation in modeling. Brownian motion, Lévy processes and beyond. PLoS One **5**, e15102 (2010) (p. 51, 175)
43. Böttcher, B.: An overshoot approach to recurrence and transience of Markov processes. Stoch. Process. Appl. **121**, 1962–1981 (2011) (p. 160, 161, 162, 164)

44. Böttcher, B.: On the construction of Feller processes with unbounded coefficients. Electron. Commun. Probab. **16**, 545–555 (2011) (p. 91, 97, 172, 173)
45. Böttcher, B., Schilling, R.L.: Approximation of Feller processes by Markov chains with Lévy increments. Stoch. Dyn. **9**, 71–80 (2009) (p. 54, 172, 174)
46. Böttcher, B., Schnurr, A.: The Euler scheme for Feller processes. Stoch. Anal. Appl. **29**, 1045–1056 (2011) (p. 76, 175)
47. Böttcher, B., Butko, Y.A., Smolyanov, O.G., Schilling, R.L.: Feynman formulas and path integrals for some evolution equations related to τ-quantization. Russ. J. Math. Phys. **18**, 387–399 (2011) (p. 175)
48. Böttcher, B., Schilling, R.L., Wang, J.: Constructions of coupling processes for Lévy processes. Stoch. Process. Appl. **121**, 1201–1216 (2011) (p. 147, 152, 153, 155, 156)
49. Bouleau, N., Hirsch, F.: Dirichlet Forms and Analysis on Wiener Space. De Gruyter, Berlin (1991) (p. 22, 80)
50. Boyarchenko, S., Levendorskiĭ, S.Z.: Option pricing for truncated Lévy processes. Int. J. Theor. Appl. Financ. **3**, 549–552 (2000) (p. 74)
51. Boyarchenko, S., Levendorskiĭ, S.Z.: Non-Gaussian Merton–Black–Scholes Theory. World Scientific, Singapore (2002) (p. 74)
52. Breiman, L.: Probability. SIAM, Philadelphia (1992) (Unabridged, corrected republication of the 1968 edition, Addison–Wesley, Reading (MA)) (p. 41)
53. Bretagnolle, J.L.: Processus à accroissements indépendants. In: Badrikian, A., Hennequin, P.L. (eds.) Ecole d'Été de Probabilités: Processus Stochastiques. Lecture Notes in Mathematics, vol. 307, pp. 1–26. Springer, Berlin (1973) (p. 40)
54. Butko, Y.A., Smolyanov, O.G., Schilling, R.L.: Feynman formulae for Feller semigroups. Dokl. Math. **82**, 679–683 (2010) (p. 175)
55. Butko, Y.A., Smolyanov, O.G., Schilling, R.L.: Hamiltonian Feynman–Kac and Feynman formulae for dynamics of particles with position-dependent mass. Int. J. Theor. Phys. **50**, 2009–2018 (2011) (p. 175)
56. Butko, Y.A., Smolyanov, O.G., Schilling, R.L.: Lagrangian and Hamiltonian Feynman formulae for some Feller semigroups and their perturbations. Infinite Dimens. Anal. Quantum Probab. Relat. Top. **15**, 1250015 (26 pp.) (2012) (p. 175)
57. Butzer, P.L., Berens, H.: Semi-Groups of Operators and Approximation. Springer, Berlin (1967) (p. 28)
58. Cancelier, C.: Problèmes aux limites pseudo-differentiels donnant lieu au principe du maximum. Commun. Partial Differ. Equ. **11**, 1677–1726 (1986) (p. 88)
59. Chen, M.-F.: Eigenvalues, Inequalities, and Ergodic Theory. Springer, London (2005) (p. 141, 147)
60. Chen, Z.-Q.: Symmetric jump processes and their heat kernel estimates. Sci. China Ser. A Math. **52**, 1423–1445 (2009) (p. xii, 143)
61. Chen, M.-F., Li, S.: Coupling methods for multidimensional diffusion processes. Ann. Probab. **17**, 151–177 (1989) (p. 151)
62. Chen, Z.-Q., Kumagai, T.: Heat kernel estimates for stable-like processes on d-sets. Stoch. Process. Appl. **108**, 27–68 (2003) (p. 86)
63. Chen, X., Wang, J.: Weighted Poincaré inequalities for nonlocal Dirichlet forms. Preprint [arXiv: 1207.7140] (2012) (p. 146)
64. Chen, X., Wang, J.: Functional inequalities for nonlocal Dirichlet forms with finite range jumps or large jumps. Stoch. Process. Appl. **124**, 123–153 (2014). doi:10.1016/j.spa.2013.07.001 (p. 146)
65. Chen, Z.-Q., Kim, P., Song, R.: Dirichlet heat kernel estimates for $\Delta^{\alpha/2} + \Delta^{\beta/2}$. Ill. J. Math. **54**, 1357–1392 (2010) (p. xii)
66. Chen, Z.-Q., Kim, P., Song, R.: Dirichlet heat kernel estimates for fractional Laplacian under gradient perturbation. Ann. Probab. **40**, 2483–2538 (2012) (p. xii)
67. Chung, K.L.: Lectures from Markov Processes to Brownian Motion. Springer, Berlin (1982) (p. xii)
68. Chung, K.L., Zhao, Z.: From Brownian Motion to Schrödinger's Equation. Springer, Berlin (2001) (p. xii, 108, 109, 110)

69. Ciesielski, Z., Kerkyacharian, G., Roynette, B.: Quelques espaces fonctionnels associés à des processus gaussienes. Studia Math. **107**, 171–204 (1993) (p. 136, 139, 140)

70. Çinlar, E., Jacod, J.: Representation of semimartingale Markov processes in terms of Wiener processes and Poisson random measures. In: Seminar on Stochastic Processes, pp. 159–242. Birkhäuser, Boston (1981) (p. 63, 67, 77)

71. Çinlar, E., Jacod, J., Protter, P., Sharpe, M.J.: Semimartingales and Markov Processes. Z. Wahrscheinlichkeitstheor. verw. Geb. **54**, 161–219 (1980) (p. 63, 66, 67, 77)

72. Cont, R., Tankov, P.: Financial Modelling with Jump Processes. Chapman & Hall/CRC, Boca Raton (2004) (p. 172)

73. Courrège, P.: Générateur infinitésimal d'un semi-groupe de convolution sur $\mathbb{R}^n$, et formule de Lévy–Khinchine. Bull. Sci. Math. **88**, 3–30 (1964) (p. 33)

74. Courrège, P.: Sur la forme intégro–différentielle des opérateurs de C_k^∞ dans C satisfaisant au principe du maximum. In: Séminaire Brelot–Choquet–Deny. Théorie du potentiel, tome 10, exposé no. 2, pp. 1–38 (1965/1966) (p. 47)

75. Cranston, M., Greven, A.: Coupling and harmonic functions in the case of continuous time Markov processes. Stoch. Process. Appl. **60**, 261–286 (1995) (p. 149, 157)

76. Cranston, M., Wang, F.-Y.: A condition for the equivalence of coupling and shift coupling. Ann. Probab. **28**, 1666–1679 (2000) (p. 149)

77. Cuppens, R.: Decomposition of Multivariate Probabilities. Academic, New York (1975) (p. 44)

78. Davies, B.: One-Parameter Semigroups. Academic, London (1980) (p. 7, 17, 23, 24)

79. Davies, E.B.: Heat Kernels and Spectral Theory. Cambridge University Press, Cambridge (1990) (p. 141)

80. Davis, M.H.A.: Markov Models & Optimization. Chapman & Hall/CRC, London (1993) (p. 26)

81. Davies, E.B., Simon, B.: Ultracontractivity and the heat kernel for Schrödinger operators and Dirichlet Laplacians. J. Funct. Anal. **59**, 335–395 (1984) (p. 141)

82. Dautray, R., Lions, J.-L.: Mathematical Analysis and Numerical Methods for Science and Technology, vol. 1. Springer, Berlin (1990) (p. 18)

83. de Acosta, A.: Large deviations for vector-valued Lévy processes. Stoch. Process. Appl. **51**, 75–115 (1994) (p. 178)

84. Dellacherie, C., Meyer, P.-A.: Probabilités et potentiel (tome 4 – chapitres XII à XVI). Hermann, Paris (1987) (p. 6)

85. Demuth, M., van Casteren, J.A.: Stochastic Spectral Theory for Selfadjoint Feller Operators. A Functional Integration Approach. Birkhäuser, Basel (2000) (p. 109, 110)

86. Dieudonné, J.: Foundations of Modern Analysis (Enlarged and Corrected Printing). Academic, New York (1969) (p. 33)

87. Dorroh, J.R.: Contraction semi-groups in a function space. Pac. J. Math. **19**, 35–38 (1966) (p. 100)

88. Dudley, R.M., Norvaiša, R.: An Introduction to p-Variation and Young Integrals. MaPhySto Lecture Notes, vol. 1. University of Aarhus, Aarhus (1998) (p. 132)

89. Dudley, R.M., Norvaiša, R.: Differentiability of Six Operators on Nonsmooth Functions and p-Variation. Lecture Notes in Mathematics, vol. 1703. Springer, Berlin (1998) (p. 132)

90. Dudley, R.M., Norvaiša, R.: Concrete Functional Calculus. Springer, New York (2011) (p. 132)

91. Duffie, D., Filipović, D., Schachermayer, W.: Affine processes and applications in finance. Ann. Appl. Probab. **13**, 984–1053 (2003) (p. 5, 20)

92. Duistermaat, J.J., Kolk, J.A.C.: Distributions. Theory and Applications. Birkhäuser/Springer, New York (2010) (p. 46)

93. Dunford, N., Schwartz, J.T.: Linear Operators, vol. 1. Wiley-Interscience, New York (1957) (p. 1)

94. Duoandikoetxea, J.: Fourier Analysis. American Mathematical Society, Providence (2001) (p. 37)

95. Dupuis, C.: Mesure de Hausdorff de la trajectoire de certains processus à accroissements indépendants et stationnaires. In: Dellacherie, C., Meyer, P.A., Weil, M. (eds.) Séminaire de Probabilités VIII. Lecture Notes in Mathematics, vol. 381, pp. 40–77. Springer, Berlin (1974) (p. 129)

96. Dynkin, E.B.: Die Grundlagen der Theorie der Markoffschen Prozesse. Springer, Berlin (1961) (English translation: Theory of Markov Processes, Dover, Mineola (NY), 2006) (p. 28)

97. Dynkin, E.B.: Markov Processes, vol. 1. Springer, Berlin (1965) (p. 15, 26, 27, 177)

98. Edwards, R.E.: Fourier Series. A Modern Introduction, vol. 2, 2nd edn. Springer, Berlin (1982) (p. 37)

99. El Karoui, N., Lepeltier, J.-P.: Représentation des processus ponctuels multivariés à l'aide d'un processus de Poisson. Z. Wahrscheinlichkeitstheor. verw. Geb. **39**, 111–134 (1977) (p. 178)

100. Ethier, S.N., Kurtz, T.G.: Markov Processes: Characterization and Convergence. Wiley, New York (1986) (p. 1, 12, 16, 17, 21, 22, 24, 25, 89, 100, 107, 168, 173)

101. Evans, K.P., Jacob, N.: Feller semigroups obtained by variable order subordination. Rev. Mat. Complut. **20**, 293–307 (2007) (p. 105)

102. Evans, S.N., Sowers, R.B.: Pinching and twisting Markov processes. Ann. Probab. **31**, 486–527 (2003) (p. 99)

103. Falconer, K.: Fractal Geometry. Mathematical Foundations and Applications, 2nd edn. Wiley, Chichester (2003) (p. 124)

104. Farkas, W., Jacob, N., Schilling, R.L.: Feller semigroups, L^p-sub-Markovian semigroups, and applications to pseudo-differential operators with negative definite symbols. Forum Math. **13**, 51–90 (2001) (p. 73, 83, 85)

105. Farkas, W., Jacob, N., Schilling, R.L.: Function spaces related to continuous negative definite functions: ψ-Bessel potential spaces. Diss. Math. **393**, 1–62 (2001) (p. 37, 73, 85)

106. Feller, W.: Zur Theorie der stochastischen Prozesse (Existenz- und Eindeutigkeitssätze). Math. Ann. **113**, 113–160 (1936) (p. 86)

107. Feller, W.: On the integro-differential equations of purely discontinuous Markoff processes. Trans. Am. Math. Soc. **48**, 488–515 (1940) (p. 86)

108. Feller, W.: The parabolic differential equations and the associated semigroups. Ann. Math. **55**, 468–519 (1952) (p. 86)

109. Franke, B.: The scaling limit behaviour of periodic stable-like processes. Bernoulli **12**, 551–570 (2006) (p. 164, 185)

110. Franke, B.: Correction to: "The scaling limit behaviour of periodic stable-like processes" [109]. Bernoulli **13**, 600 (2007) (p. 164)

111. Freidlin, M.I., Wentzell, A.D.: Random Perturbations of Dynamical Systems, 2nd edn. Springer, New York (1998) (p. 178)

112. Fristedt, B.: Sample functions of stochastic processes with stationary, independent increments. In: Ney, P., Port, S. (eds.) Advances in Probability and Related Topics, vol. 3, pp. 241–396. Marcel Dekker, New York (1974) (p. xii, 40, 111, 129)

113. Fukushima, M.: Dirichlet Forms and Markov Processes (in Japanese). Kinokuniya, Tokyo (1975) (p. 82)

114. Fukushima, M.: On an L^p-estimate of resolvents of Markov processes. Publ. RIMS Kyoto Univ. **13**, 277–284 (1977) (p. 29)

115. Fukushima, M.: On a decomposition of additive functionals in the strict sense for a symmetric Markov process. In: Ma, Z.-M., Röckner, M., Yan, J.-A. (eds.) Dirichlet Forms and Stochastic Processes. Proceedings of the International Conference Held in Beijing, China, 1993, pp. 155–169. De Gruyter, Berlin (1995) (p. xii, 85)

116. Fukushima, M., Kaneko, H.: (r, p)-capacities for general Markovian semigroups. In: Albeverio, S. (ed.) Infinite-dimensional Analysis and Stochastic Processes (Bielefeld 1983). Research Notes in Mathematics, vol. 124, pp. 41–47. Pitman, Boston (1985) (p. 29)

117. Fukushima, M., Uemura, T.: Jump-type Hunt processes generated by lower bounded semi-Dirichlet forms. Ann. Probab. **40**, 858–889 (2012) (p. 81, 82, 96)

118. Fukushima, M., Oshima, Y., Takeda, M.: Dirichlet Forms and Symmetric Markov Processes. De Gruyter, Berlin (1994) (p. 79, 80, 82, 83, 84, 102, 110, 141, 144)
119. Fukushima, M., Oshima, Y., Takeda, M.: Dirichlet Forms and Symmetric Markov Processes, 2nd edn. De Gruyter, Berlin (2011) (p. 110)
120. Garroni, M.G., Menaldi, J.L.: Green functions for second order parabolic integro-differential problems. Research Notes in Mathematics, vol. 275. Longman, Harlow, Essex (1992) (p. 72)
121. Garsia, A.M., Rodemich, E., Rumsey, H.: A real variable lemma and the continuity of paths of some Gaussian processes. Indiana Univ. Math. J. **20**, 565–578 (1970) (p. 136)
122. Gentil, I., Maheux, P.: Super-Poincaré and Nash-type inequalities for subordinated semigroups. Preprint [arXiv: 1105.3095v2] (2011) (p. 142)
123. Getoor, R.K.: Transience and recurrence of Markov processes. In: Azéma, J., Yor, M. (eds.) Séminaire de Probabilités XIV 1978/1979. Lecture Notes in Mathematics, vol. 784, pp. 397–409. Springer, Berlin (1980) (p. 160)
124. Gomilko, A., Haase, M., Tomilov, Y.: Bernstein functions and rates in mean ergodic theorems for operator semigroups. J. Anal. Math. **118**, 545–576 (2012) (p. 103)
125. Gong, F.-Z., Wang, F.-Y.: Functional inequalities for uniformly integrable semigroups and application to essential spectrums. Forum Math. **14**, 293–314 (2002) (p. 142)
126. Grafakos, L.: Classical Fourier Analysis, 2nd edn. Springer, Berlin (2008) (p. 37)
127. Gross, L.: Logarithmic Sobolev inequalities. Am. J. Math. **97**, 1061–1083 (1975) (p. 141, 143)
128. Günter, N.M.: La théorie du potential et ses applications aux problèmes fundamentaux de la physique mathématique (Collection Borel). Gauthier–Villars, Paris (1934) (p. 18, 186)
129. Günter, N.M.: Die Potentialtheorie und ihre Anwendung auf Grundaufgaben der mathematischen Physik (extended German translation of [128]), 2nd edn. B.G. Teubner, Leipzig (1957) (p. 18)
130. Hartman, P., Wintner, A.: On the infinitesimal generators of integral convolutions. Am. J. Math. **64**, 273–298 (1942) (p. 155)
131. Hasegawa, M.: A note on the convergence of semi-groups of operators. Proc. Jpn. Acad. **40**, 262–266 (1964) (p. 169)
132. Hawkes, J.: Potential theory of Lévy processes. Proc. Lond. Math. Soc. **38**, 335–352 (1979) (p. 11)
133. Herren, V.: Lévy type processes and Besov spaces. Potential Anal. **7**, 689–704 (1997) (p. 136, 139)
134. Herz, C.S.: Théorie élémentaire des distributions de Beurling. Publ. Math. Orsay no. 5, 2ème année 1962/1963, Paris (1964) (p. 47)
135. Hoh, W.: The martingale problem for a class of pseudo differential operators. Math. Ann. **300**, 121–147 (1994) (p. 93)
136. Hoh, W.: Pseudo differential operators with negative definite symbols and the martingale problem. Stoch. Stoch. Rep. **55**, 225–252 (1995) (p. 74)
137. Hoh, W.: A symbolic calculus for pseudo-differential operators generating Feller semigroups. Osaka J. Math. **35**, 798–820 (1998) (p. 44, 51, 72)
138. Hoh, W.: Pseudo-Differential Operators Generating Markov Processes. Habilitationsschrift Universität Bielefeld, Bielefeld. http://www.mathematik.uni-bielefeld.de/~hoh/pdo_mp.ps (1998) (p. 11, 47, 51, 54, 57, 62, 72, 73, 74, 90, 93, 94, 96, 107)
139. Hoh, W.: Perturbations of pseudo differential operators with negative definite symbols. Appl. Anal. Optim. **45**, 269–281 (2002) (p. 74, 107)
140. Hoh, W., Jacob, N.: On the Dirichlet problem for pseudodifferential operators generating Feller semigroups. J. Funct. Anal. **137**, 19–48 (1996) (p. 11, 57)
141. Hörmander, L.: The Analysis of Linear Partial Differential Operators II, 1st edn. Springer, Berlin (1983) (p. 61)
142. Hörmander, L.: The Analysis of Linear Partial Differential Operators I, 2nd edn. Springer, Berlin (1990) (p. 32, 46)
143. Hörmander, L.: The Analysis of Linear Partial Differential Operators III. Springer, Berlin (1994) (p. 51, 72)

144. Hu, Z.-C., Ma, Z.-M., Sun, W.: Extensions of Lévy-Khinchine formula and Beurling–Deny formula in semi-Dirichlet forms setting. J. Funct. Anal. **239**, 179–213 (2004) (p. 81)
145. Ikeda, N., Watanabe, S.: On some relations between the harmonic measure and the Lévy measure for a certain clas of Markov processes. J. Math. Kyoto Univ. **2**, 79–95 (1962) (p. 65)
146. Ikeda, N., Watanabe, S.: Stochastic Differential Equations and Diffusion Processes, 2nd edn. North–Holland/Kodansha, Amsterdam/Tokyo (1989) (p. 75, 107)
147. Ikeda, N., Nagasawa, M., Watanabe, S.: A construction of Markov processes by piecing out. Proc. Jpn. Acad. **42**, 370–375 (1966) (p. 99)
148. Imkeller, P., Willrich, N.: Solutions of martingale problems for Lévy-type operators and stochastic differential equations driven by Lévy processes with discontinuous coefficients. Preprint [arXiv: 1208.1665] (2012) (p. 90)
149. Itô, K.: Lectures on Stochastic Processes. Tata Institute of Fundamental Research/Springer, Bombay/Berlin (1961/1984) (p. 18)
150. Itô, K.: Semigroups in probability theory. In: Komatsu, H. (ed.) Functional Analysis and Related Topics, 1991. Proceedings of the International Conference in Memory of Professor Kosaku Yosida held at RIMS, Kyoto University, Japan, 1991, pp. 69–83. Lecture Notes in Mathematics, vol. 1540. Springer, Berlin (1993) (p. 17)
151. Itô, S.: Diffusion Equations. American Mathematical Society, Providence (1992) (p. 4)
152. Jacob, N.: Feller semigroups, Dirichlet forms, and pseudo differential operators. Forum Math. **4**, 433–446 (1992) (p. 52)
153. Jacob, N.: A class of Feller semigroups generated by pseudo differential operators. Math. Z. **215**, 151–166 (1994) (p. 74)
154. Jacob, N.: Non-local (semi-)Dirichlet forms generated by pseudo differential operators. In: Ma, Z.-M., Röckner, M., Yan, J.-A. (eds.) Dirichlet Forms and Stochastic Processes. Proceedings of the International Conference Held in Beijing, China, 1993. De Gruyter, Berlin (1995) (p. 84)
155. Jacob, N.: Pseudo Differential Operators and Markov Processes. Akademie, Berlin (1996) (p. 51)
156. Jacob, N.: Characteristic functions and symbols in the theory of Feller proceses. Potential Anal. **8**, 61–68 (1998) (p. 57, 58)
157. Jacob, N.: Pseudo Differential Operators and Markov Processes I: Fourier Analysis and Semigroups. Imperial College Press/World Scientific, London (2001) (p. 11, 22, 40, 44, 51, 60, 61, 73, 80, 82)
158. Jacob, N.: Pseudo Differential Operators and Markov Processes II: Generators and Their Potential Theory. Imperial College Press/World Scientific, London (2002) (p. 51, 70, 71, 72, 73, 87, 106, 107)
159. Jacob, N.: Pseudo Differential Operators and Markov Processes III: Markov Processes and Applications. Imperial College Press/World Scientific, London (2005) (p. 16, 51, 89, 160)
160. Jacob, N., Schilling, R.L.: Estimates for Feller semigroups generated by pseudodifferential operators. In: Function Spaces, Differential Operators and Nonlinear Analysis. Proceedings of the International Conference Paseky nad Jizerou, September 1995, pp. 27–49. Prometheus Publishing House, Prague (1996) (p. 104, 105)
161. Jacob, N., Schilling, R.L.: Subordination in the sense of S. Bochner – an approach through pseudo differential operators. Math. Nachr. **178**, 199–231 (1996) (p. 104)
162. Jacob, N., Schilling, R.L.: An analytic proof of the Lévy-Khinchin formula on $\mathbb{R}^n$. Publ. Math. Debrecen **53**, 69–89 (1998) (p. 33, 92)
163. Jacob, N., Schilling, R.L.: Lévy-type processes and pseudo differential operators. In: Barndorff-Nielsen, O.E., Mikosch, T., Resnick, S.I. (eds.) Lévy Processes: Theory and Applications, pp. 139–168. Birkhäuser, Boston (2001) (p. 104)
164. Jacob, N., Schilling, R.L.: Function spaces as Dirichlet spaces (about a paper by Maz'ya and Nagel). Z. Anal. Anw. **24**, 3–28 (2005) (p. 155)
165. Jacob, N., Schilling, R.L.: Towards an L^p potential theory for sub-Markovian semigroups: kernels and capacities. Acta Math. Sinica **22**, 1227–1250 (2006) (p. 30, 37, 85)

166. Jacob, N., Schilling, R.L.: On a Poincaré-type inequality for energy forms in L^p. Mediterr. J. Math. **4**, 33–44 (2007) (p. 30)

167. Jacob, N., Schilling, R.L.: Extended L^p Dirichlet spaces. In: Laptev, A. (ed.) Around the Research of Vladimir Maz'ya – vol. 1: Function Spaces, pp. 221–238. Springer, Berlin (2009) (p. 30)

168. Jacob, N., Knopova, V., Landwehr, S., Schilling, R.L.: A geometric interpretation of the transition density of a symmetric Lévy process. Sci. China Ser. A Math. **55**, 1099–1126 (2012) (p. xiii)

169. Jacod, J.: Calcul Stochastique et Problèmes de Martingales. Lecture Notes in Mathematics, vol. 714. Springer, Berlin (1979) (p. 178)

170. Jacod, J., Shiryaev, A.N.: Limit Theorems for Stochastic Processes, 2nd edn. Springer, Berlin (2003) (p. 40, 64, 89, 90, 168)

171. Jonsson, A., Wallin, H.: Function Spaces on Subsets of $\mathbb{R}^n$ (Mathematical Reports, vol. 2, Part 1). Harwood Academic, Chur (1984) (p. 135)

172. Kallenberg, O.: Foundations of Modern Probability, 2nd edn. Springer, New York (2004) (p. 89, 168)

173. Kato, T.: Perturbation Theory for Linear Operators. Springer, Berlin (1995) (p. 74)

174. Kazumi, T., Shigekawa, I.: Measures of finite (r, p)-energy and potentials on separable metric spaces. In: Azéma, J., Meyer, P.-A., Yor, M. (eds.) Séminaire de Probabilités 26, pp. 415–444. Lecture Notes in Mathematics, vol. 1526. Springer, Berlin (1992) (p. 85)

175. Keller-Ressel, M.: Affine processes—contributions to theory and applications. Ph.D. Thesis, TU Wien (2008) (p. 5)

176. Keller-Ressel, M., Schachermayer, W., Teichmann, J.: Affine processes are regular. Probab. Theory Relat. Fields **151**, 591–611 (2011) (p. 5, 20)

177. Keller-Ressel, M., Schachermayer, W., Teichmann, J.: Regularity of affine processes on general state spaces. Electron. J. Probab. **18**, 1–17 (2013) (p. 20)

178. Khintchine, A.I.: Sur la croissance locale des processus stochastiques homogènes à accroissements independants (Russian, French summary). Izvest. Akad. Nauk SSSR, Ser. Math. **3**, 487–508 (1939) (p. 129)

179. Khoshnevisan, D., Xiao, Y.: Lévy process: capacity and Hausdorff dimension. Ann. Probab. **33**, 841–878 (2005) (p. 125)

180. Khoshnevisan, D., Xiao, Y., Zhong, Y.: Measuring the range of an additive Lévy process. Ann. Probab. **31**, 1097–1141 (2003) (p. 124)

181. Kikuchi, K., Negoro, A.: On Markov process generated by pseudodifferential operator of variable order. Osaka J. Math **34**, 319–335 (1997) (p. 73)

182. Knopova,V., Schilling, R.L.: On the small–time behaviour of Lévy-type processes (2013) (p. 179)

183. Knopova, V., Schilling, R.L.: Transition density estimates for a class of Lévy and Lévy-type processes. J. Theor. Probab. **25**, 144–170 (2012) (p. 122)

184. Knopova, V., Schilling, R.L.: A note on the existence of transition probability densities for Lévy processes. Forum Math. **25**, 125–149 (2013) (p. 155)

185. Kochubei, A.N.: Parabolic pseudodifferential equations, hypersingular integrals and Markov processes. Math. USSR Izv. **33**, 233–259 (1989) (p. 87)

186. Kolmogoroff, A.N.: Über die analytischen Grundlagen der Wahrscheinlichkeitsrechnung. Math. Ann. **104**, 415–458 (1931) (English translation in [187]). (p. 86)

187. Kolmogorov, A.N.: Selected Works of A. N. Kolmogorov. Volume II: Probability Theory and Mathematical Statistics. Kluwer Academic, Dordrecht (1992) (p. 86, 188)

188. Kolokoltsov, V.N.: Semiclassical Analysis for Diffusions and Stochastic Processes. Lecture Notes in Mathematics, vol. 1724. Springer, Berlin (2000) (p. 88)

189. Kolokoltsov, V.N.: Symmetric stable laws and stable-like jump-diffusions. Proc. Lond. Math. Soc. **80**, 725–768 (2000) (p. 87)

190. Kolokoltsov, V.N.: Nonlinear Markov Processes and Kinetic Equations. Cambridge University Press, Cambridge (2010) (p. 96)

191. Kolokoltsov, V.N.: Markov Processes, Semigroups and Generators. De Gruyter, Berlin (2011) (p. 88)
192. Kolokoltsov, V.N., Schilling, R.L., Tyukov, A.E.: Transience and non-explosion of certain stochastic Newtonian systems. Electron. J. Probab. **7**, 1–19 (2002) (p. 163, 165)
193. Komatsu, T.: Continuity estimates for solutions of parabolic equations associated with jump type Dirichlet forms. Osaka J. Math. **25**, 697–728 (1988) (p. 86)
194. Kumano-go, H.: Pseudo-Differential Operators. MIT, Cambridge (1981) (p. 51, 72, 89)
195. Kunita, H.: Stochastic Flows and Stochastic Differential Equations. Cambridge University Press, Cambridge (1990) (p. 75)
196. Kunita, H.: Stochastic differential equations based on Lévy processes and stochastic flows of diffeomorphisms. In: Real and Stochastic Analysis. New Perspectives, pp. 305–373. Birkhäuser, Boston (2004) (p. 75)
197. Kurtz, T.G.: Equivalence of stochastic equations and martingale problems. In: Crisan, D. (ed.) Stochastic Analysis 2010 (7th ISAAC Congress, London, July 2009), pp. 113–130. Springer, Berlin (2011) (p. 90)
198. Kyprianou, A.E.: Introductory Lectures on Fluctuations of Lévy Processes with Applications. Springer, Berlin (2006) (p. 40)
199. Langer, H.: A class of infinitesimal generators of one-dimensional Markov processes. J. Math. Soc. Jpn. **28**, 242–249 (1976) (p. xiii)
200. Langer, H., Partzsch, L., Schütze, D.: Über verallgemeinerte gewöhnliche Differentialoperatoren mit nichtlokalen Randbedingungen und die von ihnen erzeugten Markov–Prozesse. Publ. Res. Inst. Math. Sci. **7**, 659–702 (1971/1972) (p. xiii)
201. Larsen, R.: An Introduction to the Theory of Multipliers. Springer, Berlin (1971) (p. 37)
202. Lescot, P., Röckner, M.: Perturbations of generalized Mehler semigroups and applications to stochastic heat equations with Lévy noise and singular drift. Potential Anal. **20**, 317–344 (2004) (p. 143)
203. Lin, H.N., Wang, J.: Successful couplings for a class of stochastic differential equations driven by Lévy processes. Sci. China Ser. A Math. **55**, 1737–1748 (2012) (p. 159)
204. Lindvall, T.: Lectures on the Coupling Method. Wiley, New York (1992) (p. 147, 157)
205. Lindvall, T., Rogers, L.C.G.: Coupling of multidimensional diffusions by reflection. Ann. Probab. **14**, 860–872 (1986) (p. 151)
206. Lindvall, T., Rogers, L.C.G.: On coupling of random walks and renewal processes. J. Appl. Probab. **33**, 122–126 (1996) (p. 148, 150)
207. Liptser, R.S., Pukhalskii, A.A.: Limit theorems on large deviations for semimartingales. Stoch. Stoch. Rep. **38**, 201–249 (1992) (p. 178)
208. Loève, M.: Probability Theory II, 4th edn. Springer, New York (1978) (p. 177)
209. Lumer, G.: Perturbation de générateurs infinitésimaux du type changement de temps. Ann. Inst. Fourier **23**, 271–279 (1973) (p. 100)
210. Ma, Z.-M., Röckner, M.: An Introduction to the Theory of (Non-Symmetirc) Dirichlet Forms. Springer, Berlin (1992) (p. xii, 74, 79, 80, 82)
211. Ma, Z.-M., Overbeck, L., Röckner, M.: Markov processes associated with semi-Dirichlet forms. Osaka J. Math. **32**, 97–119 (1995) (p. 82, 83)
212. Ma, Z.-M., Röckner, M., Zhang, T.-S.: Approximation of arbitrary Dirichlet processes by Markov chains. Ann. Inst. Henri Poincaré **34**, 1–22 (1998) (p. 171)
213. Ma, Z.-M., Röckner, M., Sun, W.: Approximation of Hunt processes by multivariate Poisson processes. Acta Appl. Math. **63**, 233–245 (2000) (p. 171)
214. Mackevičius, V.: Weak convergence of random processes in spaces $D[0, \infty)(X)$. Lith. Math. J. **14**, 620–623 (1974) (p. 168)
215. Malliavin, P.: Stochastic Analysis. Springer, Berlin (1997) (p. 29)
216. Mandl, P.: Analytical Treatment of One-Dimensional Markov Processes. Springer, Berlin (1968) (p. xiii)
217. Manstavičius, M.: p-variation of strong Markov processes. Ann. Probab. **32**, 2053–2066 (2004) (p. 132)

218. Manstavičius, M.: A non-Markovian process with unbounded p-variation. Electron. Commun. Probab. **10**, 17–28 (2005) (p. 132)

219. Masamune, J., Uemura, T.: Conservation property of symmetric jump processes. Ann. Inst. H. Poincaré Probab. Statist. **47**, 650–662 (2011) (p. 57)

220. Masamune, J., Uemura, T., Wang, J.: On the conservativeness and the recurrence of symmetric jump-diffusions. J. Funct. Anal. **263**, 3984–4008 (2012) (p. 57)

221. Meyer, P.-A.: Renaissance, recollements, mélanges, ralentissement de processus de Markov. Ann. Inst. Fourier **25**, 464–497 (1975) (p. 99, 107)

222. Meyer, P.A., Smythe, R.T., Walsh, J.B.: Birth and death of Markov processes. In: Le Cam, L.M., Neyman, J., Scott, E.L. (eds.) Proceedings of the Sixth Berkeley Symposium on Mathematical Statistics and Probability (University of California, Berkeley, 1970/1971). Volume III: Probability Theory, pp. 295–305. University of California Press, Berkeley (1972) (p. 99)

223. Meyn, S.P., Tweedie, R.L.: Stability of Markovian processes I: Criteria for discrete-time chains. Adv. Appl. Probab. **24**, 542–574 (1992) (p. 160)

224. Meyn, S.P., Tweedie, R.L.: A survey of Foster-Lyapunov techniques for general state space Markov processes. In: Proceedings of the Workshop on Stochastic Stability and Stochastic Stabilization, Metz, France, 1993 (1993) (p. 160)

225. Meyn, S.P., Tweedie, R.L.: Stability of Markovian orocesses II: Continous-time processes and sampled chains. Adv. Appl. Probab. **25**, 487–517 (1993) (p. 160)

226. Meyn, S.P., Tweedie, R.L.: Stability of Markovian processes III: Foster-Lyapunov criteria for continuous-time processes. Adv. Appl. Probab. **25**, 518–548 (1993) (p. 26, 160, 162)

227. Meyn, S.P., Tweedie, R.L.: Markov Chains and Stochastic Stability, 2nd edn. Cambridge University Press, Cambridge (2009) (p. 160)

228. Millar, P.W.: Path behavior of processes with stationary independent increments. Z. Wahrscheinlichkeitstheor. verw. Geb. **17**, 53–73 (1971) (p. 124)

229. Mouhot, C., Russ, E., Sire, Y.: Fractional Poincaré inequalities for general measures. J. Math. Pure Appl. **95**, 72–84 (2011) (p. 146)

230. Nagasawa, M.: Note on pasting of two Markov processes. In: Meyer, P.A. (ed.) Séminaire de Probabilités X. Lecture Notes in Mathematics, vol. 511, pp. 532–535. Springer, Berlin (1976) (p. 99)

231. Nash, J.: Continuity of solutions of parabolic and elliptic equations. Am. J. Math. **80**, 931–954 (1958) (p. 86)

232. Nelson, E.: A functional analytic approach using singular Laplace integrals. Trans. Am. Math. Soc. **88**, 400–413 (1958) (p. 103)

233. Nelson, E.: The free Markov Field. J. Funct. Anal. **12**, 211–227 (1973) (p. 141)

234. Oshima, Y.: Dirichlet Spaces – Lecture Notes, Universität Erlangen–Nürnberg, Summer Term 1988. Erlangen (1988) (p. 79, 160)

235. Oshima, Y.: Semi-Dirichlet Forms and Markov Processes. De Gruyter, Berlin (2013) (p. 79)

236. Pazy, A.: Semigroups of Linear Operators and Applications to Partial Differential Equations. Springer, New York (1983) (p. 8, 17, 23, 84, 107)

237. Peetre, J.: Réctification à l'article "Une caractérisation abstraite des opérateurs differentiels". Math. Scand. **8**, 116–120 (1960) (p. 27)

238. Port, S., Stone, C.: Brownian Motion and Classical Potential Theory. Academic, New York (1978) (p. xii)

239. Potrykus, A.: A symbolic calculus and a parametrix construction for pseudodifferential operators with non-smooth negative definite symbols. Rev. Mat. Complut. **22**, 187–207 (2009) (p. 74)

240. Potrykus, A.: Pseudodifferential operators with rough negative definite symbols. Integral Equ. Oper. Theory **66**, 441–461 (2010) (p. 74)

241. Priola, E., Zabczyk, J.: Liouville theorems for non-local operators. J. Funct. Anal. **216**, 455–490 (2004) (p. 157)

242. Priola, E., Zabczyk, J.: Densities for Ornstein–Uhlenbeck processes with jumps. Bull. Lond. Math. Soc. **41**, 41–50 (2009) (p. 157)

243. Protter, P.: Stochastic Integration and Differential Equations, 2nd edn. Springer, Berlin (2004) (p. 16, 40, 75, 132)
244. Pruitt, W.E.: The Hausdorff dimension of the range of a process with stationary independent increments. Indiana J. Math. **19**, 371–378 (1969) (p. 124)
245. Pruitt, W.E.: The growth of random walks and Lévy processes. Ann. Probab. **9**, 948–956 (1981) (p. 112, 125)
246. Pustyl'nik, E.I.: On functions of a positive operator. Math. USSR Sbornik **47**, 27–42 (1984) (p. 103)
247. Rao, M.M., Ren, Z.D.: Theory of Orlicz Spaces. Marcel Dekker, New York (1991) (p. 11)
248. Reed, M., Simon, B.: Functional Analysis, vol. 1, 2nd edn. Academic, San Diego (1980) (p. 8, 39, 41)
249. Revuz, D.: Markov Chains, 2nd edn. North-Holland, Amsterdam (1984) (p. 11)
250. Revuz, D., Yor, M.: Continuous Martingales and Brownian Motion, 3rd edn. Springer, Berlin (2005) (p. 6, 15, 16)
251. Röckner, M., Wang, F.-Y.: Weak Poincaré Inequalities and L^2-Convergence Rates of Markov Semigroups. J. Funct. Anal. **185**, 564–603 (2001) (p. 141)
252. Röckner, M., Wang, F.-Y.: Harnack and functional inequalities for generalized Mehler semigroups. J. Funct. Anal. **203**, 237–261 (2003) (p. 143)
253. Rogers, C.A.: Hausdorff Measures. Cambridge University Press, Cambridge (1970) (p. 124)
254. Rogers, L.C.G., Williams, D.: Diffusions, Markov Processes, and Martingales (vol. 2: Itô Calculus). Wiley, New York (1987) (p. 102)
255. Rogers, L.C.G., Williams, D.: Diffusions, Markov Processes, and Martingales (vol. 1: Foundations), 2nd edn. Wiley, New York (1994) (p. 2, 22)
256. Rosenbaum, M.: First order p-variations and Besov spaces. Stat. Probab. Lett. **79**, 55–62 (2009) (p. 140)
257. Roynette, B.: Mouvement brownien et espaces de Besov. Stoch. Stoch. Rep. **43**, 221–260 (1993) (p. 136)
258. Rudin, W.: Real and Complex Analysis, 3rd edn. McGraw Hill, New York (1986) (p. 7)
259. Rudin, W.: Functional Analysis, 2nd edn. McGraw Hill, New York (1991) (p. 32, 60)
260. Runst, T., Sickel, W.: Sobolev Spaces of Fractional Order, Nemytskij Operators, and Nonlinear Partial Differential Equations. De Gruyter, Berlin (1996) (p. 135, 137)
261. Ruzhansky, M., Turunen, V.: Pseudo-Differential Operators and Symmetries. Background Analysis and Advanced Topics. Birkhäuser, Basel (2010) (p. 51)
262. Saloff-Coste, L.: Aspects of Sobolev-Type Inequalities. London Mathematical Society Lecture Notes Series, vol. 289. Cambridge University Press, Cambridge (2002) (p. 141)
263. Sandrić, N.: Recurrence and transience property for a class of Markov chains. Bernoulli (2014) (p. 164)
264. Sandrić, N.: Long-time behavior of stable-like processes. Stoch. Process. Appl. **123**, 1276–1300 (2013) (p. 162, 164)
265. Sandrić, N.: Recurrence and transience property for two cases of stable-like Markov chains. J. Theor. Probab. (2013). doi:10.1007/s10959-012-0445-0 (p. 164)
266. Sasvári, Z.: Multivariate Characteristic and Correlation Functions. De Gruyter, Berlin (2013) (p. 41)
267. Sato, K.: Lévy Processes and Infinitely Divisible Distributions. Cambridge University Press, Cambridge (1999) (p. 18, 19, 33, 40, 44, 111, 121, 129)
268. Sato, K., Yamazato, M.: Operator-selfdecomposable distributions as limit distributions of processes of Ornstein-Uhlenbeck type. Stoch. Process. Appl. **17**, 73–100 (1984) (p. 4, 19)
269. Savov, M.: Small time two-sided LIL behavior for Lévy processes at zero. Probab. Theory Relat. Fields **144**, 79–98 (2009) (p. 129)
270. Sawyer, S.A.: A formula for semigroups, with an application to branching diffusion processes. Trans. Am. Math. Soc. **152**, 1–38 (1970) (p. 99)
271. Schilling, R.L.: When does a càdlàg process have continuous sample paths? Expo. Math. **12**, 255–261 (1994) (p. 28)

272. Schilling, R.L.: On the domain of the generator of a subordinate semigroup. In: Král, J., Lukeš, J., Netuka, I., Veselý, J. (eds.) Potential Theory – ICPT 94. Proceedings of the International Conference on Potential Theory, Kouty (CR) 1994, pp. 449–462. De Gruyter, Berlin (1996) (p. 103)

273. Schilling, R.L.: On Feller processes with sample paths in Besov spaces. Math. Ann. **309**, 663–675 (1997) (p. 136, 139, 140)

274. Schilling, R.L.: Conservativeness and extensions of Feller semigroups. Positivity **2**, 239–256 (1998) (p. 9, 55, 56, 57, 114)

275. Schilling, R.L.: Conservativeness of semigroups generated by pseudo differential operators. Potential Anal. **9**, 91–104 (1998) (p. 56, 58)

276. Schilling, R.L.: Feller processes generated by pseudo-differential operators: on the Hausdorff dimension of their sample paths. J. Theor. Probab. **11**, 303–330 (1998) (p. 127)

277. Schilling, R.L.: Growth and Hölder conditions for the sample paths of Feller processes. Probab. Theory Relat. Fields **112**, 565–611 (1998) (p. 39, 61, 65, 112, 117, 125, 126, 129, 131)

278. Schilling, R.L.: Subordination in the sense of Bochner and a related functional calculus. J. Aust. Math. Soc. Ser. A **64**, 368–396 (1998) (p. 103)

279. Schilling, R.L.: Function spaces as path spaces of Feller processes. Math. Nachr. **217**, 147–174 (2000) (p. 138)

280. Schilling, R.L.: Sobolev embedding for stochastic processes. Expo. Math. **18**, 239–242 (2000) (p. 136)

281. Schilling, R.L.: Dirichlet operators and the positive maximum principle. Integral Equ. Oper. Theory **41**, 74–92 (2001) (p. 22, 80, 84)

282. Schilling, R.L.: A note on invariant sets. Probab. Math. Stat. **24**, 47–66 (2004) (p. 81)

283. Schilling, R.L.: Measures, Integrals and Martingales. Cambridge University Press, Cambridge (2005) (p. 3, 12)

284. Schilling, R.L., Partzsch, L.: Brownian Motion. An Introduction to Stochastic Processes. De Gruyter, Berlin (2012) (p. 4, 17, 18, 21, 22, 26, 27, 28, 36, 75, 76, 123, 132)

285. Schilling, R.L., Schnurr, A.: The symbol associated with the solution of a stochastic differential equation. Electron. J. Probab. **15**, 1369–1393 (2010) (p. 55, 56, 59, 61, 63, 64, 76, 126)

286. Schilling, R.L., Uemura, T.: On the Feller property of Dirichlet forms generated by pseudo-differential operators. Tohôku Math. J. **59**, 401–422 (2007) (p. 63, 81, 85, 96)

287. Schilling, R.L., Wang, J.: On the coupling property of Lévy processes. Ann. Inst. Henri Poincaré Probab. Stat. **47**, 1147–1159 (2011) (p. 147, 148, 150)

288. Schilling, R.L., Wang, J.: Lower bounded semi-Dirichlet forms associated with Lévy type operators. Preprint [arXiv: 1108.3499] 18 pp. (2012) (p. 81, 96)

289. Schilling, R.L., Wang, J.: On the coupling property and the Liouville theorem for Ornstein–Uhlenbeck processes. J. Evol. Equ. **12**, 119–140 (2012) (p. 142, 147, 149, 150, 156, 158)

290. Schilling, R.L., Wang, J.: Strong Feller continuity of Feller processes and semigroups. Infinite Dimens. Anal. Quantum Probab. Relat. Top. **15**, 1250010 (28 pp.) (2012) (p. 11, 12, 59)

291. Schilling, R.L., Wang, J.: Some theorems on Feller processes: transience, local times and ultracontractivity. Trans. Am. Math. Soc. **365**, 3255–3286 (2013) (p. 67)

292. Schilling, R.L., Sztonyk, P., Wang, J.: Coupling property and gradient estimates of Lévy processes via the symbol. Bernoulli **18**, 1128–1149 (2011) (p. 147, 155)

293. Schilling, R.L., Song, R., Vondraček, Z.: Bernstein Functions: Theory and Applications, 2nd edn. De Gruyter, Berlin (2012) (p. 35, 45, 103, 104, 107, 142)

294. Schnurr, A.: The symbol of a Markov semimartingale. Ph.D. thesis, TU Dresden, Aachen (2009) (p. 61, 63, 65, 77)

295. Schnurr, A.: On the semimartingale nature of Feller processes with killing. Stoch. Process. Appl. **122**, 2758–2780 (2012) (p. 61, 63, 65)

296. Schnurr, A.: Generalization of the Blumenthal–Getoor index to the class of homogeneous diffusions with jumps and some applications. Bernoulli (2013, to appear) (p. 65)

297. Sharpe, M.: Zeroes of infinitely divisible densities. Ann. Math. Stat. **40**, 1503–1505 (1969) (p. 153)
298. Sharpe, M.: General Theory of Markov Processes. Academic, London (1988) (p. xii, 102)
299. Shieh, N.-R., Xiao, Y.: Hausdorff and parking dimensions of the images of random fields. Bernoulli **16**, 926–952 (2010) (p. 127)
300. Shiozawa, Y., Takeda, M.: Variational formula for Dirichlet forms and estimates of principal eigenvalues for symmetric α-stable processes. Potential Anal. **23**, 135–151 (2005) (p. 178)
301. Silverstein, M.L.: Symmetric Markov Processes. Lecture Notes in Mathematics, vol. 426. Springer, Berlin (1974) (p. 82)
302. Situ, R.: Theory of Stochastic Differential Equations with Jumps and Applications: Mathematical and Analytical Techniques with Applications to Engineering. Springer, New York (2005) (p. 75)
303. Skorokhod, A.V.: Limit theorems for stochastic processes. Theory Probab. Appl. **1**, 261–290 (1956) (p. 89, 168)
304. Skorokhod, A.V.: Studies in the Theory of Random Processes. Addison–Wesley, Reading (1965) (p. 177)
305. Skorokhod, A.V.: Random Processes with Independent Increments. Kluwer, Dordrecht (1991) (p. 129)
306. Song, R., Vondraček, Z.: Harnack inequalities for some classes of Markov processes. Math. Z. **246**, 177–202 (2004) (p. 85)
307. Stein, E.M.: Topics in Harmonic Analysis Related to the Littlewood–Paley Theory. Annals of Mathematics Studies, vol. 63. Princeton University Press, Princeton (1970) (p. 30)
308. Strang, G.: Approximating semigroups and the consistency of difference schemes. Proc. Am. Math. Soc. **20**, 1–7 (1969) (p. 167, 168)
309. Stroock, D.W.: Diffusion processes associated with Lévy generators. Probab. Theory Relat. Fields **32**, 209–244 (1975) (p. 72, 90)
310. Stroock, D.W.: Diffusion semigroups corresponding to uniformly elliptic divergence form operators. In: Azéma, J., Meyer, P.A., Yor, M. (eds.) Séminaire de Probabilités XXII. Lecture Notes in Mathematics, vol. 1321, pp. 316–347. Springer, Berlin (1988) (p. 4, 5)
311. Stroock, D.W.: Markov Processes from K. Itô's Perspective. Princeton University Press, Princeton (2003) (p. 78, 93, 94)
312. Stroock, D.W., Varadhan, S.R.S.: Multidimensional Diffusion Processes. Springer, Berlin (1979) (p. 56, 57, 75, 76, 89)
313. Taira, K.: Diffusion Processes and Partial Differential Equations. Academic, Boston (MA) (1988) (p. xiii, 72)
314. Taira, K.: On the existence of Feller semigroups with discontinuous coefficients. Acta Math. Sinica **22**, 595–606 (2006) (p. 71)
315. Takeda, M.: On a large deviation for symmetric Markov processes with finite lifetime. Stoch. Stoch. Rep. **59**, 143–167 (1996) (p. 178)
316. Takeda, M.: A large deviation principle for symmetric Markov processes with Feynman–Kac functional. J. Theor. Probab. **24**, 1097–1129 (2011) (p. 178)
317. Taylor, M.E.: Pseudodifferential Operators, 2nd edn. Princeton University Press, Princeton (1981) (p. 51)
318. Taylor, S.J.: Sample path properties of processes with stationary independent increments. In: Kendall, D.G., Harding, E.F. (eds.) Stochastic Analysis, pp. 387–414. Wiley, London (1973) (p. 40, 111)
319. Thorisson, H.: Shift-coupling in continuous time. Probab. Theory Relat. Fields **99**, 477–483 (1994) (p. 147)
320. Thorisson, H.: Coupling, Stationarity, and Regeneration. Springer, New York (2000) (p. 147)
321. Triebel, H.: Theory of Function Spaces. Birkhäuser, Basel (1983) (p. 135)
322. Triebel, H.: Theory of Function Spaces II. Birkhäuser, Basel (1992) (p. 135)
323. Triebel, H.: Interpolation Theory, Function Spaces, Differential Operators, 2nd edn. Johann Ambrosius Barth, Heidelberg (1995) (p. 135)
324. Triebel, H.: Theory of Function Spaces III. Birkhäuser, Basel (2006) (p. 135, 139)

325. Trotter, H.F.: Approximation of semi-groups of operators. Pac. J. Math **8**, 887–919 (1958) (p. 169)

326. Tsuchiya, M.: Lévy measure with generalized polar decomposition and the associated SDE with jumps. Stoch. Stoch. Rep. **38**, 95–117 (1992) (p. 78, 94)

327. Tweedie, R.L.: Topological conditions enabling use of Harris methods in discrete and continuous time. Acta Appl. Math. **34**, 175–188 (1994) (p. 161)

328. Uemura, T.: On some path properties of symmetric stable-like processes for one dimension. Potential Anal. **16**, 79–91 (2002) (p. 96, 164)

329. Uemura, T.: On symmetric stable-like processes: some path properties and generators. J. Theor. Probab. **17**, 541–555 (2004) (p. 96, 164)

330. Uemura, T., Shiozawa, Y.: Explosion of jump-type symmetric Dirichlet forms on $\mathbb{R}^d$. J. Theor. Probab. (2012). doi:10.1007/s10959-012-0424-5 (p. 57)

331. van Casteren, J.A.: Markov Processes, Feller Semigroups and Evolution Equations. World Scientific, New Jersey (2011) (p. 52, 96, 110)

332. Varopoulos, N.T., Saloff-Coste, L., Coulhon, T.: Analysis and Geometry on Groups. Cambridge University Press, Cambridge (1992) (p. 141)

333. Waldenfels, W.v.: Eine Klasse stationärer Markowprozesse. Kernforschungsanlage Jülich, Institut für Plasmaphysik, Jülich (1961) (p. 47)

334. Waldenfels, W.v.: Positive Halbgruppen auf einem n-dimensionalen Torus. Arch. Math. **15**, 191–203 (1964) (p. 47)

335. Waldenfels, W.v.: Fast positive Operatoren. Z. Wahrscheinlichkeitstheor. verw. Geb. **4**, 159–174 (1965) (p. 47)

336. Wang, F.-Y.: Functional inequalities for empty essential spectrum. J. Funct. Anal. **170**, 219–245 (2000) (p. 142)

337. Wang, F.-Y.: Functional inequalities, semigroup properties and spectrum estimates. Infinite Dimens. Anal. Quantum Probab. Relat. Top. **3**, 263–295 (2000) (p. 142)

338. Wang, F.-Y.: Functional inequalities for the decay of sub-Markov semigroups. Potential Anal. **18**, 1–23 (2003) (p. 141)

339. Wang, F.-Y.: Functional inequalities on abstract Hilbert spaces and applications. Math. Z. **246**, 359–371 (2004) (p. 142)

340. Wang, F.-Y.: Functional Inequalities, Markov Processes and Spectral Theory. Science Press, Beijing (2005) (p. 30, 141, 143)

341. Wang, F.-Y.: Functional inequalites for Dirichlet forms with fractional powers. Chin. Sci. Tech. Online **2**, 1–4 (2007) (p. 142)

342. Wang, F.-Y.: Coupling for Ornstein–Uhlenbeck processes with jumps. Bernoulli **17**, 1136–1158 (2011) (p. 147, 149, 150)

343. Wang, J.: Criteria for ergodicity of Lévy type operators in dimension one. Stoch. Process. Appl. **118**, 1909–1928 (2008) (p. 164)

344. Wang, J.: Regularity of semigroups generated by Lévy type operators via coupling. Stoch. Process. Appl. **120**, 1680–1700 (2010) (p. 179)

345. Wang, J.: Stability of Markov processes generated by Lévy type operators. Chin. J. Contemp. Math. **32**, 1–20 (2011) (p. 57, 97)

346. Wang, J.: On the existence and explicit estimates for the coupling property of Lévy processes with drift. J. Theor. Probab. (2012). doi:10.1007/s10959-012-0463-y (p. 159)

347. Wang, J.: On the exponential ergodicity for Lévy driven Ornstein–Uhlenbeck processes. J. Appl. Probab. **49**, 990–1104 (2012) (p. 179)

348. Wang, J.: A simple approach to functional inequalities for non-local Dirichlet Forms. ESAIM Probab. Stat. (2013). doi:10.1051/ps/2013048 (p. 144, 146)

349. Wang, J.: Sub-Markovian C_0-semigroups generated by fractional Laplacian with gradient perturbation. Integral Equ. Oper. Theory **76**, 151–161 (2013) (p. 30)

350. Wang, F.-Y., Wang, J.: Coupling and strong Feller for jump processes on Banach spaces. Stoch. Process. Appl. **123**, 1588–1615 (2013) (p. 157)

351. Wang, F.-Y., Wang, J.: Functional inequalities for stable-like Dirichlet forms. J. Theor. Probab. (2013). doi:10.1007/s10959-013-0500-5 (p. 146)

352. Wentzell (Vent'cel), A.D.: On boundary conditions for multi-dimensional diffusion processes. Theory Probab. Appl. **4**, 164–177 (1959) (p. xiii)
353. Xiao, Y.: Random fractals and Markov processes. In: Lapidus, M.L., Frankenhuijsen, M.v. (eds.) Fractal Geometry and Applications: A Jubilee of Benoît Mandelbrot. Proceedings of Symposia in Pure Mathematics, vol. 72.2, pp. 261–338. American Mathematical Society, Providence (2004) (p. 40, 111, 125)
354. Yosida, K.: Functional Analysis, 6th edn. Springer, Berlin (1980) (p. 17, 84)
355. Zhang, X.C.: Derivative formula and gradient estimates for SDEs driven by α-stable processes. Stoch. Process. Appl. **123**, 1213–1228 (2013) (p. 160)

Index

B. Böttcher et al., *Lévy Matters III*, Lecture Notes
in Mathematics 2099, DOI 10.1007/978-3-319-02684-8,
© Springer International Publishing Switzerland 2013

LECTURE NOTES IN MATHEMATICS  Springer

Edited by J.-M. Morel, B. Teissier; P.K. Maini

Editorial Policy (for the publication of monographs)

1. Lecture Notes aim to report new developments in all areas of mathematics and their applications - quickly, informally and at a high level. Mathematical texts analysing new developments in modelling and numerical simulation are welcome.

   Monograph manuscripts should be reasonably self-contained and rounded off. Thus they may, and often will, present not only results of the author but also related work by other people. They may be based on specialised lecture courses. Furthermore, the manuscripts should provide sufficient motivation, examples and applications. This clearly distinguishes Lecture Notes from journal articles or technical reports which normally are very concise. Articles intended for a journal but too long to be accepted by most journals, usually do not have this "lecture notes" character. For similar reasons it is unusual for doctoral theses to be accepted for the Lecture Notes series, though habilitation theses may be appropriate.

2. Manuscripts should be submitted either online at www.editorialmanager.com/lnm to Springer's mathematics editorial in Heidelberg, or to one of the series editors. In general, manuscripts will be sent out to 2 external referees for evaluation. If a decision cannot yet be reached on the basis of the first 2 reports, further referees may be contacted: The author will be informed of this. A final decision to publish can be made only on the basis of the complete manuscript, however a refereeing process leading to a preliminary decision can be based on a pre-final or incomplete manuscript. The strict minimum amount of material that will be considered should include a detailed outline describing the planned contents of each chapter, a bibliography and several sample chapters.

   Authors should be aware that incomplete or insufficiently close to final manuscripts almost always result in longer refereeing times and nevertheless unclear referees' recommendations, making further refereeing of a final draft necessary.

   Authors should also be aware that parallel submission of their manuscript to another publisher while under consideration for LNM will in general lead to immediate rejection.

3. Manuscripts should in general be submitted in English. Final manuscripts should contain at least 100 pages of mathematical text and should always include

   - a table of contents;
   - an informative introduction, with adequate motivation and perhaps some historical remarks: it should be accessible to a reader not intimately familiar with the topic treated;
   - a subject index: as a rule this is genuinely helpful for the reader.

   For evaluation purposes, manuscripts may be submitted in print or electronic form (print form is still preferred by most referees), in the latter case preferably as pdf- or zipped ps-files. Lecture Notes volumes are, as a rule, printed digitally from the authors' files. To ensure best results, authors are asked to use the LaTeX2e style files available from Springer's web-server at:

   ftp://ftp.springer.de/pub/tex/latex/svmonot1/ (for monographs) and
   ftp://ftp.springer.de/pub/tex/latex/svmultt1/ (for summer schools/tutorials).

Additional technical instructions, if necessary, are available on request from lnm@springer.com.

4. Careful preparation of the manuscripts will help keep production time short besides ensuring satisfactory appearance of the finished book in print and online. After acceptance of the manuscript authors will be asked to prepare the final LaTeX source files and also the corresponding dvi-, pdf- or zipped ps-file. The LaTeX source files are essential for producing the full-text online version of the book (see http://www.springerlink.com/openurl.asp?genre=journal&issn=0075-8434 for the existing online volumes of LNM). The actual production of a Lecture Notes volume takes approximately 12 weeks.

5. Authors receive a total of 50 free copies of their volume, but no royalties. They are entitled to a discount of 33.3 % on the price of Springer books purchased for their personal use, if ordering directly from Springer.

6. Commitment to publish is made by letter of intent rather than by signing a formal contract. Springer-Verlag secures the copyright for each volume. Authors are free to reuse material contained in their LNM volumes in later publications: a brief written (or e-mail) request for formal permission is sufficient.

Addresses:
Professor J.-M. Morel, CMLA,
École Normale Supérieure de Cachan,
61 Avenue du Président Wilson, 94235 Cachan Cedex, France
E-mail: morel@cmla.ens-cachan.fr

Professor B. Teissier, Institut Mathématique de Jussieu,
UMR 7586 du CNRS, Équipe "Géométrie et Dynamique",
175 rue du Chevaleret
75013 Paris, France
E-mail: teissier@math.jussieu.fr

For the "Mathematical Biosciences Subseries" of LNM:

Professor P. K. Maini, Center for Mathematical Biology,
Mathematical Institute, 24-29 St Giles,
Oxford OX1 3LP, UK
E-mail: maini@maths.ox.ac.uk

Springer, Mathematics Editorial, Tiergartenstr. 17,
69121 Heidelberg, Germany,
Tel.: +49 (6221) 4876-8259

Fax: +49 (6221) 4876-8259
E-mail: lnm@springer.com